GENETICS AND ETHICS IN GLOBAL PERSPECTIVE

INTERNATIONAL LIBRARY OF ETHICS, LAW, AND THE NEW MEDICINE

VOLUME 17

The titles published in this series are listed at the end of this volume.

GENETICS AND ETHICS IN GLOBAL PERSPECTIVE

by

Dorothy C. Wertz †

Research Professor of Psychiatry, University of Massachusetts Medical School and Shriver Center, Waltham, MA, USA

and

John C. Fletcher

Professor Emeritus of Biomedical Ethics and Internal Medicine, University of Virginia School of Medicine, Charlottesville, VA, USA

KLUWER ACADEMIC PUBLISHERS

DORDRECHT / BOSTON / LONDON

A C.I.P. Catalogue record for this book is available from the Library of Congress.

ISBN 1-4020-1768-5

Published by Kluwer Academic Publishers,
P.O. Box 17, 3300 AA Dordrecht, The Netherlands.

Sold and distributed in North, Central and South America
by Kluwer Academic Publishers,
101 Philip Drive, Norwell, MA 02061, U.S.A.

In all other countries, sold and distributed
by Kluwer Academic Publishers,
P.O. Box 322, 3300 AH Dordrecht, The Netherlands.

Printed on acid-free paper

Printed and bound in Great Britain by Antony Rowe Limited.

TABLE OF CONTENTS

PREFACE

Every book contains the story of how it came to be written. Over the years, many people have asked us how we arrived in the rather unusual position of doing social science surveys of ethics and human genetics. John Fletcher has already described his odyssey in the introduction to our previous book, *Ethics and Human Genetics: a Cross-Cultural Perspective* (1989): his deaf parents, his years in the Episcopal ministry, his founding of an interfaith theological seminary that provided pastoral training for Protestants, Catholics, and Jews, his research on families having prenatal diagnosis, which culminated in *Coping with Genetic Disorders: A Guide for Clergy and Parents* (1982), the first book of its kind, his 11 years as Chief and sole member of the Bioethics Program of the National Institutes of Health, followed by leadership of the Bioethics Program at the University of Virginia Medical School. Now it is time for me to tell my side of the story.

I was interdisciplinary from the beginning, a route that, for employment reasons, I always urge students not to take. At Radcliffe College/Harvard University I majored in Social Relations, an undergraduate program that combined Sociology, Social Psychology, and some Cultural Anthropology. I audited almost as many courses as I took for credit, mostly in Philosophy. After a rather unproductive year studying anthropology at the London School of Economics, I returned to Harvard to study existentialism, at the invitation of philosopher John Wild. As the Philosophy Department taught mainly logical positivism in the late 1950s, I had to enter an interdisciplinary program called History and Philosophy of Religion. After Wild left Harvard, I returned to my original interest in social science and specialized in Religion and Society, still under the same broad interdisciplinary committee. My thesis was on social conflict resolution in the fourteenth-century medieval morality plays; it taught me that writing a thesis and convincing a committee to accept it was more a work of salesmanship than scholarship. Harvard teaches its students to think independently and on their feet. Although several famous names appear on my oral examinations, including social psychologist Gordon Allport, sociologist Talcott Parsons, psychiatrist John P. Spiegel, and social ethicist James Luther Adams, I never had a mentor. Most women didn't in those days. Being a woman in religion was somewhat of an oddity. My first professional conference consisted of what seemed like 500 men and myself, the "invisible person." I quickly found out, on entering the job market, that most schools were not eager to hire women, especially in religion. Relying on my undergraduate major, I shifted to sociology, then a rapidly expanding field, and, over the course of 16 years at nine different colleges and universities, taught virtually every course in the sociology curriculum, plus a number in anthropology and psychology.

Along the way, I came to be co-author, with my husband Richard, of *Lying-In: A History of Childbirth in America* (1977, Expanded Edition 1989). The book emerged, not from our personal experience (we had no children), but from Dick's enduring interest in technology and society in American life, a course that he teaches to this day. In 1972, members of the Boston Women's Health Book Collective spoke in his history of medicine course at MIT and described the technologization of birth, which should be a natural event requiring little intervention. I was then commuting from Boston to Connecticut to teach (an early example of commuter marriage), and Dick asked me to go to the Yale University library on my day off and see when episiotomies and routine forceps came into use. I returned so full of fascinating information that we decided to write an entire book on the history of birth in America, as an example of the complex interactions between culture, gender, and technology. As we progressed through our ten revisions of the manuscript, the emerging Women's Movement of the early 1970s played a greater role in our thinking.

In 1979, seeking an escape from undergraduate teaching and an opportunity to support myself by research and writing, I approached Boston University School of Public Health, book in hand, and asked if I could apply for a fellowship. The School of Public Health at B.U. was not doing anything related to birth at that time (they have since instituted an entire program in Maternal and Child Health); their closest area to my interest was a national survey of genetic counseling recently completed by sociologist James R. Sorenson, who agreed to act as my supervisor. After an initial false try (the reviewers did not like Jim's calling me "a lady and a scholar"), in 1981 I received three years of fellowships from the National Science Foundation, the longest-term stability in my career to that date. Jim taught me how to write articles for scientific journals (quite a different endeavor from writing a book), about how to approach a large data set, and how to present tables. Together we wrote a number of articles based on his genetic counseling survey, which remains the largest study of its kind. At the end of September, 1984, a month after my fellowship ran out, Jim received a call from John Fletcher, who had just returned from a six-month sabbatical visiting genetics centers in Europe. John was convinced that geneticists around the world were coming to a consensus about the resolution of ethical problems that occur in clinical practice, and wanted Jim's help in doing an international survey of geneticists to document this consensus. Jim, having little time available, enlisted my help and gradually turned the project over to me. There began a collaboration with John that has lasted 16 years.

In 1984–86, we did a survey of 683 geneticists in 19 nations, and found considerably less consensus than John had anticipated. We first presented the results at the 1986 International Congress of Human Genetics in Berlin and subsequently published them in *Ethics and Human Genetics: A Cross-cultural Perspective* (Berlin: Springer-Verlag, 1989).

As time went on, we were asked to share our findings with a variety of national and international ethics bodies. The Science Council of Canada invited me to participate in a project on ethical provision of genetics services. Subsequently, the Royal Commission on New Reproductive Technologies (Canada) commissioned me to write a paper on "Prenatal Diagnosis and Society." I became a member of the various ethics committees of the American Society of Human Genetics, Council of Regional Networks of Genetics Services, New England Regional Genetics Group, and finally, the international body of researchers, the Human Genome Organization (HUGO). Together with John, I was instrumental in developing three major documents on ethics and genetics for the World Health Organization.

Some geneticists thought that our survey should be repeated every 10 years, in order to identify trends. The first survey was supported by small grants from private foundations and a lot of volunteer labor. This time we resolved to do the work with adequate support. Felix de la Cruz of the Mental Retardation and Developmental Disorders Branch of the National Institute of Child Health and Human Development (National Institutes of Health) encouraged us to negotiate a contract for the North American portion of the study. His staff members, Dorothy McKelvin and Harvey Shifrin, were attentive throughout the project. Eric Juengst of the Ethical, Legal, and Social Implications (ELSI) Branch of the National Human Genome Research Institute (NHGRI) helped with the international aspects of the project, under a peer-reviewed grant to the Eunice Kennedy Shriver Center for Mental Retardation in Waltham, MA.

This time we resolved to include United States primary care physicians in the study, because they will likely be doing much of the counseling and testing in the future. We also included US and Canadian patients, a group whose voices had been rarely heard. Internationally, we sought to include countries that had withdrawn from the earlier study, such as China and Russia, and also to include much of Latin America. In all, 36 nations participated. Unfortunately, we still failed to get adequate representation from the Muslim world.

Our purpose in doing surveys is not to "do ethics by the numbers." Majorities, even large majorities, can be wrong from an ethical point of view. Our purpose is to present a basis for national and international discussion. People often think that they already know how their professional colleagues will respond to a situation, but they may have little basis for such suppositions. The survey responses – given by individuals privately, not in the public glare of a professional conference – have shocked many geneticists around the world. Usually people are most shocked by the responses from within their own country. Until these private ethics come before the public, it is difficult to have meaningful discussions of what really should occur.

Our questions are based mostly on clinical practice, not research. Since we completed the survey, a number of research issues have come to the forefront, including cloning, enhancement of intelligence, use of embryonic stem cells, and research use of collections of stored human cells. Although some of these issues have aroused media attention, none has yet affected clinical practice. Geneticists dislike answering hypothetical questions or questions about procedures that have not yet arrived; hence we omitted most hypotheticals, with the exceptions of pre-conception sex selection and testing for genetic predisposition to behavioral conditions such as alcoholism. If we were to do the survey today, we would include more questions of a research nature.

The first eight chapters describe the survey's methods and results, in considerable detail. In order to spare the reader from verbal statistics, we have combined a largely non-numerical text with detailed tables. The text summarizes and comments on the tables and can be read independently by those who are interested in the general picture without a plethora of numbers. Those interested in seeing how a particular country responded, or how countries compare, can turn to the tables. We considered consolidating the responses of individual countries into "culture areas," as for example, Latin America, an approach that would have considerably shortened the tables. However, there was so much difference between countries in a culture area (as for example, between Brazil and Chile, or Japan and China) that we think it would be misleading to try consolidation, except in the few instances where there was close agreement. International readers are usually interested in how a particular country responded, so that they can use the information in ethical and policy discussions. We consider the detail a strength of this report.

There follow sections on Genetic Services, Laws Affecting Genetics, and Geneticists Speak Out. The latter reports personal observations or pleas for change from our international authors. In the previous book we had a chapter written by a geneticist from each of the 19 countries surveyed. In this book, we could not do the same for 37 nations. Instead, we have selected what we consider the most interesting six chapters for inclusion, and have incorporated portions of the remainder into the three sections above.

The final two chapters describe some of the ethical complexities involved in the survey questions, in the context of varying economic, cultural, and religious backgrounds, and attempt to delineate what should be done in genetic services. The outline of the concluding chapter parallels that of the introductory chapters.

We would like to thank our international collaborators who distributed and collected the survey questionnaires, and whose names appear on page, our authors (who appear in the "About the Authors" section), and those who helped us to design the survey questions, including Betsy Anderson, Joan O. Weiss, Richard W. Erbe, Marvin Natowicz, Philip R. Reilly, Thaddeus E. Kelly, and the providers and consumers who participated in the field tests. This project could not have succeeded without the help of research assistants Wanda Hunt and Robin Gregg. Mary Jo Rosenblatt assisted with telephone follow-ups, and Lee Picard oversaw financial concerns. Leslie Cabral provided secretarial assistance in the early stages, and Donna Caira typed the final manuscript.

We hope that others will undertake similar pursuits in the future and that our efforts will serve to guide them.

Dorothy C. Wertz†
University of Massachusetts Medical School
Shriver Division
Waltham, MA

†Dr. Wertz died on April 29, 2003.

PART I

THE SURVEY AND ITS RESULTS

CHAPTER 1

THE 36-NATION SURVEY

INTRODUCTION: A GLOBAL ENTERPRISE

Modern genetics is a global enterprise. Although the United States likes to think of itself as the leader in the Human Genome Project (HGP), Britain, Japan, Germany, France, and other nations are also playing major roles. The concept of "genetic disease" has expanded to include not only heart disease, cancer, asthma, and diabetes, but also infectious diseases such as malaria, HIV, and leprosy, which affect millions of people in the developing world. Genetic studies of the immune system may eventually lead to genetically-based vaccines for these and other infectious diseases. Millions of human DNA samples from around the world will be amassed and stored for this research. Sample donations from people in one country may help people halfway around the world.

Clinical genetics, the area where research and medicine converge, poses ethical problems both old and new, which provoke responses governed in part by culture, religion, and economics. Through widely-accepted networks of referrals, both professionals and patients travel across national boundaries to provide or receive services. International referral is a generally accepted alternative when legal or ethical conditions prohibit an action in one's own country. The actions of geneticists in one country cannot be completely independent of the actions of geneticists in other countries. Therefore, global discussion of the issues becomes necessary.

RATIONALE

One way – perhaps the only practical way – to begin a global discussion of ethics is to put everybody's views on the table so that they can be examined openly. This was the purpose of our two international surveys of geneticists' ethical views. The first, in 19 nations, took place in 1984–86 (Wertz and Fletcher, 1989) and the second, in 36 nations, in 1993–95, is reported here. Often, only the views of the most outspoken geneticists in a country are heard, at conferences, in journals, or in the media. The outspoken do not necessarily represent prevailing opinion or practice. An anonymous survey, like a ballot, has the advantage of soliciting the views of those not usually heard.

Why solicit the views of geneticists, a profession limited in numbers? Geneticists possess specialized knowledge that is rare in many countries. They are in a unique position to inform government policymakers and those who reimburse for genetic services. Enlightened policy depends on the help of experts. Even if many prefer to avoid the political arena, some will inevitably be called forward to provide expert testimony. Their views should carry more weight than the views of non-specialist physicians, whom geneticists should help to educate, both medically and ethically.

Medical genetics differs from the rest of medicine in several important ways (Table 1), notably its possible effect on family members who may share mutations and its potential for inadvertently revealing the "nonbiological" nature of some social relationships, especially paternity. Some geneticists would argue that in genetics the true patient is the family, not the individual (Berg, 1989). Presymptomatic genetic tests can predict, with a far greater certainty than ordinary biochemical testing, whether a person who is now healthy will later develop a disorder.

Finally, at least in the nations where genetic counseling began (UK, US, and Canada), there is a tradition of nondirectiveness that differs markedly from the treatment-and-advice-oriented ethics of general medicine. This nondirective approach is the product of the peculiar history of doctor–patient interactions in genetics. Genetic counseling originated, in the years immediately after World War II, as an educational specialty for couples who wondered whether they should take the risk of having a child with a "birth defect." There were no treatments or preventive measures. Prenatal diagnosis did not appear in clinical practice until the 1970s. All that couples could do was to decide whether or not to have children. In order to make intelligent decisions, they needed information they could understand. Placing people at the level of students, rather than patients (patient, or "sufferer" is a somewhat infantilized status) automatically helped to equalize the "provider–patient" relationship. People were treated as autonomous agents who made their own decisions on the basis of information provided. In the absence of treatment, telling people how to govern their reproductive lives was considered an intrusion on individual freedom, at least in the English-speaking nations where counseling began. Today, as genetics moves into adult medicine, and prevention and treatment – genetic, biochemical, and lifestyle – become available, genetics will likely become more directive, like most other areas of medicine.

METHODS

The Questionnaires

The anonymous questionnaire is the general standard in survey research. Ours combined closed-ended questions (a checklist of possible actions in a situation) and open-ended questions asking people to write, in their own words, why they had chosen a particular course of action. The questions themselves were devised to reflect a broad range of ethical problems in clinical practice, including confidentiality, individual autonomy in decision-making; directiveness in counseling; attitudes toward abortion; requests for prenatal diagnosis when not medically indicated (e.g., for sex selection), spouses', families', and adoptees' access to genetic information; and priorities in national health budgets. In addition, we asked some general questions about perceptions of persons with disabilities, the role of the state, rights to referral, and responsibilities of individuals and families. We also included questions on ethically and politically important issues that do not fall within the practice of most medical geneticists, such as testing in the workplace and DNA fingerprinting for identification.

Most questions were presented as case vignettes. This helps concretize the issue for medical professionals who are used to thinking in terms of cases, and also engages their attention. Although some cases were rare, such as disclosure to a woman found to have XY chromosomes, most were likely to occur at least once during a lifetime of practice. Such cases were included because they illustrate basic issues such as disclosure of psychologically and socially sensitive information.

Questions were clinical rather than research-oriented. For the most part, we tried to avoid hypothetical or futuristic situations. Professionals dislike answering such questions, because they know that when a test or treatment becomes a reality, it almost inevitably has drawbacks in terms of accuracy, predictive value, generalizability across ethnic groups, or cost. Nevertheless, we did include a few questions on behavioral genetics. In all, the questionnaire had 50 questions on ethics with a total of 491 subparts.

Looking back, there are some questions we wish we had included, notably about research. The course of science is difficult to predict. At the urging of our sponsor at the National Institutes of Health (NIH), we were prepared to add questions on prenatal diagnosis through isolation of fetal cells from maternal blood, a technique thought to be in the near future for the last 10 years, but which is still almost as far from clinical practice as in 1990. On the other hand, no one would have guessed that cloning, a supposedly "impossible" technique, would suddenly come to the forefront of ethical discussion, or that stem cells taken from human blastocysts (early pre-embryos) would be used in therapy. Research on banks of stored human blood or tissue samples – a non-issue several years ago – has come to the forefront of bioethical discussion. Some researchers discuss germline gene therapy as inevitable (Silver, 1998). In retrospect, it would have been useful to include questions on these topics, but these must await another survey.

In the United States, the geneticist questionnaires were part of a set of related surveys for primary care physicians, genetics patients, and the general public (Table 2). The geneticist questionnaire served as the template. The primary care physician survey translated some genetic terms into more common medical parlance (e.g., 45, X became Turner syndrome) and omitted a few situations that would not be seen by a nonspecialist. It added a series of questions on doctors' knowledge about the functional (daily life) aspects of relatively common genetic conditions such as Down syndrome or cystic fibrosis.

The patient questionnaire translated selected questions (about two-thirds of the total) from the geneticist questionnaire into lay terms. These questionnaires were administered both before and after counseling in order to see whether people's ethical views changed after counseling (they didn't). They also included sections on what was and was not discussed in counseling, and whether the session met the respondent's expectations.

The public survey drew 12 greatly simplified questions, with subparts, from the patient questionnaire, the maximum that could be accommodated as part of a larger "Limobus" survey on a variety of other issues conducted by Roper-Starch Worldwide, a US commercial survey firm.

During development, all questionnaires received input from genetics professionals, parents, genetic support groups, and primary care physicians. All were field-tested three times on up to nine people (the maximum allowed under US Office of Management and Budget regulations). Unfortunately, it was not feasible to test outside Canada and the US, though our colleagues in each nation offered comments. The geneticist and primary care physician questionnaires took, on average, one to two hours to complete, the patient questionnaire one-half hour, and the public questionnaires 15 minutes.

Sample Populations

We included all nations known to have at least ten practicing medical geneticists, 36 nations in all. Large parts of the world do not meet this criterion, notably sub-Saharan Africa (with the exception of South Africa) and most of the Muslim world. Our attempts to include those Muslim nations with active genetics professions failed, either because of strict censorship (Saudi Arabia), political turmoil (Iran), or low response rate (Egypt), leaving Turkey as the only, and possibly atypical, Muslim representative. On the other hand, we were able to include 18 nations not represented in the 1984–86 survey, including China, Thailand, Russia, Poland, the Czech Republic, Belgium, The Netherlands, Portugal, Spain, Finland, South Africa, and seven additional nations in Latin America (Argentina, Chile, Colombia, Cuba, Mexico, Peru, and Venezuela). New Zealand, though widely known for its pioneering work with embryos, actually has a genetics community too small for inclusion. In contrast, Costa Rica has by now reached the minimum ten for inclusion.

In each nation except the US, we asked a well-respected geneticist colleague to develop a list of currently practicing medical geneticists. (In China and Sweden one colleague was a philosopher. In Denmark and Germany, one colleague was a social scientist specializing in social issues in genetics.) Determining who should be on the list to receive a survey was sometimes difficult. Most countries (except the US, Canada, UK, Germany, and Hungary) did not have specialty boards. Most countries do have genetics professional societies with membership lists. However, these may include researchers as well as clinicians, and sometimes do not include important types of practitioners. Our colleagues therefore had to use their own judgment in determining who was actually providing genetic services or acting as expert administrators of such services. Ultimately the samples they selected were not homogeneous. The medical profession is structured differently in different nations. For example, in Poland, midwives do most of the prenatal care, including the prenatal diagnosis counseling. Their inclusion explains what seems like an overly large sample of Polish geneticists (Table 3). In South Africa and the United Kingdom, genetic nurses play a role. In France, most geneticists are pediatricians; prenatal diagnosis is done by another specialty group. In Germany, professional society lists made it virtually impossible to separate scientific researchers from medical practitioners, though our colleagues attempted to do so. Our Argentine colleague used a narrow definition, including mostly well-known specialists and leading to underrepresentation of Argentina's large genetic community. In the US and Canada a separate professional group called genetic counselors is responsible for much of the person-to-person interaction. In the United States, counselors receive a Master's degree after two years of scientific, counseling, and clinical training in one of 25 genetic counseling programs. They may then take a specialty board examination (formerly from the American Board of Medical Genetics, since

1993 from their own independent American Board of Genetic Counseling). In Canada, most genetic counselors are nurses, with their own professional association. In the UK, the training program is part-time, designed for people already working in the field, mostly genetic nurses. Australia and South Africa are also training genetic counselors. Counselors were included in the sample in all five countries.

In the United States, we included all persons certified by the American Board of Medical Genetics as of 1992 (the most recent list available at the time). The National Society of Genetic Counselors provided an additional list of members eligible for and awaiting Board certification. In Canada, our colleagues used the membership lists of the Canadian College of Medical Geneticists and the Canadian Association of Genetic Counselors.

Primary care physicians were randomly sampled though the American Medical Association's Physician Masterfile, which includes all licensed physicians, regardless of whether they are members of the AMA. Using our criteria that participants be certified by their specialty board and currently practicing, the AMA selected 300 pediatricians, 300 obstetricians, and 300 family practitioners. (Knowing what we know now about the inadequate knowledge of these groups, we realize that we should also have included nonboard-certified physicians, who comprise 25–40% of the total.)

The patient surveys took place at 12 general genetics (not prenatal diagnosis) clinics in the US and two in Canada. Our contacts at each clinic were asked to include all adults (including parents of minor children) visiting the clinic for the first time. Limitation to "new" patients was intended to avoid possible prior effects of genetic counseling or people's ethical views. Questionnaires were provided in Spanish as well as English. Persons unable to read either language were excluded.

The public survey was an in-person quota survey of 1000 members of the adult US public, conducted door-to-door by Roper-Starch Worldwide.

In Germany, our colleagues, Irmgard Nippert (University of Muenster) and Gerhard Wolff (University of Freiburg) surveyed patients at two clinics, using a German translation of the US patient questionnaire. The German version was shorter and was given out only before counseling. In France, our colleague Ségoléne Aymé distributed a French translation of the entire questionnaire to patients at five clinics, before counseling.

Administration of Questionnaires

Our international colleagues were asked to use three waves of mailings: a questionnaire, followed by a reminder, followed by a second questionnaire. Postcards, returned separately from the questionnaire, enabled them to monitor who had already returned the questionnaire. In the US, UK, Germany, and possibly some other nations, a telephone follow-up completed attempts to raise response rates.

Telephone follow-ups proved impractical for the primary care physicians, who were almost impossible to reach. Instead, we offered payment of $25 on receipt of a form that they returned separately from the questionnaire.

Patients received the questionnaires by mail from the clinic prior to their appointment. They were asked to fill them out and bring them to their appointment, at which time they were asked to place them in sealed envelopes addressed to us, and to put them in a special box. Clinic staff then mailed the envelopes. Patients were given post-counseling questionnaires matched by number to the pre-counseling questionnaires (with clinic personnel holding the code, which was not available to us, the researchers) and a stamped envelope addressed to us at the Shriver Center. After we received a batch of ten post-counseling questionnaires we sent the numbers to the clinic, which either paid people $25 directly or sent us a list of names and addresses (not matched by numbers) so that we could pay them.

In order to make the public survey comparable to the other surveys, the interviewer handed out printed questionnaire booklets and waited while people completed them. This preserved a measure of anonymity on sensitive questions.

Data Analysis

Data were entered into a Statistical Package for the Social Sciences (SPSSX) program on the Boston University Computer. We chose SPSS because of its labeling capabilities and easily readable print-out for distribution to our international colleagues. We did cross-tabulations of each question by country, age, sex, attendance at religious services, and in the US and Canada by type of degree (Master's-level genetic counselor, M.D., or Ph.D.). Multivariate analyses such as stepwise logistic regressions or factor analyses are not reported here, as they tend to become rather complex and are usually of interest only within particular countries.

We applied a quantitative shorthand for data entry of respondents' qualitative (open-ended) explanations of why they had chosen particular courses of action, using an expanded version of the codebook of ethical reasoning that we employed in 1985 (Wertz and Fletcher, 1989).

Respondents' Characteristics

Response rates (Tables 2 and 3) were generally good, especially in view of the surveys' lengths. There were a few exceptions, such as Cuba, and some countries where a major center refused to participate (e.g., Belgium, which is missing the French-speaking University of Louvain). The numbers asked to participate, in Tables 2 and 3, are based on persons who are presumed to have received the questionnaires. Even recent lists had substantial numbers who had moved without a forwarding address, died, or left the field, thus reducing the original samples considerably.

Respondents' professional characteristics appear in Table 4. Although the majority were M.D.'s, they represented a smaller portion of respondents (56%) than in 1985 (81%), probably because of the sizable representation of US genetic counselors, who comprised over half the US sample (555 out of 1084). One-quarter had received training outside their own countries, usually in the US, UK, Canada, Netherlands, Germany, Japan, or Russia. This was a somewhat younger population than in 1985, both in age (42 vs. 45 years, Table 5) and in years of experience (9 vs. 14 years, Table 4). This means that over half the respondents were not in the field when we did the initial survey ten years previously. Most worked in large cities or suburbs of large cities, though some practiced in small cities or towns (15%), rural areas (1%), or entire regions (2%). Most practiced in university (39%) or hospital (38%) settings; a few worked in commercial labs (4%) or office practice (4%). Most (83%) saw patients. Outside the US, the median number of patients per week (1 to 5) was the same as in the 1985 survey, but in the US the median had increased to 6 to 10, probably because of the high representation of counselors. In the US, Master's level counselors, who earned $35,000–40,000 a year, as compared to over $100,000 for physicians, brought down the median income. In many countries, especially in Eastern Europe, Latin America, and Asia, geneticists' salaries are appallingly low by the standards of the developed world, approximating the income of a waiter or waitress in the US. Those who did prenatal diagnosis – a lucrative field in many countries – earned considerably more than their colleagues.

Respondents' personal characteristics appear in Table 5. Far more (62%) were women than in 1985 (35%). Women comprised more than 50% in 22 nations. In some nations, including most of the Middle East, medicine is considered a good career for women and an extension of women's "helping" role. The pay is too low to be attractive to most men, who prefer to go into business. There are exceptions, however. In Japan, genetics is largely a man's world. There was also a preponderance of men in Australia, Sweden, and Switzerland. Most US Master's-level counselors (95%) were women.

Women MDs reported fewer years in practice (9 years), were less likely to have a university affiliation, and spent more time in direct patient care and (in the US) less time in research than men. Fewer women were married (73% vs. 87% for men) or had children (87% vs. 95% for men). In the US, Master's-level counselors' marital status (64% married) and number of children (59% had children) reflect their younger median age (34 years). Differences in income are particularly striking. A consistent income gap between men and women remains, regardless of years in the profession. In the US, 50% of women M.D.s, and 71% of men had incomes over $100,000. Outside the US, 87% of women and 42% of men M.D.s earned less than $50,000. There were no gender differences in frequency of attendance at religious services, importance of religion in life, personal experience with disability, or experience with patient support groups.

Switzerland reported the oldest median age (55), followed by Israel (53), China (50), and Sweden (50). In the Netherlands, Norway, Portugal, Hungary, Turkey, Cuba, and Thailand, the median age was 40 or below. Most were married and had two or more children, including 49% in China (many perhaps born before implementation of the "one-child policy"). In Mexico, which has implemented family limitation programs, one of our collaborators was proud that geneticists have set an example, with only 34% reporting more than three children.

Although close to half said that religion was at least "moderately important" in their lives, this appeared to be a private, personal interpretation of religion, rather than adherence to organized religion. As in the 1985 survey, the median number of religious observances attended per year was one to two. There were exceptions. In Chile, where there is a strong Catholic influence on the genetics profession, respondents attended mass almost once a week. In South Africa, Spain, Poland, and Peru, they attended services once a month, and in the US, Greece, Italy, Argentina, Mexico, and Thailand about five times a year. In the US, France, and India, which had the highest rates in 1985, attendance had dropped markedly. Geneticists as a whole appeared to be a nonobservant lot, though many had inner religious feelings.

About two-thirds reported personal experience with disability, outside their professional roles. Usually the disability was in a relative. A minority (about 30%) had had some experience with genetic support groups, and fewer still (about 13%) said that they had had “a lot” or “very extensive” experience. The low figures for the United States, which reported less experience than in many other countries, are surprising given the large numbers of genetic counselors and the counseling profession’s avowed standards of working closely with genetic support organizations. Many counselor training programs still provide an inadequate introduction to life with disability (Brown and Wertz, 1999), though training is improving. The last column in Table 5 represents only the two-thirds of respondents who reported personal experience with disability. This means that among the entire sample, only 20% had some experience with support groups and about 8–9% (5% in the US) had a lot of experience. In many countries, support groups are just beginning, but they will likely become a worldwide trend. Since we completed the survey, support groups for families with Down syndrome have appeared in Japan and Hong Kong, and a group for families with mental retardation has begun in India. The newly-formed World Federation for the Prevention of Birth Defects, which works with the US-based March of Dimes, is acting as an international umbrella organization for support groups. Except in North America and Western Europe, however, most groups are still too new and too small to have political force.

The three groups of US primary care physicians (pediatricians, obstetricians, and family practitioners) responded almost equally. Unlike the geneticists, the majority (72%) were men. They had been in practice a median of 14 years and had a median age of 44. They attended religious services more frequently than US geneticists, about once a month as compared with five times a year, and earned more money, a median of $100,000–150,000. Seventeen percent were foreign medical graduates, and 85% of the total were white. Two-thirds were in office practice, including 17% in solo practice. Unlike geneticists, some practiced in small towns (27%) or rural areas (10%). Three-quarters said that they had referred at least one patient to a genetic specialist during the last year, but only 8% had had experience with genetic support groups.

Patients’ characteristics appear in Table 6. Most North American “patients” were actually parents of small children (median age 5) seeking diagnosis, evaluation, or counseling about whether a subsequent child might have the same condition or seeking information about “tests on an unborn baby.” About 15% were having prenatal diagnosis only for advanced maternal age (35 or over). Their most common reason for coming to a counseling clinic, just as in a survey of 1369 US counseling sessions conducted 20 years ago (Sorenson, Swazey, and Scotch, 1981), was “my doctor told me to come.” Most were women, white, and visiting US (not Canadian) clinics. The sample differs in important respects from the college-educated, experienced consumers sometimes reported in studies of “genetic discrimination” (Lapham, 1997). Ours was basically a working-class sample, with most women and their husbands employed in clerical, sales, service, or factory production work. Their median family income of $25,000–45,000 is close to the US median ($44,000 for a family of four), and their median education of 13 years represented one year of college, business, or technical training after high school. They attended church considerably more often than US geneticists and included a larger percentage of Catholics (44% vs. 19%) and a smaller percentage of Jews (3% vs. 22%). Most of the clinics they visited were in older industrial cities of the Northeast or Midwest, sometimes called the “Rust Belt.”

Respondents to the public survey were almost equally divided between men and women; they included 11% who were African–Americans, barely enough to make some general comparisons between their views and those of whites (Wertz, 1999). In the data analysis, respondents were divided into regions (Northeast, West, Midwest, and South) and education and income groups.

In Germany, 906 patients at two genetics centers (Muenster and Freiburg) received questionnaires; 593 (65%) returned completed questionnaires. Of respondents, 90% were women, with a median age of 33 years; 97% were German nationals, 79% were married; 47% had no children; 28% had one child, 18% had two, and 6% more than two. Twenty-one percent lived in a large city or suburb, 63% in a medium-sized or small town, and 16% in a rural area. They included both general genetics patients and prenatal patients.

Among 773 French patients asked to respond, 394 (51%) returned questionnaires. Most were women (88%), with a median age of 32, married (68%), with 12 years (56%) or 16 years (36%) education, Catholic (71%), who attended church less than once a year (79%), an income of $5,001–25,000 (30%) or $25,001–45,000 (60%), residing in large metropolitan areas (42%), small cities (30%) or rural areas (28%). Five percent had no children, 31% had one, 32% two, and 32% more than two. Most had come for evaluation of a child or information on prenatal diagnosis.

Although the rest of this book focuses on the international survey of geneticists, we have included these sociodemographic data on respondents to all surveys so that readers will be able to characterize the people behind the numbers in forthcoming tables.

Limitations of the Survey Approach

Surveys have some disadvantages. There is no way of proving that people actually do what they say they would do, unless one were to observe their actions on a day-to-day basis, hardly a practical method of validating large-scale research. Especially when a respondent has not yet encountered a particular situation in real life, as was the case for many of our respondents, it is extremely difficult to predict what that individual will do when faced with the actual situation. Prediction of individual actions was not the intention of the survey, however. The point was to provide a general impression of how genetics specialists in a country think about an ethical situation. For example, survey respondents overwhelmingly rejected forced hospitalization of a pregnant woman with a genetic condition called phenylketonuria (PKU), whose refusal to stay on a special diet is endangering her fetus. What the survey does is to document both strong moral opposition to medical control of a pregnant woman and strong support for women's individual autonomy among genetics professionals, the group who should be most familiar with the effects of maternal PKU. Documentation of this professional view could provide guidance for non-specialist physicians, legal experts, and policymakers.

Is there any way of judging how people's actions might differ from their survey responses? Perhaps. In some cases, there are general norms in genetic practice, for example, "nondirectiveness" in counseling. The ideal of nondirectiveness combines provision of unbiased information, refusal to tell people what to do, and empathy for the person being counseled. In North America especially, genetics professionals try to adhere to this goal. If their survey answers are biased, the bias is more likely to be toward than away from the professional norm. In other words, counselors may be more directive in practice than they say they would. It is unlikely that they would be less directive than they say they would, since being nondirective is the professional norm. Similarly, maintaining individual confidentiality – even from family members at genetic risk who might benefit from the information – has been the traditional norm in medicine in North America and Western Europe. It is possible that more respondents would actually tell family members against the wishes of their patients than say they would. Despite possible divergences from actual practice, the tables in the next seven chapters give a general estimate of people's views and provide what we intended – a forum for discussion.

Summary of Results

The international survey's major findings – as summarized in a 10-minute presentation – appear in Table 7. In the following seven chapters, we will describe these under the major headings of Equity, Autonomy, Directiveness in Counseling, Eugenics, Family Relationships, Privacy, and Ethical Reasoning. These chapters are descriptive and make no judgements about what ought to be done. We have reserved the "shoulds" for the final chapter.

GROUPS OF NATIONS

The tables are presented in country-by-country detail because many readers will want to see how their country answered. The tables are also organized into regional or language groups. The designation "Eastern Europe" is political rather than geographical, and includes countries in the former Communist bloc, although some are in Central Europe. Sometimes groups of nations presented responses that differed, as a group, from other groups. This was often the case with family-related questions, such as access to genetic information for a spouse or blood relatives, or testing of minor children at parents' request. On the other hand, countries within a group were sometimes widely divergent with regard to counseling after prenatal diagnosis or perceptions of disability. Some countries stood out from their regional groups. For example, Japan presented Western-like responses in regard to patient autonomy, in contrast to China, India, and Thailand. Japanese geneticists have long monitored Western ethical developments and incorporated them into professional society statements, for example on genetic testing and prenatal diagnosis, and have sometimes added original and useful provisions, such as requiring approval of an ethics board before an individual's desire for confidentiality can be overridden in order to warn relatives at genetic risk.

In Latin America, Cuba stands out for its acceptance of abortion and its pessimistic view of disability, a view that comes close to that of the other socialist nation in the survey, China. At the other end of the spectrum, Chile stands out for its anti-abortion view, affected by strong Catholic beliefs.

In Eastern Europe, Poland stands out from its formerly socialist neighbors on abortion issues. About half of the Polish respondents had some anti-abortion sentiments, while the other half, in their written comments, opposed the resurgence of the Catholic Church.

In Southern Europe, Spain, with its greater Catholic influence in the profession, was more hesitant about abortion than its neighbor, Portugal, even though prenatal diagnosis has a longer history in Spain.

Belgium and Greece stand out as the European nations most likely to offer pessimistic counseling about disability. Our Greek collaborator thought that everyone else counseled this way until informed otherwise. (See his explanation in Chapter 9, "Geneticists Speak Out.")

Two of our groupings are somewhat arbitrary. We included South Africa under English-speaking nations because there were no other African nations in the survey and also because the genetics professionals there are largely of English-speaking background and English-trained. Although their responses often differ from others in the English-speaking group, notably in being less autonomy-oriented, geneticists' background moral traditions and culture are probably closer to the English-speaking than to any other group. Israel and Turkey, though both in the Near East, do not constitute a culture group and are placed together largely for ease of identification for readers who do not want to run through 36 nations alphabetically.

The responses of US primary care physicians and public are placed after responses of US geneticists, for easy comparison. Describing these is the task of another book. Meanwhile, readers should notice that patients and genetics professionals in the US often saw issues very differently. While genetic professionals emphasized privacy and nondirectiveness, patients were more interested in getting help for their families, whatever that required. The views of primary care physicians were often closer to the patients' views than to those of genetics professionals.

We have also included German and French patients' responses, immediately below German and French professionals. Readers will notice considerable similarity between North American, French and German patients. The totals in the tables include only the genetics professionals. They do not include patients or primary care physicians.

CHAPTER 2

GLOBAL INEQUITY

The basic underlying ethical issue in genetics – as in medicine generally – is equity. Until or unless affordable genetically-engineered vaccines for common tropical diseases are developed, genetics will be a specialty serving the haves rather than the have-nots. There are disparities both between countries and within countries. Table 8 displays some economic and social characteristics of nations in the survey. Some nations, such as Switzerland ($2252) spend large sums per capita on public health; others, such as China ($10), India ($3), Colombia ($12), Peru ($17), and Thailand ($20) spend very little. In China, children with cleft lip or cleft palate are rarely treated. In the absence of government health insurance (which now covers only government employees), few parents can afford an operation. Entire orphanages are filled mostly with children with cleft lip/palate, mostly girls, the undesired sex. In India, children with thalassemia, a blood disease, are not treated, unless the parents are wealthy enough to pay the considerable expenses out-of-pocket. Prospective parents are told, after prenatal diagnosis, that they will have a sick child for about ten years before inevitable death. It is hardly surprising, under these economic circumstances, that counseling about what Westerners consider successfully treatable conditions is pessimistic, often directively so.

Population per physician (Table 8, col. 3) is not always a good indicator of medical quality. Western European nations average between 300 and 400, but the UK, with excellent services, has 639. When a nation has only one physician for several thousand people, however, as is the case in Thailand (4,432), India (2,563), and South Africa (1,875), one may assume that many receive little medical care. When physicians in general are rare, genetic services would be like the frosting on the cake, a luxury received by few. Russia (220) and Cuba (255), which has long loaned its surplus physicians to other developing nations, have the lowest physician-population ratios.

Nations with the highest infant mortality under one year, India (77 per 1000), Peru (66 per 1000), Brazil (51 per 1000), and South Africa (48 per 1000) have five to 16 times the mortality rate of nations with the lowest rates, Finland, Sweden, and Japan (all with 5 per 1,000). Life expectancy at birth is highest in Japan, at 79 years. It hovers between 75 and 77 years in most English-speaking nations, Western and Southern Europe, and Israel, and is several years less in much of Eastern Europe and Latin America. It is lowest in India (60) and South Africa (63). Late-onset genetic disorders such as Huntington disease, a progressively incapacitating neurological disorder that usually appears in the forties and leads to death in the mid-fifties, may be seen as less devastating in a nation with a life expectancy of 60 than in nations where people live into their late seventies.

Literacy rates – especially women's literacy rates, because women are responsible for their children's health care – are an important predictor of health. Literate people are better able to interact with physicians on a more equal basis, and are more likely to ask questions or volunteer important information. Simply educating women – not necessarily about health – has been shown to increase their children's life expectancy. Women's and men's literacy rates are shown separately in Table 8. Although rates are equal, and also high, in most nations, in some countries women's literacy is not only low, but considerably lower than men's. India (38%), China (73%), and Turkey (72%) stand out.

Contraceptive use never comes close to 100%, but in most nations is at least 50%. In English-speaking nations and Europe, it is about 75–80%. Norway (84%) and China (83%) have the highest rates of use; Russia (32%), India (43%), and Chile (43%) have the lowest.

A rough ranking of socioeconomic standing among the world's nations, based on gross national product per capita, education, health expenditures, infant mortality, and life expectancy, appears in the far right-hand column of Table 8. Nations in the survey ranged from rank "1" for Switzerland to 125 for India.

Genetics, like other medical services, is unequally distributed around the world. In Table 11, we have tried to estimate the population per geneticist in each survey country by dividing the number of geneticists asked to participate in the survey by the country's population. This table is divided into developed, Eastern European, and developing nations. Genetics is a specialty of the developed world, with one geneticist per 233,372 people, as compared with the developing world, with one per 2,704,470. Cuba, which has an oversupply of doctors, was an exception. The 122 (mostly developing) nations not in the survey, because they had fewer than ten geneticists, accounted for 2,038,980,000 people, about one-third of the world's population. (Sivard, 1996).

Genetic services are also unequally distributed within nations. In Sweden, with national health insurance, good transportation, and 100% literacy, rural people utilize fewer services than urban dwellers. In the United Kingdom, with similar social parameters, people in the lower socioeconomic groups are less likely to receive or utilize prenatal and other services than people in the higher groups (Harris, 1998). In the United States, members of some minority groups receive fewer services than the white majority. All of our collaborators reported similar inequities in their countries. In many developing nations, genetic services are almost a luxury, limited to the few who can afford them.

We asked respondents whether the highest quality services were available to all, regardless of ability to pay. Surprisingly, majorities said yes, except in the United States, Turkey, Argentina, Brazil, Chile, Colombia, Peru, China, India, and Thailand (Table 9). Their answers to the next survey question stand in direct contradiction. In most countries, except Canada, United Kingdom, United States, Denmark, Finland, Czech Republic, and Cuba, geneticists estimated that fewer than 50% of those who needed them actually received genetic services in 1995. In eight countries, estimates were 15% or fewer (Table 9). Geneticists in all countries, except South Africa, which is in political turmoil, estimated that the percent receiving needed services would increase in ten years (2005). For some countries, even the increased figure is appallingly low, for example 10% in Peru and 15% in Thailand.

Geneticists' rankings of a list of priorities in health care budgets were generally similar around the world (Table 10). Very sensibly, they gave top ranking to the measures most likely to help the most people: prenatal care and sex education, including contraception. Measures likely to help the few, such as gene therapy and research and treatment for infertility, ranked at the bottom. As in the 1985 survey, genetic testing for susceptibility to occupational disease also received a low ranking (9). In contrast, patients (Table 10, col. 2) gave gene therapy a high ranking (3); they ranked carrier testing higher, and public education about genetics lower than did geneticists, but, like geneticists, put prenatal care and sex education at the top and research and treatment for infertility close to the bottom of the scale.

Given the vast economic inequalities among nations, and the costs of most genetic services, it appears that the benefits of genetics may accrue mostly to well-off people in rich nations. The increasing role of commercial entities, which have to make profits for their stockholders, may increase these inequalities. Since 1992–93, private companies have spent more than governments on genetic research (Cook-Deegan *et al.*, 2001). Companies correctly believe that they are most likely to recoup their considerable costs by selling products in North America, Europe, Australia, and Japan, rather than in other areas of the world. This could lead to steering research toward the needs of the most profitable areas. Patenting is a potential further source of inequality. Although some nations, including India and China, have not recognized foreign patents until recently and still may not enforce them, some genetic technologies (such as testing for breast cancer mutations) are so expensive that scientists in developing nations may choose not to set up their own labs.

Inequities in genetic services are part of a matrix of inequitable health care, and do not differ from other kinds of health inequities. Genetics affects future generations. However, environment also has generational effects, through a weakened or malnourished physiology that causes low birth weight or prematurity. Generational studies in the United Kingdom have shown that women who grew up poor in the Great Depression of the 1930s but subsequently joined the middle class nevertheless had more miscarriages, stillbirths, and low birthweight babies than women who were middle-class from birth (Baird, 1970). There are indications that the effects of poverty may continue into a third generation. Environmental inequalities amplify genetic inequalities; in the developing world, infections, malnutrition, toxic exposures, and absence of basic health care or services for people with disabilities all make life more difficult for people with genetic conditions.

Inequalities of access to contraception or safe abortion severely limit choices of families at risk for genetic disorders. Contraceptive use is about 54% in developing nations as compared with 72% in developed nations (UNICEF, 1988). Half of the 45 million abortions worldwide each year are unsafe (illegal), and 90% of these occur in developing nations (United Nations Population Fund, 1998). In Latin America, where abortion is illegal except to save the mother's life or for rape or incest, there are 41 unsafe abortions per 1000 women each year, as compared with 2 per 1000 in Europe (Allan Guttmacher Institute, 1994). Although abortions after prenatal diagnosis account for a small percentage of overall abortions (possibly 1%, according to records in a few European countries), there are nations in the developing world where access to a safe, (although illegal) abortion of a fetus with a genetic condition is limited to those with financial means. Limited access to contraception means that many women in developing countries continue to have children well into their forties, an age at which the incidence of Down syndrome increases markedly. The percent of all births to women over 35 averages 11–15% in developing nations as compared with 9% in developed nations (United Nations Population Fund, 1998). Although when adjusted for maternal age, the birth rates of children with Down syndrome are similar worldwide, the older ages at which women give birth in developing nations mean a higher birth prevalence of children with Down syndrome (possibly 3 per 1000, as compared 1 per 1000 in developed nations) (Penchaszadeh, 2000). Parents in these countries are cheated in three ways: unequal access to reliable contraceptives, unequal access to prenatal diagnosis and safe abortion, and inadequate health and education services for children with Down syndrome. There is evidence that most people, worldwide, will use genetics services to avoid the birth of a child with a genetic condition, if such services are available (Penchaszadeh, 2000). In countries where abortion is legal, including both developed and developing nations, prenatal diagnosis is usually regarded as the most practical current use of genetic services (Harris, 1997; Penchaszadeh, 2000).

Hemoglobinopathies (thalassemia and sickle cell anemia) pose the largest health problems worldwide caused by single genes, especially in Africa, India, the Middle East, the Mediterranean, and Southeast Asia, and among descendants of people from these areas. About 250,000 children are born each year with a major hemoglobinopathy, 95% of them in the developing world. As a whole, single-gene disorders have the same prevalence (3.5 per 1000 births) worldwide. However, some conditions that are common in Europe and North America, such as cystic fibrosis and PKU, are rare in developing nations, especially in Asia. In many developing nations, the medical profession does not take genetic services seriously, because these services are perceived as expensive, applicable only to rare diseases, diagnostic rather than therapeutic, and associated with abortion.

Global patterns of disease are changing, and with these changes genetics will assume a greater role. Already, noncommunicable diseases – almost all of which have a genetic component – far surpass infections, maternal and perinatal conditions, and nutritional deficiencies combined as causes of death in Latin America, the Caribbean, China, the Middle East, and Southeast Asia (Murray and Lopez, 1996). Only in India (51%) and sub-Saharan Africa (65%) do infections and nutrition still exceed noncommunnicable diseases such as heart disease, cancer, and cerebrovascular disease as causes of mortality. Heart disease is now the leading cause of death worldwide, followed by cerebrovascular disease. (The most common fatal cancer, lung cancer, rates as number 10.) Heart disease, vascular disease, and cancer are all genetically influenced. Future strategies for prevention and treatment will be based on genetic research.

At least six of the world's ten leading causes of disability, including alcoholism, chronic obstructive pulmonary disease (asthma, bronchitis, emphysema), bipolar disorder (manic depression), congenital anomalies, osteoarthritis, and schizophrenia, have some genetic components (Murray and Lopez, 1996).

By 2020, noncommunicable diseases are projected to account for 73% of deaths worldwide (up from 55% in 1990), with most of the increase occurring in the developing world. The aging population and increasing use of tobacco will be major factors in this increase (Murray and Lopez, 1996). Surely developing nations have as great a need for genetic services, merged into general health care, as developed nations. By 2020, more people worldwide will probably have access to genetic services, as predicted by our respondents, but in many countries this may still be only a fraction of the population, usually the well-educated and well-off. By then, genetic services may have a broader range that encompasses the common multifactorial diseases (caused by interactions of several genes plus environment and health behavior) of adulthood. In the poorer nations especially, the preferred approach to single-gene disorders will be prevention, through counseling, contraception, prenatal diagnosis and abortion, rather than treatment or support that the country's health system cannot afford. Treatment could become a luxury, affordable mainly for industrialized nations. Greater utilization of genetic services, including prenatal diagnosis, by well-off, educated people in developed nations to avoid births of children with genetic conditions could further increase health inequities between rich and poor in these nations. Geneticist Lee Silver (1997) envisages a special class of people who have availed themselves of all genetic services, including future genetic therapies for enhancement, whom he calls the "Generich." Current data on global inequalities tend to support this futuristic vision. Unless governments, international organizations, and corporations take drastic actions, genetic inequalities will remain for generations to come.

CHAPTER 3

THE RISE OF AUTONOMY

In the context of genetics, autonomy usually involves appeals to human rights: rights to know or not know information, rights to respect, rights to decide, rights to services, rights to referral, or rights to privacy. Among those with means in rich nations, autonomy may also mean frivolous consumerism, as people pick and choose among medical services of dubious necessity, or demand access to any service they desire and can pay for out of pocket. Originally the word meant independence or self-governance for a political entity, such as a city-state in ancient Greece (Gouldner, 1965). In countries influenced by the eighteenth century Enlightenment, the concept of autonomy became attached to individuals. Now the word autonomy applies mainly to individuals as separate, independent entities who own their genetic information and may even seek a share in intellectual property rights derived from it.

Philosopher Gerald Dworkin (1988) describes different kinds of autonomy. Ideally, autonomy might mean actions by educated people who are fully aware of all alternatives in a situation, the outcomes of these alternatives, and the effects of their actions on others, on society, and on the moral order. Autonomy would include not only awareness but the ability to actualize each of the various alternatives that people find acceptable without hindrance from economics, politics, religion, discrimination, coercion by medical professionals, or other factors external to the individual. These conditions rarely, if ever occur. Many autonomous actions take place in the absence of full awareness and therefore may occasion harm to the individual or others. One factor of which many people are unaware is the effect of their actions on others. As our French colleagues in the survey said, paraphrasing Diderot, "My rights end where the rights of others begin."

Nevertheless, even imperfectly informed autonomy is an advance in the interactions between professionals and the laypeople they serve. Our survey explored the different dimensions of autonomy and some possible trends in nations also surveyed in 1985. The concept of autonomy has reached its fullest development in the United States, which was born during the Enlightenment and still reflects it, perhaps to greater degree than any other nation. In some parts of the world, especially East Asia, the concept is still new and untried. The dominance of the principle of autonomy in bioethics, especially in a Western context, has been widely criticized in recent years. We discuss some of this criticism in the context of genetics in Chapter 20.

Not surprisingly, the United States is the world leader in autonomy, at least according to responses to the questionnaire. Individual autonomy triumphs over other concerns such as social justice or family needs. Writing in 1835, Alexis de Tocqueville pointed to Americans' belief is that individuals should have control over their own interests and that society cannot, and should not, try to interpret the interests of individuals. De Toqueville wrote, "Every individual, private person, society, community, or nation, is the only lawful judge of its own interest, and, provided it does not harm the interests of others, nobody has the right to interfere. I think that one must never lose sight of this point." "Another principle of American society of which one must never lose sight [is that] every individual being the most competent judge of his own interest, society must not carry its solicitude on his behalf too far, for fear that in the end he might come to count on society, and so a duty might be laid on society which it is incapable of performing ..." (Schleifer 1980, pp. 124–125).

The US was not alone in preferring autonomy in responses to the survey. Russia, and to a lesser extent, Israel and some other nations appear almost as autonomy-oriented as the US in regard to some questions. Yet in Russia there is no political history of individual rights and no word for "privacy." Possible explanations are the near total breakdown of the public health infrastructure in Russia, leaving the door open to unrestricted for-profit medicine, and the absence of any discussion of ethical issues pertaining to genetics. In this void, some answers to the survey may be insufficiently considered and may reflect an "anything goes" situation rather than an ethics-based respect for autonomy.

At the other end of the spectrum, responses from China and India reflect the least autonomy. In China, a Confucian-based ethic prevails; individuals become human only as members of family and society, not as entities in themselves. Individuals owe something to those who helped them to become human. The Western concept of "rights" does not translate well into Chinese, although the word appears in English translations of some Chinese laws. As in the rest of Asia, people expect the doctor to tell them what to do rather than helping them to come to their own decisions. In India, paternalistic medicine prevails, possibly exaggerated in a country where 34% of men and 62% of women cannot read and physicians may consider it their duty to direct an "ignorant" populace.

Japan stands out among other Asian nations in its Western-appearing approach. The Japan Society of Human Genetics keeps its eyes open to the latest American Society of Human Genetics "Points to Consider" statements on particular ethical issues and usually makes similar statements, refining, adding to, and adapting them to the situation in Japan. Historically, Japan was the first Asian country to adopt Western technology on a large-scale basis, without forsaking its own religions and culture. It now appears to be adopting Western medical ethics, at least in formal approaches and professional statements.

In Latin America, Cuba stands out as less autonomy-oriented than other nations. In their responses to some questions, Cubans were more similar to Chinese and Indian geneticists than to their fellow Latin Americans. We attribute this in part to a feeling of social solidarity in Cuba that continually reminds people that an individual's welfare is inextricably linked to a social whole; the welfare of society is the welfare of the individual, and individuals must not seek for more than their fair share. Bioethical discussion is developing in Cuba; there is a Bioethics Institute at Holguin, and there have been debates on the fair distribution of genetics services. The population is virtually 100% literate, but not oriented toward printed media and with no ready access to the Internet, so public discussion proceeds orally. Medical services are limited by unavailability of key reagents or equipment as a result of the American embargo; however, the services that exist are available to all persons equally, free of charge, a situation that exists perhaps nowhere else in the world. Under these conditions, social justice often takes precedence over individual autonomy.

Respondents in European nations hold varying views on autonomy. Most, especially in Northern and Western Europe where the Enlightenment arose, are generally strong supporters of autonomy. However, on some questions there is a sharp division between Northern/Western Europe and Eastern Europe or Southern Europe, with far less emphasis on the individual in the latter two groups of nations.

GENERAL ATTITUDES: RIGHTS TO SERVICES, RIGHTS TO DECIDE, RIGHTS TO KNOW, RIGHTS TO REFERRAL

Tables 12–14 report general attitudes toward various aspects of autonomy. The questions were expressed in general form, not as case vignettes. The first column of Table 12 shows some, but not overwhelming agreement with a statement that "withholding any service requested [by patients] is paternalistic." Paternalism has acquired a bad name in Western public discussions, as meaning arbitrary and authoritative medical pronouncements that are unresponsive to people's feelings and needs. However, there is an earlier meaning of paternalism as responsibly caring for people who do not have the intricate knowledge to make the most beneficial decisions for themselves. This type of paternalism is ethically acceptable in some situations and is widely applied in much of the world. Although national figures vary, slightly over half in many English-speaking and European nations, and in some Latin American nations favored not withholding requested services. US patients, as elsewhere in these three tables, were the most autonomy-oriented of all, with 69% agreeing that withholding services was paternalistic.

The question of whether "patients are entitled to any service they request and can pay for out-of-pocket" has portent for the future, as commercial companies play a greater role in genetic research and development and as private, fee-for-service medical systems emerge in countries with national public health insurance systems. Responses to this question (Table 12, col. 2) reflect the state of a country's health care system; the greater the role of private health care, the more likely respondents were to agree. Thus, 36% of US geneticists, but only 4% of UK geneticists agreed that patients were entitled to anything they could pay for. Perhaps the most telling figure in all the tables on autonomy is the US patients' responses; 59% said people were entitled to anything they asked for, as long as they could pay for it. This may be the wave of the future, as private companies advertise tests for breast cancer mutations, private health systems emerge alongside public health systems, and "patients" turn into "consumers." Most patients (80%) also said that prenatal diagnosis should be provided "regardless of its intended use." (Table 12, col. 3). Intended uses could have included sex selection and prenatal paternity testing, since these had already been raised earlier in the questionnaires.

The only country where more than 80% of geneticists agreed that prenatal diagnosis should be available regardless of use was France (86%); most nations hovered somewhere between 20% and 40%, with some very small percents in Hungary, Turkey,

Cuba, and Japan. The majority of geneticists apparently thought that provision of prenatal diagnosis should be tied to uses that they considered "responsible." A large majority in the US (85%) but a minority (46%) elsewhere thought women "should have an unqualified right to abortion before viability" (Table 12, col. 4). The low figure from Chile reflects strongly-held Catholic belief that human life begins at conception, a belief influencing Chilean responses to all questions on abortion. Elsewhere, the lack of consensus may be a reaction to the word "unqualified," with many respondents perhaps believing that there should be guidance or limits. A similar question about whether "a woman's decision about abortion should be her own, without intervention by anyone" (Table 14, col. 3) produced more agreement (55%) outside the US than the question in Table 12, but there was still wide variation.

There was universal agreement that it was the professionals' responsibility to offer a referral if they were "unwilling to perform a procedure for moral reasons" (Table 13, col. 1), but no consensus about whether referrals to other countries or states should be offered if a procedure were illegal (Table 13, col. 2). The low percents who would offer such referrals in Germany and Norway may reflect an unwillingness to circumvent the law in these countries. The high percents in many Latin American countries are in response to the total unavailability of legal abortion (except to save the mother's life or for rape or incest) in most of Latin America (except Cuba) and the consequent need either to find illegal abortions after prenatal diagnosis or to refer outside the country for women who can afford to travel. That relatively few in China would refer may reflect the financial and legal difficulties most people face in leaving the country.

Geneticists in most nations drew the line at referrals for sex selection (Table 13, col. 3). Except in the US (55%), Russia (60%), and Mexico (54%) most would not refer for this purpose, though half of US patients thought they should.

There was general but not overwhelming support for rights to know and not know, women's rights to decide about abortion (Table 14, cols. 1–3) and also for professionals supporting patient decisions with which the professional disagrees (Table 14, col. 4). In 28 nations, including 91% in the US and 76% in Canada, majorities thought parents should be told the sex of the fetus if they ask (Table 14, col. 1). The low percents in China, India, Japan, and Turkey who would give out this information are understandable in view of the preference for boys in these nations and its potential effects on society. The low percent in Germany may point to fear of abetting requests for sex selection from immigrants from countries that prefer boys.

Although patients strongly supported rights to know (92%), rights to decide about abortion (60%), and rights to support for their decisions (79%), far fewer (41%) thought that people had a right *not* to know their test results. In their write-in comments, patients said they could not understand why anyone would take a test without wanting to know the results. They thought that people who did not want to know should not take the test in the first place. To many patients, the concept of a right not to know appears to be new. This concept may gain acceptance over time.

That majorities in all but five nations believed counselors should support all patient decisions, including those with which they disagree, evidences almost worldwide acceptance of a basic tenet of autonomy: respect for people's rights to make "wrong" decisions.

Supporting patients' decisions has long been a preferred approach in medical genetics in Western nations. Starting with its first delineation of the ethics of counseling

(Fraser, 1974), genetics, perhaps unique among medical specialties, specified that patients' decisions must be respected. This degree of respect was possible because for many years, most decisions were about reproduction, an area of deeply-held personal values. As genetics enters adult medicine in regard to multifactorial diseases, geneticists' responses to patient decisions may become more similar to responses of other physicians. If successful treatments become available, it will become difficult to support a "wrong" decision. Meanwhile, the survey results suggest a strong respect for deeply-held human rights to decide. China, the major exception (with only 14% agreeing that all patient decisions should be supported), may eventually accept the majority view, as its new generation of "only children" produced by the "one-child" policy achieves professional leadership.

There was one limit to otherwise strong support for autonomy. Most people, including autonomy-oriented US patients, did not think people should have the right to choose their children's sex (Table 14, col. 5). Only 7% worldwide agreed. This attitude is congruent with the reluctance to approve preconception sex selection described in Table 22.

DISCLOSING SENSITIVE INFORMATION: THE WOMAN WITH XY CHROMSOMES

There was clear agreement favoring full disclosure of psychologically sensitive or ambiguous test results and of colleagues' disagreements about the meaning of test results (Table 15). In the first of two cases, we described a woman undergoing a work-up for amenorrhea (absence of menstruation).

> "Tests indicated that she has a 46, XY karyotype, and a diagnosis is made of complete androgen insensitivity syndrome (testicular feminization syndrome). You are asked to consider what you would tell patients." The two alternative answers were: 1. "Tell her that the reason for her amenorrhea is well understood, explain to her the biology of gender, and then explain to her that although she has a normal female phenotype, her chromosomes are XY." 2. "Give her the reasons for her amenorrhea without telling her about the XY chromosomes."

In 1985 (Table 15, col. 4), we did not specify her age but said that she was being examined for infertility. In 1995, because this condition usually becomes apparent in the teenage years, we specified ages 15, 18, and 25. At all three ages, except for a 15-year-old whose parents do not want her to know, the majority of geneticists would tell her about the XY chromosomes. Not only is her infertility untreatable because she lacks female reproductive organs, but she has vestigial male organs in her abdominal cavity that are likely to become cancerous and must be removed. This is extremely sensitive information in many situations, because to some people it connotes that she is not a full woman and it may hurt her chances of marriage. In 1985, respondents split almost evenly (51–49%) on this question, with 64% in the US, 68% in Canada, and 41% elsewhere saying they would disclose. That in 1995, 78% outside the US, 92% in the US, and majorities in all nations except Mexico would tell her reflects a growing confidence on the part of professionals that they can explain the situation with a minimum of psychosocial harm, and also growing adherence to an ethic of full disclosure. In all nations surveyed in 1985 (except Turkey) more would disclose in 1995 than in 1985.

DISCLOSING AMBIGUOUS TEST RESULTS

The ambiguous findings case described a fetus that by chorionic villus sampling:

> "has an apparently balanced de novo translocation in 5% of the cells. This is confirmed by amniocentesis. Level III ultrasound examination is negative. The parents decline PUBS (percutaneous umbilical blood sampling) because they consider the risk of the procedure too high. Your colleagues differ about the meaning of the test results."

In this case, there is a possible risk of mental retardation, but the test results are not clear. Most respondents would disclose the uncertainty (Table 15, col. 6) and would also disclose the fact that their colleagues disagreed about the meaning of the test results. We asked about colleague disagreement in a similar but not identical case in 1985. At that time, many respondents said disclosing colleague disagreement was unnecessary and gratuitous information or that it made the profession look bad. In most nations surveyed in 1985, more would disclose in 1995 on the basis that parents had a right to or needed the information.

SEX SELECTION

Requests for prenatal diagnosis for sex selection, in the absence of a sex-linked genetic condition, strain the limits of autonomy. Is selecting sex a logical and permissible extension of parents' rights to control the number, timing, health, and spacing of their children, or is it the first step on a moral "slippery slope" toward selection of cosmetic traits such as eye color and weight?

In 1985, we posed this case:

> A couple with four daughters desire a son. The woman is pregnant. They request prenatal diagnosis solely to select the sex of the child and say they will abort a female fetus.

Such requests occur all over the world, but most frequently in countries where families desire a son, such as India, China, Turkey, and a number of Asian and Middle Eastern countries not in the surveys. Until the government of India outlawed the practice in 1995, sex selection was the most common use of prenatal diagnosis. China likewise outlawed it in 1994 (China, Law on Maternal and Infant Care). An estimated 100 million women are missing from the world's population (Sen, 1990) though mostly through preferential treatment of male children, or outright female infanticide rather than as a result of prenatal diagnosis.

In response to suggestions that geneticists might respond differently if a family sought a girl or came from a different culture, in 1995 we asked five different questions about sex selection.

1. A single woman wishes to become a mother. She wants a girl because a girl will be easier to raise. She may decide to terminate a male fetus.
2. A couple with four daughters desires a son. The woman is pregnant. They say that they will abort a female fetus.
3. A poor family has five sons. The sixth pregnancy is unwanted, but they say they will make financial sacrifices if the fetus is a girl. Otherwise, they will have an abortion.
4. A woman from a non-Western culture that prefers sons has so far given birth to two daughters. Her husband wants a small family but also wants a son. He says that if

she has another daughter, he will divorce her and send her and her daughters back to their own country and abandon them.
5. A professional couple in their early 40's have two teenaged sons. The pregnancy is unexpected. They were going to have an abortion, but have decided to wait until they find out the fetus's sex. If it is a girl, they will carry it to term.

The 1995 question also specified a noninvasive procedure such as ultrasound, with a hypothetical 100% accuracy, so as to eliminate considerations of procedural risk.

The only one of the five cases where responses differed significantly from the others was Case 5 (Table 16, col. 5). Many geneticists argued that this couple was entitled to prenatal diagnosis anyway on the basis of their age and that prenatal diagnosis might identify a genetic disorder in addition to identifying sex. In many countries there was also greater sympathy for the non-Western woman facing divorce if she fails to bear a son than for the other cases. Addition of the description "poor" to the parents who want a girl influenced some respondents, who said they felt more sympathy for this couple than for the couple (economic status undescribed) who want a boy. The single woman received the least sympathy; some respondents said she should not have a baby at all. Overall, however, there was little difference in responses to the various cases. The act of sex selection itself, rather than the particular circumstances, was the major factor influencing responses.

Outside the US, minorities (23–33%) would actually perform prenatal diagnosis, except for the couple in their forties (Table 16). Exceptions, where the majority would perform prenatal diagnosis, in all or some cases, were the Czech Republic, Hungary, Russia, Israel, Cuba, and Venezuela. The picture changes, however, when one looks at referrals. Many geneticists believed that it was their duty to offer a referral even if they disapproved of sex selection, because they thought all patients are entitled to a referral. Many hoped or expected that the referral would also refuse sex selection and that patients would run out of money, energy, and time before they found someone to accede to their requests.

When the percents offering referrals are added to those performing prenatal diagnosis, the world total increases to 49%, including 34% outside the US (Table 17). In most countries surveyed in 1985, with the exceptions of France, Sweden, Turkey, and India, the totals who would perform or refer increased. The totals who would actually perform prenatal diagnosis also increased, except in Canada, Sweden, and India.

In the United States in 1985, 62% would perform prenatal diagnosis or refer (Wertz and Fletcher, 1989). This was the most shocking and publicized finding of the earlier survey. In 1995, after ten years of bioethical discussion, mostly opposed to sex selection, 72% of US geneticists would perform or refer. How can we explain this? The entire 10% increase was in referrals, with Master's-level genetic counselors more willing than M.D. or Ph.D. geneticists to refer rather than to refuse. Increased respect for patient autonomy explains much of the increase. A total of 43% gave autonomy-based reasons (Table 18). Respondents pointed out that where abortion on request was legal, it made no sense to provide abortions without requiring any reason at all while refusing abortions to families who provided a reason. Others pointed out that couples could lie, as in Table 21, and that those who are truthful about their requests should not be penalized. A few said that sex selection might actually create happier families and lead to better acceptance of children (3%). Most drew the line at insurance payments (94%) but 8% said that patients had a right to the service if they could pay out of pocket (Table 18).

The most common reason for refusing sex selection, as in 1985, was that "sex is not a disease" (43%), a sentiment expressed in write-in comments in exactly these words. The phrase is more an immediate emotional reaction than a philosophical analysis, but it makes the points that (1) the purpose of prenatal diagnosis is to diagnose disease, however "disease" may be defined (a complex issue), and (2) children are not commodities that one buys at the supermarket. Some respondents likened the parents to people who put a coin in a gumball machine and then throw out the gumball if it is not the color they want. A few (7%) thought sex selection would lead to sex discrimination or pointed out that professionals had rights to refuse a service if, in the professional's view, it is not medically indicated (7%). Very few, however, referred to social implications, such as maintaining a balanced sex ratio (3%), avoiding a possible trend toward "cosmetic" selections (1%), or limiting population by limiting the number of women (2%). As in 1985, most reasoning about sex selection appears to be "individual-based," rather than societally based.

Subanalyses suggest that professionals with longer experience in the practice of genetics, and those who have had requests for sex selection, are more likely than others to say they would refer, rather than perform or refuse (Wertz and Fletcher, 1998). Referral seems to be a means of appearing to respond to patient requests without having to take any action. Some geneticists may have found by experience that it is easier to refer than to say no. Women were significantly less likely than men to give outright refusals (Wertz, 1997).

Outside the US, more respondents would attempt to dissuade (Table 20) than would simply provide information, but in 14 countries more "would provide information only" than would attempt to dissuade. In the US, 24% would dissuade, 47% would provide information only, and 29% would support the parents' decisions.

Sex selection is a particularly troubling issue for most geneticists. Close to half have had outright or suspected covert requests for sex selection (Table 19). Such requests call into question the role of the professional. Should the professional try to dissuade people from sex selection, thereby offering moral guidance but stepping outside the ethos of nondirectiveness, or should the professional play the role of morally neutral technical expert who provides information only? Offering moral guidance smacks of undesired paternalism. On the other hand, many people would like to think that the professionals whom they see are moral agents with ethical beliefs and values, rather than "value-neutral" technicians. Many would like to know what the professional would do if in their situation (Anderson, 1996). Yet even in this morally extreme case where many consider the couple's proposed actions reprehensible, many professionals would choose the value-neutral approach, placing the moral burden on the patient. A non-involved approach may be psychologically easier for the professional. Some, especially in the US, would follow the counseling ethos (Fraser, 1974) of supporting all patients' decisions.

In many and perhaps most cases, the request for sex selection may be veiled under a request for a medically accepted indication for prenatal diagnosis, such as a fictional family history (Table 21). We presented the following case:

> A couple aged 27 requests prenatal diagnosis. They say a previous child with Down syndrome died in another area of the country. There is not time to document the previous Down syndrome and still permit early termination, so you perform the test. After you have performed the test, and found no abnormalities, the parents show an unusual interest in learning the fetus's sex. You begin to wonder whether their story about the child with Down syndrome was true. (Assume that laws and professional rules in your country permit disclosure of fetal sex.)

In this case, majorities in every country (69% worldwide) would confront the parents directly and ask whether sex selection is their real purpose (Table 21, col. 1). However, majorities in 20 countries, including 85% in the US, would tell parents the sex within the time line for legal abortion (Table 21, col. 2). This acquiescence to requests occurred even in countries such as the UK and Denmark, where centralized records make it easy to check parents' stories.

Somewhat surprisingly, there was comparatively little support for use of preconception sex selection (Table 22), in the absence of a sex-linked genetic disorder, possibly because an accurate method is still hypothetical and its use would probably not be part of most genetic practices. In the US, sizable minorities of geneticists (38%), patients (35%), and the public (40%) thought such a method should be available to everybody, without restrictions; 12% of US geneticists and 26% of patients and the public said they would use it, though few thought it would benefit society.

THE DEAF COUPLE WHO WANT A DEAF CHILD

One case surpassed even sex selection in arousing respondents' emotions: the deaf couple who ask for prenatal diagnosis so that they can abort, not a deaf child, but a hearing child. Our case read as follows:

> A deaf couple with several hearing children want one who will be deaf like themselves. They both have a type of hereditary deafness that can be prenatally diagnosed. They say they will abort a hearing fetus.

Many deaf couples secretly hope that at least one of their children will be deaf, so that they can communicate on the same level, with sign language as a "first language." This desire is natural. So is the hope of some couples with achondroplasia (the most common form of dwarfism) that their children will not tower over them or become discipline problems because of their "normal" height. Most couples do not act on these desires, except perhaps by adopting children like themselves. Most would prefer to let nature take its course or to have so-called "normal" children, rather than using prenatal diagnosis to select in favor of a deaf or achondroplasic child. Nevertheless, requests such as described in Table 23 have occurred. They represent extreme examples of autonomy. They also present issues of equity. If a hearing couple can abort a deaf child and not incur moral blame for doing so, why should not a deaf couple who are members of a vibrant minority, the Deaf Community, be able to abort a hearing child who could not be a full member of that community? Write-in comments on this question showed outbursts of emotion surpassing all other questions in the survey. Respondents from all over the world called the parents "selfish," "perverse," "irresponsible," "unfit," and "unfair" to both the deaf child who would be born and to the hearing fetus that would be aborted. Responses stem in part from the generally pessimistic view of disability described in the next chapter.

In this case, majorities in all nations except the US and Canada would refuse prenatal diagnosis without offering a referral. The percent outside the US who would perform prenatal diagnosis (14%) is markedly lower than for any of the sex selection cases in Table 16, and fewer (7%) would refer. It appears that most saw this case as genuinely different from sex selection, because it is not only outside the traditional goals of medicine but appears to stand them on their head by producing a child that the larger community labels "unhealthy" or "abnormal." In the US and Canada, however, respondents tended to regard deaf selection as similar to sex selection, an autonomy-based issue, and said this in their comments. In both countries, the percents who would perform prenatal diagnosis for the deaf couple (18% in Canada, 35% in the US) were almost the same as those would perform it for the couple with four girls in Table 16. The totals who would perform or refer were the same in Canada (51%) and only slightly lower in the US (68% vs. 72%). Although fewer patients (47%) thought prenatal diagnosis should be performed for the deaf couple than for the couple with four daughters (59%), many said that the two cases were the same, morally, in terms of people's rights to choose the family that is best for them. Some respondents, including patients, also raised the issue of equity between "abled" (or "temporarily abled," since many will become disabled) and "disabled" families, supporting people's rights to select the child that will best be part of their community.

Prenatal Paternity Testing

Prenatal paternity testing is also controversial. Use of the method in cases of rape or incest is widely accepted, but sometimes a request comes from a woman (Table 24) who has had relations with several men. She wishes to find out which one is the father in order to plan the future. Such requests often involve psychosocial complexities beyond the expertise of the average geneticist. We posed the following case:

> A pregnant woman requests prenatal diagnosis for paternity testing. She is involved with two men: Mr. X, who wants children, and Mr. Y, who does not. If Mr. X is the father, she will keep the child. If Mr. Y is the father, she will terminate the pregnancy. Mr. X has offered to cooperate in testing. If she cannot be sure that Mr. X is the father, she will probably terminate.

Options were:

1. Perform prenatal paternity testing without comment on her personal situation.
2. Perform prenatal paternity testing with some comment on the possible negative Pconsequences of her actions.
3. Perform prenatal paternity testing only if she commits herself to long-term psychological counseling.
4. Refer to someone who might perform prenatal paternity testing.
5. Refuse to perform prenatal paternity testing and not refer.

There was somewhat greater willingness to perform prenatal diagnosis for this woman than in the sex selection cases; 56% in the US and 46% in other nations would actually perform prenatal diagnosis, though these percents included 24% and 29% respectively who would either make a negative comment about her approach to the situation or would require commitment to long-term psychological counseling as a precondition of prenatal diagnosis. In some nations – the UK, France, Germany, and the Netherlands – the percents who would perform were markedly lower than elsewhere. The requirements of national health systems and the need to conserve resources only partially explains these differences. Most geneticists would not refuse this woman outright without providing a referral. When referrals are added in, 96% in the US and 71% elsewhere, including majorities in all countries, would perform or refer. Among patients, 87% thought the doctor should perform prenatal diagnosis, including 56% who suggested negative comments or counseling. Comments on the woman's behavior tended to be pejorative. Some said she was an unfit mother who should not have a child at all. Others said she should tell the man of her choice that it is his child, without misusing genetic services.

PRENATAL DIAGNOSIS SOLELY TO RELIEVE ANXIETY

Providing prenatal diagnosis solely to relieve maternal anxiety, in the absence of any medical indications such as age or family history, raises issues of medical paternalism versus women's autonomy, and also of the fair allocation of technological resources. We raised this question in 1985 and found that majorities in most nations would perform or refer (Table 25). In some nations, such as Switzerland, anxiety was a medically accepted indication. In the intervening ten years, prenatal technologies have become much more plentiful and issues of fair use of "scarce resources" have receded. Prenatal diagnosis has also become safer. In 1995, we provided more explanatory detail about the case.

> A woman under 30 with no family history of genetic disorders or exposure to toxic substances requests amniocentesis. Non-invasive testing through maternal serum alpha-fetoprotein (MSAFP), estriol, hCG, [blood tests on the mother] and ultrasound reveal no cause for alarm. There are no medical or genetic indications for amniocentesis in this case. Nevertheless, she appears very anxious about the normalcy of the fetus, and persists in her demands for prenatal diagnosis even after being informed that in her case, the potential medical risks for the fetus, in terms of miscarriage, may outweigh the likelihood of diagnosing an abnormality.

In response to suggestions that geneticists might respond differently to women in different situations, we also asked the question for three different women:

1. A 15-year-old single mother who says her best friend had a child with Down syndrome.
2. A well-educated woman aged 25 with no first-hand knowledge of Down syndrome.
3. A woman aged 25 who works in an institution for people with retardation.

For the most part, there were no significant differences in responses, although some said the 15-year-old was at increased risk for having a child with Down syndrome because of her young age and therefore more deserving of the service. Table 25 reports only the responses for situation 2, which is most comparable to the question asked in 1985. Very few would refer the woman to a clinical psychologist or psychiatrist to evaluate her anxiety. In most nations, with the exceptions of Scandinavia and Switzerland, the percents who would perform prenatal diagnosis or refer increased; totals were 96% in the US and 71% elsewhere. As an ethical and social issue, we may consider prenatal diagnosis for maternal anxiety resolved, not through bioethical discussion, but through increased respect for autonomy in general and through proliferation of prenatal diagnostic services.

The question remains of who should pay for this and other genetic services. In the US, 78% of patients, 30% of geneticists, and 21% of primary care physicians thought insurance (not specified as public or private) should pay for prenatal diagnosis for the anxious young woman (Table 25). In thirteen nations, greater percentages than in the US thought insurance should pay, including Australia (53%), South Africa (44%), the UK (46%), Belgium (53%), Denmark (71%), Germany (61%), Sweden (50%), and Argentina (62%). The low percents (4–6%) in the four Asian nations reflect either the absence of insurance, or, in Japan, limitations on or absence of coverage for prenatal diagnosis.

Around the world, few (5–10%) thought that publicly-funded insurance should pay for prenatal diagnosis for sex selection or deaf selection, or for preconception sex selection (Table 26). More, but not a clear majority, thought publicly-financed insurance should pay for cystic fibrosis carrier testing or DNA banking on behalf of a person's descendants (Table 26). In the absence of medical indications or treatments, most would be sparing of the public purse.

"LATE" ABORTION AND ABORTION LAWS

"Late abortion" is another autonomy-related issue, complicated by laws and institutional policies. The case below describes how such a situation might occur:

> A 36-year-old woman has declined chorionic villus sampling (CVS) or amniocentesis. At 23 weeks, she decides to have ultrasound. Ultrasound shows a fetus with a nuchal membrane (excess fold of skin on posterior neck) seen in Trisomy 21 (Down syndrome), Trisomy 18, or 45,X (Turner syndrome). She has PUBS (percutaneous umbilical blood sampling), which provides definitive diagnosis. She wants to terminate the pregnancy. If she cannot get a legal termination at your institution, what would you do?

Table 27 shows a gradient in willingness to refer, based on geneticists' perceptions of the three disorders. Although majorities would refer within the country for all three disorders, the greatest percent (86% in all) would refer for Trisomy 18, which causes profound mental retardation and death in infancy. Almost as many (74% outside the US) would refer for abortion for Down syndrome. However, for 45,X, a comparatively mild condition, which causes infertility and may affect the internal organs but usually not the brain or nervous system, markedly fewer (57% outside the US) would refer within the country. About half as many would offer referrals outside the country. The low percents in Asia who would do so perhaps reflect the lack of any close place to refer.

Abortion for fetal abnormality is legal in most nations we studied (Table 28), with the major exception of Latin America, where abortion is illegal except to save the mother's life or for rape or incest. Cuba permits legal abortion. Illegal abortions – of varying degrees of safety – are widely available everywhere in Latin America. Germany and Japan, because of their past history of eugenics laws, now have no fetal indications for abortion, but in these situations abortions are easily obtained for maternal indications (raising a child with an abnormality would adversely affect the mother's physical or mental health) or for social indications (family finances, housing, etc.). Removing indications for abortion for "abnormality" in the fetus, under a law passed in 1995, has made no practical difference and has not changed the number of abortions for genetic abnormalities in Germany, but has changed the reasons recorded for such abortions. The motives behind this law are to discourage judgments, whether by professionals or parents, that appear to be eugenic in intent, and to elevate the moral status of the medical profession.

Abortion for fetal abnormality was never legal in Japan, except for a limited list of about 30 conditions that were presumed to be inherited, such as epilepsy, but was widely performed for social indications, though not without feelings of guilt (La Fleur, 1992). In some countries – Canada, South Africa, a few states in the US, Denmark, France, Norway, Sweden, Israel, Turkey, and China – there is no time limit on abortion for fetal disorders. In Portugal, the Czech Republic, and Hungary, there is no time limit for "lethal" disorders such as anencephaly. However, obtaining abortions after 22–24 weeks in many of these countries may be extremely difficult or even impossible, despite the absence of legal restrictions. In most of the countries surveyed, abortions are also legal for social or maternal indications, at least to 12 weeks, except for Latin America, Portugal, and Poland. In Turkey, the limit is 10 weeks. Geneticists in most nations agreed with their country's laws on abortions for genetic abnormalities (Table 29), but there was overwhelming dissatisfaction with the restrictive laws in Latin America that make it necessary for women to seek illegal and unsafe abortions after prenatal diagnosis. In Cuba, where abortion is legal, most geneticists approved. In Chile, the majority of geneticists adopted the "pro-life" stance that it should remain illegal. Some expressed dissatisfaction with what they considered overly restrictive laws in Poland, laws that were not restrictive enough in Spain, or the absence of laws in Thailand.

Majorities in all nations except the US, Russia, and China thought abortions should be illegal for sex selection (Table 30), as did 53% of US patients and 58% of the US public. A closer look at Table 30 shows, however, that the percents who think abortions for sex selection should be banned have decreased in all nations since the 1985 survey, including a decrease from 50% to 32% in the US. In the 1980's, there was a furor over sex selection as prenatal diagnosis became more widespread, and feminists and others thought that parents would use the method to tailor-make their families. In the United States, Pennsylvania and several other states tried to forbid sex selection as part of more general abortion laws that courts overturned as too restrictive (Planned Parenthood vs. Casey, 1992). Since then, the furor over sex selection in Western nations has died down, as most people came to realize that prenatal diagnosis is not changing the gender composition of their countries. Most parents in Western nations would not use prenatal diagnosis for sex selection, and the few who do would conceal the reason for their requests, thus making laws unnecessary or useless.

There is some support for making abortion illegal for other conditions. Outside the US, about one-third would ban abortion for severe, untreatable obesity (if this becomes prenatally diagnosable), cleft lip/palate, and predisposition to alcoholism (Table 30); about one-fourth of US patients concurred. Comparatively few geneticists would ban abortions for Huntington disease, an adult-onset single gene disorder that causes progressively severe motor and mental deterioration, leading to death after 10–15 years (7%), or for 45,X, a sex chromosome disorder that causes infertility but not mental retardation (10%).

CONTROLLING PREGNANT WOMEN

Sometimes a pregnant woman's behavior is detrimental to the fetus. In genetics, the most common such case is maternal phenylketonuria (PKU), a recessive metabolic disorder that causes a buildup of phenylalanine in the body, leading to severe mental retardation. Forty years ago, most people with PKU were institutionalized. In the 1960s, discovery that a diet low in phenylalanine, if instituted at birth, could prevent retardation, combined with mandatory screening of newborns for PKU, led to the possibility of normal life for children born with PKU. As this generation matured, married, and became parents, a new problem emerged. The children of women with PKU did not have PKU (this would require a gene from each parent), but had severe mental retardation and heart defects caused by the buildup of phenylalanine in the mother's body. Most people who have been on a strict PKU diet in childhood relax the diet somewhat, with medical permission, as teenagers or young adults, and their phenylalanine levels rise. A return to a very strict diet, including dependence on an ill-tasting special formula, before conception and throughout pregnancy may prevent severe damage to the fetus. Some women are unable or unwilling to maintain this diet. Such cases raise the issue of whether a pregnant woman's autonomy should be limited on behalf of the developing fetus and future child. We posed the following case:

> Phenylketonuria (PKU) in the untreated mother during pregnancy is known to cause fetal damage, expressed as microcephaly, moderate mental retardation (MR), congenital heart disease, and low birthweight. It is known that dietary treatment which controls blood phenylalanine levels and other biochemical abnormalities of PKU in the mother is at least partially protective of the fetus, provided that the diet begins before conception or early in the first trimester.
>
> Suppose that she becomes pregnant without having gone on the PKU diet and appears unwilling to adhere strictly to the diet. She is still early in the first trimester, so dietary intervention could offer some protection to the fetus. Your own efforts and her partner's and family's efforts to keep her on the diet have failed. What should you do?
>
> Now suppose that she has reached the second trimester without controlling her high serum phenylalanine level. It therefore appears likely that the child will be at least moderately retarded and also might have multiple physical defects. After giving her this information, how would you counsel her about the pregnancy?

Most geneticists, except in China and India, rejected the idea of forced hospitalization of the pregnant woman (Table 31). Most also rejected the use of child abuse laws to punish her if the child was born with PKU-related defects (not shown in table). If she reaches the second trimester without controlling her phenylalanine level, a minority (11% in the US, 29% elsewhere) would suggest abortion.

The preferred approaches were either to tell her about her options of carrying to term or aborting (30% in US, 20% elsewhere), or to go further and also offer to support whatever decision she makes (59% in US, 47% elsewhere). Responses to this question affirm the primacy of a woman's autonomy over the welfare of the fetus, though many respondents expressed a deep moral agony over the situation. There were exceptions. Most in China and a majority in India thought the woman should be hospitalized involuntarily to protect the child, and few would support a decision to carry a presumably damaged fetus to term. In Switzerland, Russia, China, India, and Japan, sizeable percents (17–30%) thought that the geneticist, "in view of her disregard for the fetus's health, should tell her husband to tell her to abort" (Table 31). Respondents' reasoning (from a checklist) appears in Table 32. Although more (77%)

considered a child's "right to be born healthy" important than checked "avoiding forced medical/social control of pregnant women" (55%), their choice of actions in this case belies their reasoning.

PARENTHOOD FOR PEOPLE WITH DISABILITIES

Whether people at risk for having children with disabilities should become parents is a difficult issue both in genetic counseling and in terms of fairness to society. If the parents themselves are disabled, the issue may be complicated by their own ability to support and rear a disabled child. We included four cases (Table 33). In the first case (col. 1), a blind couple both have the same autosomal recessive disorder. Any child of theirs will be blind from birth. They have rejected adoption of a sighted child or use of reproductive technologies such as donor insemination or a donor egg that would avoid blindness. Their socioeconomic situation is not described. The second (col. 2) repeats for comparison the percent supporting the woman with PKU in Table 31. In the third case, a couple, both of whom carry a gene for cystic fibrosis (CF), oppose prenatal diagnosis and abortion. Respondents were asked whether they would support "taking their chances of having a child with CF," among other possible options. The fourth case (col. 4) presents a woman with fragile X. In fragile X, the number of "repeats" in letters of the genetic code expands in succeeding generations, leading to greater retardation in the children than the parent. Each child has a 50% likelihood of receiving the fragile X from the mother. Boys are more severely affected than girls, but women may also be affected. (Fragile X does not behave like most sex-linked disorders, which affect only boys.) The children who receive the fragile X are likely to be more severely affected than the mother, because the genetic "mistake" expands.

We posed the following case:

> A 25-year-old mildly retarded woman with fragile X syndrome who lives in a residential unit for people with retardation has expressed an interest in marrying and having children. She is assumed to be competent and does not have a legal guardian. After thorough counseling, you are convinced that she does not really understand her risk of having children with retardation. She persists in her desire to have children.

Table 33 reports the percents who would seek to provide "social support so that she can carry out her wish to have children." Surprisingly, especially in view of the pessimistic prenatal diagnosis counseling discussed in the next chapter, majorities in the US, and half or close to half elsewhere, supported the parents' decisions to have children in all four cases. Majorities in 21 nations, including India, would help the woman with fragile X to have children, even though there is a 50% chance that they will have more retardation than she does. Responses to these questions indicate a sympathy with people with disabilities and a surprising openness to their decisions that demonstrates an underlying respect for individual autonomy. As always, there were some exceptions, notably in China, where the welfare of society takes precedence. In general, however, responses to these cases illustrate a shift away from population or eugenic concerns to a model of genetics focused on the individual (Juengst, 1995).

These cases raise the more general issue of fairness – to the child with a disability, to siblings, and to society – especially if the birth could have been avoided. John Stuart Mill (1855) raised these questions in his endorsement of "responsible parenthood" as a moral obligation and attached moral blame to parents who knowingly had children with disabilities. The responses in Table 34 show that many geneticists – the majority in 25 nations, including the United States (52%), and including all in Southern and Eastern Europe, the Near East, Latin America (except Argentina and Chile), and Asia (except Japan) thought bringing a child into the world with a serious genetic disability was unfair to the child (Table 34, col. 1). Almost half the patients (46%) also thought this was unfair. The variation among nations may reflect in part the lack of consensus among geneticists about what constitutes a "serious" disorder (Wertz and Knoppers). Some of the highest levels of agreement also came from the poorer nations in the survey where there are fewer resources for such children. The two nations that stood out – Germany and Japan – with only 18% agreement, were the two with "eugenics" histories that they are still struggling to overcome. In both nations, there is a reluctance in the genetics profession openly to attach moral blame to parents who choose to have children with disabilities. There was less perceived unfairness to the child's siblings than to the child, possibly because the siblings rarely appear as patients. The disability literature in general also gives less attention to siblings than they perhaps deserve.

Substantial minorities (26% in the US and 45% outside the US) thought that choosing to have a child with a disability was not fair to society (socially irresponsible in an era of prenatal diagnosis) (Table 34, col. 3). Agreement with this statement was strongest in China, India, and Cuba, but majorities also agreed in all of Eastern Europe, the Near East, most of Latin America, Greece, Portugal, and Thailand. In the US, 26% of both geneticists and patients, 46% of the public, and 55% of primary care physicians agreed. It appears that in much of the world, women who carry a child with a known genetic "defect" to term will face social blame, perhaps accompanied by social and cultural pressure to abort.

NON-TREATMENT OF GENETICALLY HANDICAPPED NEWBORNS

One final case appearing under the heading of autonomy is parental requests for non-treatment of genetically handicapped newborns. We described the following case:

> A professional couple has a newborn with a genetic disorder. The child will die without cardiac surgery. The parents refuse to give consent for the operation. They say that if the ethics committee insists upon the operation, they will leave the baby in the hospital and will abandon all parental rights and responsibilities. Putting legal issues aside, what would you do?

We also posed questions for infants with Trisomy 13, 45,X (Turner syndrome), severe, open spina bifida, and Down syndrome.

Table 35 shows a gradient in willingness to support parental decisions, based on geneticists' perceptions of the severity of the disorder. Most (83% in the US, 82% elsewhere) would support non-treatment of Trisomy 13, which causes profound mental retardation and death in infancy. Slightly fewer (79% in the US, 72% elsewhere) would support non-treatment of severe, open spina bifida, which may cause mental as well as severe physical impairment. About half would support non-treatment for a child with Trisomy 21 (Down syndrome), with notably low figures in Germany (perhaps because of the history of systematic killing of such children by the Nazis), Chile (because of a pro-life stance), and Peru. However, responses to this question varied widely. The majority would not support non-treatment for a child with 45,X, which many regarded as capable of normal life, though sizable minorities in the UK (39%), US (39%), Canada (37%), Norway (44%), and China (50%) would accede to parents' requests. Most did not consider placement for adoption a viable alternative for children with Trisomy 13 or severe spina bifida; in some countries adoption is not culturally accepted and it is difficult to place even a healthy child. For Trisomy 21 and 45,X, however, substantial percents said they would discuss the possibility of adoption, though we do not know whether they would have said this if they had been asked to describe their responses to the situation entirely in their own words, with no "prompt" from the questionnaire.

Overall, responses to autonomy-related questions indicate substantial and growing respect for individual autonomy in most nations. In view of the past history of a sometimes dominant and authoritarian medical paternalism, this is a welcome development. Whether it will go too far, toward individual selfishness at the expense of others, is a question that will play out largely in the prenatal diagnosis debates about sex selection, payment for nontraditional services, and selective abortion for the milder genetic conditions described in the next chapter.

CHAPTER 4

COUNSELING: DIRECTIVE AND NONDIRECTIVE

"Nondirectiveness" has long been the preferred approach in genetic counseling in the US, Canada, the UK, and other nations influenced by English/American medical traditions. Genetic counseling, a term coined by the American, Sheldon Reed, in 1947 to replace the earlier "genetic hygiene" or "genetic advice," began in the shadow of the Eugenics Movement and of World War II (Reed, 1947). Its founders sought to distinguish it from earlier, more directive forms of interaction in order to avoid accusations of possible coercion.

As originally outlined by a committee of the American Society of Human Genetics headed by F. Clarke Fraser in 1974, nondirectiveness meant: "(1) helping clients understand their options and the present state of medical knowledge, so they can make informed decisions, (2) helping clients adjust to and cope with their genetic problems, (3) the removal or lessening of patient guilt and anxiety, (4) helping clients achieve their parenting goals, and (5) the prevention of disease or abnormality" (Fraser, 1974).

In our 1985 survey, close to 100% said they agreed with these approaches. More recently, however, many geneticists and genetic counselors have argued that totally "nondirective" counseling is neither possible nor desirable (Clarke, 1994; Kessler, 1992; White, 1997). Nondirective counseling is impossible because counselors have their own values, and these values tend to creep into the counseling situation however mightily the counselor strives to suppress them. Furthermore, most people expect a counselor to have values. If the counselor will not discuss values openly, those being counseled may try to discover them through observations of gestures or tone of voice (Anderson, 1996).

When modern genetic counseling began, there were, and for the most part, still are no treatments. Therefore, the directions accompanying most medical encounters ("take 4 pills 3 times daily") were not appropriate. Instead, people with family histories of "inherited" diseases came to counseling to get information to help them decide whether to have a child. In the North American and British contexts, telling people not to have children appeared to be a violation of basic human rights. Some geneticists, such as Reed, also believed that people who had the common sense and responsibility to seek genetic counseling would be such "good parents" that they ought to have children even if their decisions sometimes led to their having children with disabilities (Reed, 1954). The new specialty was education-oriented rather than treatment-oriented. Relationships in an educational encounter may be more equal than in a treatment encounter and may therefore empower the parties receiving counseling.

In recent years, the profession has reinterpreted "nondirectiveness" to mean not only providing information in a manner that is as unbiased as possible, but also empathizing with the person(s) receiving counseling and offering them guidance. Both our patient survey and a series of focus groups conducted for a separate study in the Northeastern US (Wertz and Gregg, 2000) suggest that people want guidance in addition to full and accurate information. The WHO "Proposed International Guidelines" (Appendix) (1998) describe nondirective counseling as follows: "Non-directive counseling has two major elements. The first is the provision of accurate, full and unbiased information that individuals and families may use in making decisions. The second is an understanding, empathetic relationship that offers guidance and helps people to work towards their own decisions. In non-directive counseling, the professional avoids purposely slanting information that may lead people to do what the counselor thinks best. Individuals and families must depend on the counselor as a source of accurate information, and usually have no way of discovering when information is biased." (p. 5).

In the survey, we asked only about information, not empathy. Results suggest that providing "unbiased information" may be a practice largely peculiar to English-speaking nations. This was perhaps the most important finding of the entire survey. In the following pages, we begin with counselors' presentation of reproductive alternatives in preconception counseling (Tables 36 and 37), proceed to discuss the overall level of directiveness in counseling (Tables 38–41), look at the largely pessimistic direction of the directiveness (Tables 42–45), at professionals' personal attitudes toward abortion (Tables 46–48), and at their generally pessimistic attitudes toward disability (Table 49). Tables 38–44 present the most important findings in our survey and are also the most unsettling for those who think that counseling is providing unbiased information. We urge readers to peruse these tables carefully. There are many differences among nations that we cannot explain but that may be useful for both national and international discussion.

REPRODUCTIVE ALTERNATIVES

The reproductive alternatives listed in Table 36 as possibilities for a couple at 25% risk for having cystic fibrosis (CF), a recessive life-shortening lung and digestive disorder, vary in their availability and cultural acceptance. Adoption of a "healthy" child, which the majority would discuss, is not favored by prospective parents in much of Asia and is not approved by Muslim religious law (Muslims may adopt, but the child has a special status that differs from that of biological children); conversely, "healthy" infants are difficult to obtain in many developed countries. On the other hand, adoption may be gaining acceptance in India to avoid other genetic disorders (cystic fibrosis is rare in India). Donor insemination and use of donor eggs are illegal, restricted by law, or unacceptable in some countries, and preimplantation diagnosis is still in the research stage. Substantial percents would discuss all three, but there was no consensus. A question asked about these alternatives in 1985 was not comparable because it used the example of tuberous sclerosis, a dominant disorder sometimes accompanied by mental retardation.

Many found surrogacy an acceptable alternative for a woman with PKU, whose metabolic disorder could poison her fetus; 78% in the US, and 42% elsewhere would present this as an alternative (Table 37) without being asked about it. The woman's own egg could be used in this case, since the damage comes from the environment in the womb, not from the egg. Many also found surrogacy morally justified to prevent autosomal dominant or X-linked disorders; 74% of US geneticists, 62% of primary care physicians, 52% of patients, and 42% outside the US, including majorities in Canada, the Czech Republic, Poland, Russia, China and India agreed. Acceptance of surrogacy, at least to prevent genetic disorders, appears to have come a long way since the mid-1980s, when states and nations passed laws to ban it and feminists mounted protests. It is somewhat surprising that so many would approve surrogacy, when, except for the woman with PKU, a donor egg could have served as well. However, egg donation poses a risk for the donor, and *in vitro* fertilization (IVF) is expensive. Surrogacy is a technologically simple solution though not a psychosocially or ethically simple solution. Fewer would present the option of surrogacy to a CF carrier couple than in the other cases; donor insemination would serve to prevent CF, which is a recessive disorder requiring a mutated gene from each parent.

DIRECTIVENESS IN COUNSELING

The questionnaires listed 21 different genetic conditions or fetal malformations that might be found in prenatal diagnosis. Four of these were hypothetical, including severe untreatable obesity in the absence of a known genetic syndrome, and predispositions to alcoholism, mental illness, and Alzheimer disease. None of these is now diagnosable prenatally except for a few rare variants of Alzheimer disease or some mental illnesses that run in families. We also listed two maternal situations (mother's life in danger, pregnancy is the result of rape) and three fetal infections (HIV, toxoplasmosis, and rubella) in the first trimester, which may lead to early death or neurological impairment. We asked respondents how they would counsel for each of the 26 conditions or situations. Choices of responses were:

Urge parents to carry to term
Emphasize positive aspects so they favor carrying to term without suggesting it directly
Try to be as unbiased as possible
Emphasize negative aspects so they will favor termination without suggesting it directly
Urge termination
Not tell them this particular test result

Tables 38–41 show the percents who would "try to be as unbiased as possible," the gold standard of Western genetic counseling. The rest would be openly directive or provide slanted information. What people actually do in professional practice may depart considerably from what they say they would do in response to questionnaires. We suggest that the data in Tables 38–44 underestimate the amount of directive, pessimistic counseling. So much has been written for the past 20 years about nondirective counseling as the most ethical approach, starting with Fraser's 1974 article, that many professionals are reluctant to show open evidence of directiveness. Although there are wide and perhaps inexplicable variations among nations, an overall pattern emerges. Geneticists in English-speaking countries would try to present unbiased information about almost all fetal conditions, with the exception of anencephaly. Geneticists in some northern European countries, especially Denmark, Finland, Germany, the

Netherlands, Norway, and Sweden would be unbiased about some, but not all conditions, and not to the same extent. Belgium and France are less unbiased than other northern European nations. Outside English-speaking nations, most counseling is biased. Most, though not all, of the bias is pessimistic, and most of it occurs by presenting slanted information (so that parents will have an abortion without the counselor suggesting it directly) rather than by directly urging abortion. There were two conditions for which most counseling was optimistic; cleft lip/palate, and child not of the desired sex. The most consistently biased counseling occurs in Asia (with the exception of Japan, which has been influenced by Western professional views on nondirectiveness), Eastern Europe (with the exception of Poland), Cuba, and Peru. Greece stood out as more biased than other southern European nations.

Examination of the column for Trisomy 21 (Table 38), the most common chromosomal disorder compatible with life, shows not only considerable variation, but overall predominance of directiveness, with the exception of the nations already mentioned. Outside the US, the majority would counsel directively about Down syndrome and the other seven conditions in Table 38, giving credence to the allegations of some members of the disability community that counseling is indeed biased. Readers should keep in mind, however, that some disorders in this table vary in incidence and severity according to geography, and responses from some nations may reflect an absence of clinical experience with a condition. Cystic fibrosis is extremely rare in China, India, Japan, and Thailand; the mutations most frequently found in Southern Europe and the Mediterranean are associated with milder forms of the disorder than the most common mutations found in the United Kingdom. Sickle cell anemia is extremely rare in China and Japan, and infrequent in much of Latin America; the form found in India is relatively mild as compared with that found in the United States.

Counseling for the six conditions listed in Table 39 is also, on the whole, biased outside the group of English-speaking nations. In this group, as in the other tables on counseling, South Africa, with its hybrid English-Boer professional traditions, stood out as more directive and also more pessimistic.

"Severe obesity" was described as "severe, untreatable obesity in the absence of a known genetic syndrome," to distinguish it from Prader–Willi syndrome and other conditions that include both obesity and mental retardation. There is at present no prenatal test for obesity. Cleft lip/palate varies in its severity and treatability. Treatability depends to great extent on economics. Although considered successfully treatable in developed nations, cleft lip/palate may impose an insuperable economic barrier for most families elsewhere. It is not treated in rural China, and some infants (especially girls) are abandoned. The direction of counseling went both ways for this condition – optimistic in developed nations, pessimistic in China and India. We asked the question twice, for a girl and for a boy, at a widely separated interval in the list of conditions, in order to see whether counseling would be more pessimistic for a girl, whose cosmetic disability might reduce chances of marriage to greater extent than for a boy. There were no significant differences in responses, either for overall directiveness or for pessimistic/optimistic directions; at most, responses in a few nations were about two points apart. The last column in Table 39 is averaged between the two questions.

Counseling for adult-onset disorders (Table 40) appears slightly less directive than for disorders appearing in childhood. Respondents perhaps recognized that prenatal diagnosis for "predisposition" to mental illness (described in the question as schizophrenia or bipolar disorder), alcoholism, and Alzheimer disease, even if possible, might have little predictive value. About half the counseling for adult-onset conditions would be directive, but there was no consistent pattern among nations with regard to its direction.

US primary care physicians, who provide much of the initial post-test counseling for spina bifida and Trisomy 21 (though most would offer a referral to a genetic specialist), were notably more directive than US or Canadian geneticists. Their responses in Tables 38–41 are closer to the responses of geneticists in South Africa, Argentina, or Brazil than to the responses of their North American colleagues. This difference may result from the more directive and paternalistic culture of treatment-oriented medical specialists. Obstetricians (who comprised one-third of the primary care sample), in particular, tend to be more directive than pediatricians, but this difference did not appear consistently in their responses to counseling questions.

Counseling for maternal situations or fetal infections was also largely directive (Table 41). In English-speaking nations, counselors would urge the mother to save her life by terminating the pregnancy, but substantial percents were willing simply to provide information and to leave open the option of attempting to carry to term. More would be unbiased about a pregnancy resulting from rape than for other conditions in this table. It may be easier to be unbiased in this than in other situations, because most women seeking prenatal testing after a rape have already made up their minds about their course of action. Prenatal diagnosis is done after rape mainly to preserve a fetus – if it turns out to be the husband's or partner's – from what would otherwise be almost certain abortion. The test results are highly accurate and easy to interpret, and in most cases the counselor need only present the information.

The direction of counseling was split between optimistic and pessimistic in many nations. For three conditions, cleft lip/palate in a boy, cleft lip/palate in a girl, and child is not the sex desired by parents, counseling in most nations (except Russia, China and India) would be optimistically slanted or openly directed toward carrying to term. For sex, some (7%), primarily in China, the United States, Canada, and Germany would not tell parents the result. For other conditions – predispositions to mental illness, alcoholism, or Alzheimer disease, severe obesity in the absence of a known genetic syndrome, 45,X, and neurofibromatosis (a disorder causing large, sometimes dangerous tumors on head and neck, which cannot always be removed) counseling tended toward greater optimism than pessimism, with considerable variation among nations. The remaining 17 conditions, which include all genetic disorders on the list that are usually associated with mental retardation, and the two maternal situations and three fetal infections, elicited substantial percents who said they would counsel pessimistically, as reported in Tables 42–44. In these tables, the first figure given is the total who would be pessimistic, including both those who would "urge termination" and who would "emphasize negative aspects so they will favor termination without suggesting it directly." The figures in parentheses represent those who would "emphasize negative aspects," in other words, present purposely slanted information. In developing nations and Eastern Europe, especially, most of the pessimistic counseling occurs through presenting slanted information, rather than openly and directly urging termination. Most parents do not have the knowledge, experience, or boldness to question the veracity of information put forward by an expert, so the information serves in effect as propaganda for termination. (The same can be said of optimistically slanted information as propaganda for carrying to term.) In English-speaking nations and some countries in

Northern Europe, comparatively few of those giving pessimistic counseling would do so through slanted information. In most of the world, however, this was the preferred form of presentation. Chile stands out as offering little pessimistic counseling because of its anti-abortion stance.

In general, China and India were the most pessimistic nations. China, with 1.2 billion people, has modern educational settings for perhaps several hundred children with mental retardation. Pessimistic counseling was not always related to lack of support, however. It appeared in the Netherlands and Israel, which have high levels of support for people with disability or mental retardation.

US primary care physicians were more pessimistic than their US geneticist colleagues for 13 of the 14 conditions about which they were asked in Tables 42–44. The exception was familial hypercholesterolemia (high cholesterol), which is treatable with partial success. Outside the US, majorities would counsel pessimistically for seven of the conditions listed in the tables: anencephaly (79%), mother's life in danger (71%), Trisomy 13 (70%), severe, open spina bifida (65%), Hurler syndrome (60%), rubella in the first trimester fetus (56), and fetal HIV (51%), all of which lead to early death, or in the case of spina bifida and rubella, major disability. For eight other disorders, however, some, but not all of which are in varying degrees either treatable, compatible with fulfilling lives, or of late onset, substantial percents outside the US would also offer pessimistic counseling: Trisomy 21 (49%, including 29% who would provide slanted information), toxoplasmosis in the fetus (49%), cystic fibrosis (46%, including 24% who would offer slanted information), Huntington disease (39%), sickle cell anemia (37%), achondroplasia, the most common form of dwarfism (36%), PKU in the fetus (32%), and XXY (27%). XXY (Klinefelter syndrome) is a sex chromosome disorder causing infertility, absence of puberty without hormonal treatment, and sometimes behavioral or learning problems. There was a notable absence of pessimistic counseling for XXY in Northern Europe, perhaps as a result of concerted campaigns in some centers to provide unbiased counseling. Support groups may also have played a role in encouraging nondirective counseling.

We included placement for adoption as an option in the prenatal diagnosis questions, except for anencephaly, Hurler syndrome, and Trisomy 13, which lead to death shortly after birth or in infancy. Table 45 shows the percents who answered "yes" to a question about whether they would present, in counseling, the options of carrying a fetus to term and giving the child up for adoption. The question, included at the suggestion of people with disabilities and of US genetic counselors, may have encouraged more people to answer "yes" than actually would present the option in counseling. Results suggest that outside the US, this option is rarely mentioned. In many countries, placement of a child with a disability may not be possible.

For each of the 26 conditions, we asked (in a separate question) whether respondents would have abortions themselves. Choices were:

(1) I would have an abortion;
(2) I would not have an abortion, but it should be legal for others; or
(3) I would not have an abortion, and it should be illegal for others.

The question was asked separately for the first trimester and the second trimester of pregnancy. The percents who would themselves abort are shown in Tables 46–49.

(The percents who thought abortion should be illegal appear in Table 30, discussed in Chapter 3.) Most geneticists would abort for most conditions on the list, including 89% for severe spina bifida, 79% for Trisomy 21, 72% for cystic fibrosis, 62% for achondroplasia, and 49% for XXY. Exceptions were obesity, cleft lip/palate, child not the desired sex and the three hypothetical predispositions (to mental illness, alcoholism, and Alzheimer disease).

Primary care physicians, and most especially US patients and public, were much less willing to abort, though majorities believed abortion should be legal for others (except for sex selection). The percents shown for patients represent the 70% who said, on a screening question, that abortion should be legal for purposes other than to save the mother's life, rape, or incest. Those who thought that abortion should be illegal (except in these situations) were asked to skip all the prenatal diagnosis questions. Many answered them anyway. Interestingly, their answers about abortion for genetic conditions were similar to answers of those who thought abortion should be legal. It is as if the fetus with a genetic condition stands outside the parameters of many people's thinking on abortion. Perhaps, to some, these fetuses are not quite human beings.

One of us (DCW) asked similar questions in a 1987 survey of parents of children with cystic fibrosis (Wertz *et al.*, 1991). In that survey, the most shocking finding was that 11% said they would abort for severe obesity. This finding ultimately reached Broadway, in "The Twilight of the Golds," via a report in *Newsweek*. In 1995, the figure for geneticists was even more shocking at 29% – both inside and outside the US – exceeded or equaled by percents for Australia (33%), Czech Republic (54%), Hungary (29%), Russia (79%), Israel (57%), Turkey (60%), Cuba (43%), China (69%), and India (49%). Compared with these figures, the percents for US primary care physicians (13%), patients (8%), and public (16%) appear low, but represent millions of people selecting against obesity, North America's number one "cosmetic problem." It is likely, however, that many respondents perceived obesity as a dire medical as well as cosmetic problem. We did not ask for comments on the abortion questions and can only conjecture about their reasons.

US primary care physicians were less willing than their geneticist colleagues to abort for most, but not all conditions. Primary care physician questionnaires, like the patient questionnaires, contained screening questions; 17% thought abortion should be illegal except when the mother's life was in danger and were asked to skip the abortion questions.

Patients and the public were less willing to use abortion than either professional group. In some cases, the wording that we used to translate medical labels into concrete symptoms may have affected responses. For example, for sickle cell anemia (without using the name), we described a "blood disorder that may lead to episodes of severe pain." The fact that 72% of patients (mostly white) would abort for this reflects a strong parental reaction against causing pain in a child. Instead of schizophrenia or bipolar disorder, we described a mental illness that would make the child "unable to work or relate to others." That 59% would abort for this perhaps reflects their perception of being cut off from normal social discourse as a peculiar horror.

If one bases the patient responses on the entire group of 476, rather than the 348 who thought abortion should be legal, the percents who would use abortion drop considerably. Among the entire group, there were only two fetal conditions where majorities would abort: (1) a child would be severely retarded and would die within the first few months of life (Trisomy 13 in geneticist questionnnaire), and (2) a blood disorder causing severe pain (sickle cell anemia). There was no fetal condition for which the majority of the public would abort, though "child would be severely retarded and die in the first few months of life" came close at 48%. Majorities of the public thought abortion should be legal for others, except for sex selection.

We expected that fewer would abort in the second than the first trimester, both because of the increased difficulty and risk of the procedure and the "increasing moral status" of the developing fetus. However, there was little difference between responses for first and second trimesters, usually two to three percentage points for geneticists, about five (at most eight) points for patients. Apparently respondents thought in terms of being pregnant at all, rather than in terms of stage of the fetus' development.

Geneticists' personal attitudes toward abortion were significantly related ($p < 0.05$) to the direction of their counseling, except in countries where most would try to be unbiased. Another factor related to pessimistic counseling in some countries was the geneticist's age, with older professionals more likely to be both directive and pessimistic (Cohen and Wertz, 1997).

The amount of directiveness raises the question of how professionals view the goals of counseling, and what gives them satisfaction. We asked five questions about factors related to geneticists' satisfaction and dissatisfaction with counseling sessions (Table 50). Most (95%, Table 50) felt "satisfied with a counseling session if the patient understands the medical/genetic information provided." Fewer (49%) felt satisfied "if the patient gets supportive counseling, even if the patient does not remember the genetic information." In English-speaking nations (except South Africa) and most of Western Europe (except Belgium, France, and Norway) most would *not* be dissatisfied if women refuse prenatal diagnosis or "decide to carry a seriously affected fetus to term." Outside the US, however, almost half (48%), including majorities in Greece, Portugal, Spain, all nations in Eastern Europe, Turkey, Brazil, Cuba, Mexico, Peru, China, and India felt "dissatisfied if patients refuse prenatal diagnosis if it is medically indicated," and a substantial minority (37%), including majorities in Belgium, Czech Republic, Hungary, Russia, Turkey, Cuba, China, and India said they "feel dissatisfied with counseling if patients decide to carry a seriously affected fetus to term." Everywhere, most (86%) "would be dissatisfied if legal abortion were not available after prenatal diagnosis."

In sum, geneticists tended to rank the educational success of a session as more important to their own self-image than the empathetic/supportive content. Many felt that it was their job to convince women to use prenatal technology, and to prevent the birth of children with genetic conditions, through selective abortion if necessary. These responses raise the possibility of "private eugenics" and covert coercion in counseling. In English-speaking nations and many Western European nations, however, most geneticists do not base their satisfaction on numbers of prenatal diagnoses or births of affected children prevented. Nevertheless, some geneticists still do so. In the US, 27% would be dissatisfied if a woman did not use prenatal diagnosis, and 15% would be dissatisfied if she did not abort a severely affected fetus.

In the next chapter, we discuss whether the ethics of prevention, aided by directive counseling, amounts to private eugenics.

CHAPTER 5

IS MODERN GENETICS EUGENICS?

The word "eugenics" has many different meanings. It can mean simply having a healthy baby – this is what it means in China and Greece today. This is what it meant to the American farm families who exhibited themselves at fairs, along with their prize cows, at the Eugenics Society booths. At the other extreme, eugenics can mean a government sponsored, coercive program of sterilizing the genetically "unfit." Many Western nations, including the US (Allen, 1989) Canada (Mclaren, 1990) Brazil (Stepan, 1991), Sweden, France (Adams, 1990), and Germany had eugenic laws in the 1930s that required sterilization of people with mental retardation, mental illness, deafness, epilepsy, alcoholism, or criminal or other socially deviant behavior. Over 60,000 people in the US were eugenically sterilized (Reilly, 1991). The Nazis sterilized perhaps as many as 200,000 persons (Burleigh, 1994).

Most people today think of eugenics as a coercive social program enforced by the state for the good of society (American Society of Human Genetics, 1999), although many people prominent in the Eugenics Movement of the nineteenth and early twentieth centuries favored an individual, voluntary eugenics based on the free choices of enlightened parents. As Paul and others have pointed out, however, "freedom of choice" may be compromised by economic and cultural forces, including societal views about disability (Paul, 1992; Wertz, 1993; Wertz and Fletcher, 1993).

EUGENICS AS AN OPENLY STATED GOAL

The word "eugenics" elicits favorable responses in parts of the word today, usually in terms of having and rearing healthy babies and preventing the births of unhealthy babies. In China, all 252 respondents declared, in their "write-in comments" explaining reasons for their choices of responses, that eugenics was the major goal of genetics. They defined eugenics as "improving the population quality and reducing the population quantity" (Mao and Wertz, 1997). Most Chinese geneticists are unaware of the history of eugenics in the West and of the negative connotations of the word for Westerners.

China was not alone. Openly stated approvals of eugenics as a goal of genetics also appeared in comments from India, the Czech Republic, Hungary, Thailand, Russia, and Greece. Some commentators said that geneticists should revive the use of the word and not let the Nazi experience ruin its original meaning of "having a healthy baby." They equated "prevention," which is a professionally accepted goal in most of the world, with "eugenics." Clearly the profession of genetics in many countries has an open or thinly-disguised eugenic goal.

INDIVIDUAL EUGENICS: DIRECTIVE GENETIC COUNSELING

In the modern world, eugenics is sometimes described as the result of individual choices after genetic counseling. Philip Kitcher (1996) speaks of a "laissez faire eugenics" based on the free and informed choices of individuals and couples, most of whom presumably want a baby that is as healthy as possible. Kitcher regards this form of eugenics as not only benign, but ethically desirable, because it would reduce the burden of family suffering arising from genetic disease. However, Kitcher's "free choices" depend upon the full and honest presentation of information in genetic counseling, on optimal social supports for people with disabilities, and also on a fairly high level of knowledge among the public. In the absence of adequate and unbiased information, these "free choices" become impossible, and genetic counseling becomes a means of furthering underlying goals of genetic professionals or of society. As explained in the previous chapter, around the world most genetic counseling after prenatal diagnosis is directive, usually pessimistically so. Many geneticists apparently believe that directively pessimistic counseling accords with generally accepted views in most nations that the goal of genetics is prevention of "birth defects," whether by prenatal nutrition (folic acid to prevent spina bifida), contraception, prenatal diagnosis and abortion, or diagnosis and treatment of newborns. The problem of eugenics arises when one considers who has the power to decide what should be prevented. Who defines what constitutes a "birth defect"? Who decides what, if any, prevention measures should be used? Some individual professionals, public policy makers, ethnic groups, and individual families may regard a particular genetic condition as worth avoiding by termination of pregnancy; others may regard that same condition as acceptable. Although many professionals apparently believe that they have a moral obligation, based on the principle of beneficence, to steer families toward termination of pregnancy, practices placing the ultimate power of decision-making in the hands of anyone other than the family may be considered eugenic. In genetics, information is power.

In sum, "individual eugenics" is prevalent, not as free choice, but as part of the explicit (China) or implicit goals of professionals. These goals may reflect felt social needs. National patterns ran throughout responses to all questions related to government involvement, responsible parenthood, and disability. The nations with the most pessimistic counseling were all (except Greece and Belgium) from the developing world or from formerly Communist Eastern Europe. Some (China, India, and until recently, Mexico) have major overpopulation problems.

There was widespread support, however, for people's rights to choose their mates, even if these choices increased their personal risks of having children with a genetic disorder, at least if they opposed prenatal diagnosis and abortion. Majorities in most nations (except the Czech Republic, China, India, Colombia, Peru, and Venezuela) disagreed with the statement that "people who carry a recessive gene should not mate with another carrier of the same recessive gene" (Table 52). Nations where geneticists supported rights of carriers to marry each other included those with the strongest opposition to prenatal diagnosis and abortion (Chile, Poland, Spain). There was no consensus in Northern Europe or English-speaking nations about whether people should know their own and their partner's carrier status before marriage (Table 52). Majorities in all nations of Southern and Eastern Europe, the Near East, Asia (except Japan), and Latin America (except Chile) agreed that they should, as did 66% of US primary care physicians and 65% of patients. Fewer (10% in the US, 37% outside the US) thought that "people at high risk for [having children with] serious disorders should not have children unless they use prenatal diagnosis and selective abortion" (Table 52).

"RESPONSIBLE PARENTHOOD"

Before Galton (1883) coined the term eugenics, from the Greek words *eu* ("good") and *genic* ("born"), John Stuart Mill had already described "responsible parenthood" as taking steps to have children who were as healthy as possible (Mill, 1855). Many geneticists (and patients) agree with this view.

Geneticists in most nations do not believe in restricting marriage, but most believe that responsible parents should have prenatal diagnosis if medically indicated. Majorities in all nations except Australia, Canada, the United Kingdom, the United States, Germany, the Netherlands, Switzerland, and Japan thought that "women should have prenatal diagnosis if indicated by age or family history" (Table 52). English-speaking nations rejected this statement because of beliefs about women's autonomy. The "should have" was too strong a statement. Instead, respondents believed that women should be able to make a choice, and should be able to refuse prenatal diagnosis. In both Switzerland and Germany, there has been popular opposition to prenatal diagnosis as a "eugenic" practice, and in Japan there is professional concern about eugenics. Most US primary care physicians (78%) and patients (81%) thought women with a family history should have prenatal diagnosis. It appears that prenatal diagnosis has become part of medical culture in most of the world, without any conscious connection with eugenics. If followed by the type of directive counseling previously described, however, it could have a eugenic effect.

In sum, there is widespread belief among genetics professionals that parents have a social responsibility in choosing what kinds of children they will bring into the world. Even in North America and Europe, where geneticists tend to see the counseling situation as an individual professional–patient relationship and may claim that they are not influenced by economics or society, there is nevertheless some belief that parents have a social responsibility to avoid bringing children with serious disorders into the world. In China, the child becomes a person by virtue of belonging to and being accepted by society. Society therefore should ensure that children who are born will be of a "quality" that will be acceptable to the society as a whole. In China, some geneticists commented that children who will not be accepted (including the children of single mothers) should not be born.

However, many geneticists around the world, including 27% in China and 58% in India, believed that the right (or more likely duty) to reproduce was so integral to humanity that they would support the choices of prospective parents with inheritable mental disabilities (Table 33).

The economic influence on "individual eugenics" is illustrated in Table 53. Large majorities in English-speaking nations, Northern / Western Europe, and the Near East would support the decision of a blind couple, 100% of whose children would be blind, to have a child. Many respondents said that blind people could have fulfilling lives. Exceptions were the Czech Republic, Russia, China, India, Thailand, and Peru. The economic status of the blind couple was not specified. However, when we described a blind woman on welfare with three blind children, who wants to have another child, responses were dramatically different. Our case read as follows:

> A single blind woman on public welfare has a dominant genetic disorder that leads to legal blindness. She has three sons. Each has a different father. All three fathers are absent. All three sons are 90% blind in spite of early attempts at treatment. The mother tells the geneticist, "Now I think I'll have a little girl." There is a 50% risk that the child will be blind. The geneticist thinks privately, "Don't you think you have brought enough suffering into the world?" For each statement below, indicate the extent to which you agree or disagree.
>
> 1. I agree with the geneticist's private sentiment
> 2. The geneticist should convey this sentiment to the woman
> 3. The geneticist should try to persuade her not to have another child
> 4. The geneticist should suggest sterilization
> 5. There should be a law requiring her to be sterilized.

A majority outside the US (62%) as well as primary care physicians (69%) and patients (75%) would try to persuade this woman not to have a child. Many said they agreed with the geneticist's private sentiment. This case produced outbursts of emotion exceeded only by the deaf couple and sex selection cases. Prejudices about social class and sexual morality spoke more strongly than beliefs about disability. For many respondents, parenthood for those with disabilities was reserved for the well-off.

GOVERNMENT INVOLVEMENT IN EUGENIC PRACTICE

Government involvement in genetics may take several forms, including mandatory sterilization, required carrier tests before marriage, and mandatory newborn screening for the primary purpose of identifying and counseling carrier parents before they have another child.

There are few instances of state-coerced eugenics in the modern world. For details on the 1994 Chinese Law on Maternal and Infant Health Care, which has aroused so much international concern, see the chapters by Qiu and Mao (Chapters 12 and 13). There is a question as to whether this law actually prescribed sterilization or prenatal diagnosis or whether the law used the word "shall" in an ethical sense as "ought" and not in a legal sense as "should". The law carries no penalties and is not being enforced, to the knowledge of our Chinese colleagues. Prenatal diagnosis is rare in China. Taiwan has had a similar law in effect for 14 years with no evidence that it has ever been enforced. Singapore has had a Presidential decree providing economic incentives for educated women to have more children and for uneducated women to have sterilization operations (Chan, 1985). There is no evidence about enforcement of this decree. In 1996, Japan revised its Eugenic Protection Law of 1948 to remove a list of disorders that approximated such lists in the US from the 1930s. The 1996 Japanese law (the Mother's Body Protection Law) is somewhat similar to the German abortion law in that it does not permit abortions for fetal disabilities but does permit them for the mother's psychosocial health. Most abortions after prenatal diagnosis in Germany and Japan are performed for "social" reasons. In sum, there is little evidence of governments using legally sanctioned coercion to effect eugenic practices.

PREMARITAL SCREENING

Mandatory carrier screening is a type of eugenic practice, because it is aimed only at reproductive choices, rather than to improve the health of the individuals involved. In eight countries (Table 54), majorities of respondents thought that "governments should require carrier tests for common genetic disorders such as cystic fibrosis and sickle cell anemia, before marriage." These countries were Greece, Russia, Turkey, China, India, Thailand, Cuba, and Peru. Greece has previously experimented with laws requiring premarital carrier testing for thalassemia. The experiment was unsuccessful, because many people (including doctors) lied in order to avoid the stigma attaching to carrier status. Today premarital screening for thalassemia carrier status in Cyprus is a requirement of the Greek Orthodox Church, not the government (Angastiniotis *et al.*, 1986). The church will officiate at a marriage of carriers, as long as they are aware of each other's carrier status.

Substantial minorities of US primary care physicians (36%) and patients (31%) favored government required premarital tests. They may have considered such tests an extension of the state-required premarital blood tests for syphilis that have long been standard in the US. Few US or Northern European geneticists approved of mandatory tests, however.

NEWBORN SCREENING

In most nations, newborn screening programs are provided by the government, usually on a mandatory basis. The original purpose of newborn screening was to help the newborn by providing early diagnosis and treatment, especially when treatment must begin almost immediately after birth, as with PKU (phenylketonuria, a metabolic disorder that causes severe mental retardation, unless a special diet is begun soon after birth). The rationale for newborn screening is shifting, however. Large majorities in all nations except Japan agreed that "an important goal of newborn screening is to identify and counsel parental carriers before the next pregnancy." Most patients (72%) in the US said that they would want their child screened for this purpose (Table 80). In other words, newborn screening could be instituted for conditions where early diagnosis does not benefit the newborn but parental carriers could be identified and possibly prevented (through counseling) from conceiving another child. As newborn screening is usually a government-mandated program, this would amount to a government program of screening instituted to influence reproduction. Geneticists in most nations were not willing to go this far, however. When asked whether governments should require screening for Duchenne muscular dystrophy (which would not benefit the newborn), geneticists in most nations said no. Exceptions were Greece, Portugal, Russia, China, India, and Thailand. US patients also balked at the idea of a government requirement of screening for untreatable conditions; only 26% favored it. In the future, however, inexpensive multiplex tests will make it attractive for governments to add to their lists of newborn screens, and some conditions may be included mainly to identify parental carriers. In many nations, as noted earlier, the counseling will be directive.

STERILIZATION

Although few geneticists supported mandatory sterilization (see Table 55), even in the extreme case of a single blind woman on public welfare who has already had three blind children by three different men (all absent from the household) and who wants to continue having children with a 50% risk of blindness, there were some notable exceptions. Favoring legally required sterilization in this case were 82% in China, 50% in India, 28% in Thailand, 23% in Russia, 19% in Chile, and 21% of patients in the US.

Thus it appears that most professionals in most countries reject one of the worst abuses attached to eugenics, coercive sterilization of the socially and medically "unfit." However, the percentages in favor serve as a reminder that belief in coercive eugenics remains alive in countries with economic problems compounded by problems of expanding population. There is also a likely undercurrent of eugenic thought among the public in developed nations, especially in the socio-economic groups (such as represented by our US working class patient respondents) that are most resentful of people on public welfare.

"CLEANING UP THE GENE POOL"

Most people would probably agree that means taken to directly reduce the number of asymptomatic carriers of recessive genes is a truly coercive "eugenic" goal. From a scientific viewpoint, it is also ridiculous.

For a recessive condition with a disease incidence of 1 in 2000 and a gene frequency about 1 in 45, if all people who actually have the condition refrain from reproducing, it would take 18 generations (500 years) to reduce the disease incidence by half, because most cases result from the matings of "silent" carriers who have no symptoms. For rarer disorders, a similar reduction would take far longer. Mandatory population screening, followed by mandatory sterilization of carriers, could theoretically eliminate a mutation, except in the many disorders with high rates of new mutations. (Neurofibromatosis, for example, has about a 50% "new mutation" rate, meaning that neither parent has the mutation.) If ever implemented, mandatory sterilization programs would remove a sizable percentage of the population from the reproductive arena, given that 5% of Europeans carry a cystic fibrosis mutation, and about one-third of Nigerians carry a gene for sickle-cell anemia. A program that screened for carriers of genes for a number of different recessive or X-linked disorders could quickly lead to sterilization of the majority of the population.

Nevertheless, the majority of geneticists in 12 out of 36 nations agreed with the statement: "An important goal of genetic counseling is to reduce the number of deleterious genes in the population" (Table 56). These nations and the percentages of agreement were China (100%), India (87%), Turkey (73%), Peru (71%), Spain (67%), Poland (66%), Russia (58%), Greece (58%), Cuba (57%), Mexico (52%), Venezuela (50%), and Thailand (50%). Over one-third in another five countries (the Czech

Republic, Hungary, Portugal, Brazil, and Columbia) also agreed with the statement. Most countries where geneticists favored this goal were developing nations or Eastern European nations.

"Eugenic thought," if interpreted as a desired eugenic outcome based upon individual decisions, underlies genetic practice in much of the word today. Individual eugenics is achieved through presentation of purposely slanted information in counseling, sometimes accompanied by openly directive advice. The "non-directive" counseling found in English-speaking nations is an aberration from most approaches found elsewhere.

The "individual eugenics" found in many countries today is not the benign "free choice" or "laissez faire" eugenics envisaged by Kitcher (1996), because most people (especially in developing nations) do not have (1) the expertise or status to question or challenge professionals' often slanted counseling, or (2) the economic or cultural support to take their chances of having a child with a disability or to carry such a child to term. For many people, the only "realistic" choice is to prevent the birth of a child with a disability, whatever the personal or familial costs of this prevention.

Government coercion appears to play a minor role, or no role at all, in modern eugenics, at least as perceived by most members of the genetics profession. On the other hand, governments may not need to play a eugenic role, if professionals are already doing so. There may be a fine line between individual professional–patient encounters and state intervention if professionals are paid by the state, and if the state audits counseling and prenatal diagnosis for their cost-effectiveness. In some countries, private insurance companies may play the cost-accounting role of the state in encouraging what amounts to eugenic practice.

Nations with the most pessimistic counseling and the greatest interest in government involvement are mostly developing nations. In the future, it is possible that these nations will be better able to support persons with disabilities and will become less coercive and less concerned about "birth quality" as a social issue.

Another scenario is possible in the future. Developed nations may encounter economic difficulties and may become more interested in a form of eugenics that uses economic measures to influence people's choices. Both culturally-coerced and government coerced eugenics flourish in a climate of economic depression (Allen, 1989). Eugenics thrives in hard times. Ideas of individual choice may be difficult to sustain.

CHAPTER 6

GENETICS AS A FAMILY AFFAIR

Genetics is a family matter, to greater extent than other areas of medicine. We share our genes with family members. One person's diagnosis may contain important information for siblings, children, parents, and other relatives about their risks of developing disease, including common diseases such as cancer. Sometimes the information may be useful for taking preventive measures, planning lives, or deciding whether to have children. Sometimes family members would rather not know, because the information serves mainly to arouse anxiety. Sometimes the individual initially diagnosed does not want relatives to know. And sometimes the information does more harm than good, by disrupting family relationships.

In this chapter, we discuss five family issues that arise in the practice of clinical genetics:

(1) Telling relatives at genetic risk, against a patient's wishes (Tables 57–60)
(2) Spouse or partner's access to information (Tables 61–63)
(3) False paternity (Table 64)
(4) Testing children (Tables 65 and 66)
(5) Adoptees' rights (Table 67).

On all of these issues, except adoptees' and birthparents' rights, there were considerable differences of opinion in the US among geneticists, primary care physicians, and patients. On some issues, such as testing children, primary care physicians' views were closer to patients' views than were the views of geneticists.

Except for adoption and disclosure of false paternity, there was no worldwide consensus on any of these issues. People perceived individual rights and family obligations differently. In most of the world, except North America and Northern/Western Europe, the "unit of privacy" is the family, not the individual.

GENERAL CONSIDERATIONS: PRIVACY VS. CONFIDENTIALITY

"Privacy" – as Westerners in developed nations now think of it – differs from medical confidentiality. Confidentiality – described as "the medical secret" by many respondents – is an age-old tradition known and respected worldwide. Medical confidentiality includes three types of secrets: role-related secrets, natural secrets, and contractual secrets. Role-related secrets are descended from the days when healers were priests or shamans who guarded secrets known only to God. Medical secrets are similar to secrets of the Catholic sacrament of Confession, which the priest is forbidden to disclose under pain of expulsion from the priesthood, with a few extraordinary exceptions. Natural secrets, if disclosed, would disrupt the natural order; they are too "awful" to be spread abroad. False paternity would be an example. Contractual secrets are those the parties have agreed not to tell; secrecy is based on the integrity of the parties making the agreement, not on the content of the secret or the roles of the parties. In most of the world, there is an implied agreement between doctor and patient not to disclose to third parties without the patient's permission. However, family members and spouses are not always considered third parties.

"Privacy" is more encompassing and more individualistic than medical confidentiality. It means surrounding oneself with personal possessions and personal space that one controls and that is not automatically shared with others, not even with family members. It appeared with the rise of the bourgeoisie in Northern Europe in the 17th century. Historian Simon Schama describes the development of bedrooms with doors and separate entrances off a hallway so that other members of the family did not have to walk through other people's bedrooms in order to reach their own (Schama, 1987). Even in royal palaces before this time, bedrooms stood open to long, common corridors frequented by courtiers.

The concept of privacy reached its full development in Western Europe and North America in the 18th century as part of the Enlightenment, which emphasized individual rights and freedoms. British philosopher John Locke said that "every man has his property in his own body" (Locke, 1977), a belief that comes to us today in disputes over ownership of cells and rights to individuals' genetic information. Today, privacy has reached its fullest extent in countries influenced by the Enlightenment, especially in the US, UK, and Northern/Western Europe. The Enlightenment penetrated the Iberian peninsula to lesser extent than in the North. "Daughter countries" in Latin America reflect a somewhat different view of privacy than is found in North America. In some countries, such as Russia and China, there is no word that is fully equivalent to privacy. These cultural differences inform discussion of ownership of genetic information in the next two chapters.

There are other ethics of family relationships. According to the Confucian ethic of "humanness" (*ren*), described by Qiu in Chapter 12, persons are integral parts of a whole, to which they owe their basic humanity. The family is the origin of humanness. Genetic information will be handled differently than in a Western, privacy-oriented context. In Japan, where Confucian tradition remains influential despite strong US influence, the family, rather than the individual, may be the first to hear about a diagnosis, especially if the prognosis is poor, as it may be for cancer. Physicians may prefer not to cause anxiety to the patient, and instead, they tell the spouse, adult children, or other close relatives. With the spread of Western ideas about autonomy, individual-oriented medicine is replacing these traditions in Asia.

DISCLOSURE TO RELATIVES, AGAINST AN INDIVIDUAL'S WISHES

Sometimes a patient refuses to share information about a genetic diagnosis or genetic test with blood relatives who may share the same mutation and who might benefit from the information. Usually, the individual agrees to tell relatives after intensive counseling, sometimes spread over a period of time. Occasionally, however, a person remains adamant. This refusal may result from family conflict, loss of contact with relatives, belief that relatives do not want to know or that the knowledge will only upset them, or feelings of shame about the diagnosis itself. The latter sometimes occurs in the early stages of Huntington disease, and may be a part of its symptoms.

In 1983, the US President's Commission on Ethical Problems in Medicine and Biomedical and Behavioral Research recommended that confidentiality could be overridden if the following four conditions were met:

(1) All efforts to persuade the patient to disclose voluntarily have failed.
(2) There is a high risk of serious harm to the relatives.
(3) The information would be used to avoid or minimize harm.
(4) Only information directly germane to the relatives' medical/genetic status would be conveyed (US President's Commission, 1983, p. 44).

The Institute of Medicine of the US National Academy of Sciences repeated these recommendations in 1994. The American Society of Human Genetics endorsed them in a "Points to Consider" statement in 1998, but limited disclosure to disorders where treatment is possible. Although the statement could be construed to allow disclosure if information is useful for the relatives' reproductive planning, there was not full consensus on this issue. Nevertheless, this is a landmark professional statement that departs from the traditional Western ethic of individual confidentiality, in the name of preventing harm.

The cases described in Tables 58 and 59 give examples of the kinds of cases where overriding confidentiality might be considered. Table 57 presents the following case:

> The mother and grandmother of a 40 year old woman both died of breast cancer. Given her family history of the disease, she, her daughters, sisters and female relatives on her mother's side are at increased risk. Suppose that scientists have found a gene in some women with a family history of breast cancer. If a woman has the gene, her risk of getting breast cancer is more than 50%. If she has the gene, her daughters would have a 50% chance of having the gene. Suppose that a simple and accurate test will find the gene. This woman decides to have the test because she would be relieved if the tests were negative. This woman's test results are positive. The entire family belongs to the same HMO (Health Maintenance Organization, a type of practice and insurance group found in the US), so relatives can be located easily. Would you tell her 42-year-old (healthy) sister that she is at risk for cancer, even if the patient will not tell her?

This question was substituted for a question about a nondisclosing test for Huntington disease after the Huntington gene was identified, and was asked only in the US, Germany, Japan, and Hungary. There was no consensus among M.D.-Ph.D. geneticists in any of these nations. Master's-level genetic counselors in the US, however, would adhere to the more traditional ethic of preserving confidentiality, as was their position in all the questions about disclosure to relatives. We are not sure why. They may have had greater awareness of the complexities of family situations and possible harms to family integrity. It may also be more difficult for members of a relatively new profession to step outside the boundaries of traditional ethics than it is for medical doctors. Profession, rather than gender, was the differentiating factor. Women professionals at the doctoral level were more willing than their male colleagues to override confidentiality (Wertz, 1997).

Tables 58 and 59 report responses to a generic question about disclosure, presented as follows:

> A patient recently diagnosed with a genetic or partly genetic disorder refuses to permit disclosure of the diagnosis and relevant genetic information to relatives who may be at genetic risk of developing or transmitting the same disorder. The patient is mentally competent at the time of refusal. All efforts to persuade the patient to consent to disclosure have failed.

The list of possible responses were as follows. Respondents were asked to check only one.

I would:

1) Respect the patient's desire for confidentiality.
2) Provide the information to relatives whether or not they ask for it, provided that they can be found. (Assume that professional rules in your country permit this.)
3) Provide the information to relatives *only if they ask for it*, taking care to insure that only information directly relevant to the relatives' risks is provided.
4) Send the information to the patient's referring doctor and let that doctor decide what to do.

We asked the question for five different conditions: Huntington disease, hemophilia A, familial hypercholesterolemia, schizophrenia, and Li–Fraumeni syndrome, a germ-line mutation well known to cause serious malignancies at an early age. The first two were included in the 1985 questionnaire. There was no consensus at that time. In 1995, respondents answered similarly for all five conditions, though slightly more would maintain confidentiality for schizophrenia, as more stigmatizing. Table 58 reports responses for Li–Fraumeni syndrome. There was no consensus. In all, 47% in the US and 29% outside the US would respect confidentiality; 30% and 33% respectively would tell relatives if they ask; 12% and 21% would tell relatives even if they do not ask; 11% and 18% would let the referring doctor decide. Passing the dilemma to the referring physician is not necessarily an abdication of responsibility. Many respondents wrote in that the family physician knew the patient and family situation best and could make a better-informed decision than the geneticist. Responses revealed some interesting perceptions of privacy. Although asked to check only one answer, some, especially in Latin America and Southern or Eastern Europe, checked three boxes: respect confidentiality, tell relatives if they ask, and send information to the referring doctor. In their write-in comments, they said they were respecting patient confidentiality by not telling relatives unless the relatives asked. They saw no contradiction between telling relatives who asked and maintaining patient confidentiality. The only nations where even a small majority favored maintaining confidentiality were Australia, Denmark, Germany, and Sweden.

Table 59 compares 1985 and 1995 responses for Huntington disease. In all nations included in 1985, except Switzerland and Hungary, there was an increase in the percents who would respect confidentiality and a concomitant decrease in the percents who would tell relatives unasked. There was an increase in 12 of the 19 nations included in 1985 in the percents who would tell relatives if they asked. The overall picture, however, is absence of consensus. As in 1985, respondents found these questions some of the most difficult to decide.

Access to cells is another occasion for balancing rights and needs of various family members. The two cases in Table 60 are typical of the "messy" family situations geneticists often face. Although both cases describe Huntington disease and are therefore outdated because the HD mutation has been identified and individuals can be tested directly, the generic issue remains. There are still neurological disorders where several family members must be tested and linkage analysis performed to make a diagnosis. Although technology will eventually do an end-run around the family dilemmas as more and more mutations are identified directly, variations of these two cases are likely to be around for some years. The cases were as follows:

> Case 1. A man is dying of an adult-onset neurological disorder. The gene has been mapped but not cloned. Linkage testing is available. His first wife has two children, ages 14 and 16. She wants access to a blood sample from her husband for DNA studies so that when the children become adults they will be able to learn whether they have the gene. Their father is currently dying in a nursing home and is no longer legally competent. His second wife is now his legal guardian. His second wife refuses to allow his ex-wife access to his blood because the ex-wife would not permit the children to see him in the years after he developed the disorder, despite his repeated requests. The ex-wife argues that she doesn't want the children to see him in his present condition because someday they may be in his situation. To see their own future would be psychologically disturbing. Although two unaffected members of the man's family have agreed to cooperate in testing, access to the affected man's own blood is necessary. Time is running out. If this dispute is not resolved, the children will never be able to have presymptomatic tests when they become adults.
>
> Case 2. The father of two young adults (ages 20 and 21) is in the early stages of an adult-onset neurological disorder. The gene has been mapped but not cloned [identified]. Linkage testing is available. The father is still mentally competent. Both of his children want presymptomatic tests. He is willing to permit use of his DNA so that his daughter can be tested, but refuses to permit its use for his son. He believes that the boy is too unstable. In answering this question, assume that the center where you practice will provide adequate counseling support for the children.

In the first case, there was strong consensus to "require that cells be collected and stored to permit access for the man's children when they become adults." In the second case, there was consensus to test the daughter even if the patient refused to allow testing of the son. Respondents thought that testing one was better than not being able to test either.

There was no consensus about whether DNA banks should require depositors to sign pre-banking agreements about relatives' access to cells (Table 60, col. 3). Respondents in English-speaking nations did not favor this approach, regarding it as too restrictive. In the interval since we sent out the questionnaires, private DNA banks have not developed as we anticipated, but the issue of access to stored samples has emerged as a major and unresolved problem for research collections (National Bioethics Advisory Commission, 1999).

SPOUSES' ACCESS TO INFORMATION

Spouses are responsible for the care and support of children who may be born with genetic conditions and also for a spouse who may develop a genetic condition later in life. Therefore they have a morally legitimate claim to some genetic information about their partner. The question is whether this claim should override the partner's right to individual privacy. Sometimes keeping the information secret is vital to preserving the integrity of the family and may be necessary to protect a party who would otherwise be blamed and shunned.

Three-quarters of survey respondents said that in their countries, decisions were made equally by husband and wife (Table 61). However, there were nine countries where almost one-fifth or more said that the husband's view prevails, including half or more who said this in Mexico and Peru. There were six countries where about one-fifth or more said the wife's view prevails, including half in Hungary.

There were clear divisions among geographical areas with regard to spousal access to genetic information, reflecting different power balances within a marriage and different beliefs about whether the unit of privacy is the partnership or the individual. In English-speaking nations and Northern/Western Europe, few thought a spouse or partner ought to have access to an individual's genetic information without the individual's consent (Table 62), even for a balanced translocation of chromosome 21 that causes no harm to the individual but may cause Down syndrome in the couple's children. In the rest of the world, however, substantial minorities believed that spouses should have automatic access, and that there should be no genetic secrets within marriages. US patients diverged from US geneticists, with about 43% saying that spouses should have automatic access. The patients were largely white, married, working-class women. We do not know what other groups would have said.

A related question is whether a spouse or partner should have access to prenatal diagnostic test results. One might think that these results are private to the mother, since the fetus is within her body; however, the father will be legally and economically responsible for supporting the child. Table 63, column one, describes a couple in the process of divorce, where the husband wants to know the fetus's test results. The case read as follows:

> A pregnant woman undergoing divorce proceedings has prenatal cytogenetic testing. The results indicated XXY (Klinefelter syndrome, which causes infertility, failure to reach puberty without hormonal treatment, and sometimes learning and behavioral problems). She intends to carry to term and is in the process of negotiating child support. She does not want her separated husband to know the test results because he will either insist on termination or try to reduce the amount of child support. This is not the kind of son he wants. Assume that he has no clear legal right to the information from the prenatal tests.

Although most geneticists in English-speaking nations thought that this was the mother's private information, 36% outside the US, including majorities in ten countries in Southern and Eastern Europe, Latin America, and Asia, would tell the husband. Comments on this question were heated. Some accused the woman of being greedy and concealing the information to extort unwarranted amounts of child support. Some said that she was carrying the fetus to term only to obtain money from her ex-husband. There was also considerable support for the man's plight. Others, especially in the US, said that the husband's request was a gross invasion of the woman's bodily privacy.

"Genetic responsibility" for a child's condition is one of the more difficult communications in counseling. If the child's condition is attributable to the genetic makeup of only one partner, that partner may receive blame from spouse, extended families, society, and even the child itself. If the information is not disclosed, however, the couple will not be able to avail themselves of reproductive alternatives in the next pregnancy. We presented the following case in both 1985 and 1995:

> Genetic testing of a parent reveals that this parent carries a balanced translocation that has already caused Down syndrome in this couple's child. Assume that in this case, the couple has not been forewarned of this possibility before testing. What would you tell them?

In 1995, we asked respondents to answer for two situations: (1) the woman is the carrier, and (2) the man is the carrier. Responses did not differ for the two situations, and many respondents wrote in that they regarded both as equal. The format of the 1995 question confused many respondents, but there appears to be a general trend toward disclosing this information to both parties, unasked (Table 63, cols 2 and 3). This is part of a general trend toward full disclosure reported in Chapter 3.

Although English-speaking nations and Northern Europe rejected the idea that the father of a fetus should be able to make the mother have prenatal diagnosis, substantial minorities elsewhere and majorities in Russia, Israel, Turkey, Colombia, Cuba, Mexico, Peru, China, and Thailand thought he should be able to do so, on the grounds that the child was his as much as the mother's (Table 63, col. 4). The question of rights over the contents of a woman's womb remains unresolved in much of the world.

NONPATERNITY

Nonpaternity, sometimes called false or misattributed paternity, is one of the most common "accidental" results of genetic testing conducted for other purposes. For years geneticists have estimated that the biological fathers of 5% or even 10% of children were not the same as the men who thought they were the fathers. These estimates of nonpaternity were usually based on hearsay from another lab and were not confirmed by reports in the scientific literature. We finally traced the origin of the 5% figure to some blood-typing research by J.V. Neel on Caucasians in a small Michigan town shortly after World War II (Sing *et al.*, 1971). Neel thought that many of the 5% were unacknowledged adoptions and that biological nonpaternity may have been known to most of these parents. More recently, some reports have found rates of false paternity that are greater or less than the traditional 5%, ranging from 1% to 30% (Baker and Bellis, 1995; Macintyre and Soomar, 1992; Sasse *et al.*, 1994; Cerda *et al.*, 1999).

Paternity testing is the most common use of genetic testing in the US today, with over 250,000 such tests conducted annually at the request of one or both parents, or occasionally, at the request of an adult child. However, other tests conducted for diagnostic purposes in the course of medical care sometimes reveal that the social father is not the biological father. These findings are called "accidental" because they were not the purpose of the testing. The question remains of what to do with the unsought information. Generations of medical students have faced this question in courses on medical ethics. One standard ethical argument was that the husband had a right to know because, when the doctor took his blood for testing, the husband became a patient, and patients have rights to know their test results. Few doctors would tell him, however. In the 1985 survey, 96% of geneticists would not disclose nonpaternity to a husband; 83% would tell the mother alone and let her decide what to do with the information. Some would tell the couple there had been a new mutation. Since that survey, discovery of uniparental disomy (e.g., inheritance of two cystic fibrosis genes from one parent) and other rare anomalies has lent credence to some parental denials of nonpaternity and has also made additional testing necessary in order to confirm initial accidental findings. New approaches have been suggested to handle the information. One approach, popular with US Master's-level counselors in our 1995 survey, is to treat the unanticipated information as if it does not exist. The argument is that people came to the clinic seeking a certain type of information (usually a diagnosis for a child), were tested in order to obtain this information, and should receive the information they were tested for. Anything else is superfluous, unnecessary, and probably harmful information that is outside the genetics professional's role. If the information could prevent an unnecessary prenatal diagnosis in a future pregnancy, in some cases the mother might be told, but going through the extra procedure may be less harmful on balance than entering a complex family dilemma. Variations of this "they deserve only the information they asked for" approach appeared in many write-in comments in 1995, but not in 1985.

Another new approach is for clinics to adopt "public policies" about what they will do with information about nonpaternity and to make these policies known through posters at the door or brochures given to all patients before testing. Philosopher Bernard Gert has suggested this as a fair and open way of avoiding difficult dilemmas in individual cases by letting people decide, in advance, how or whether they wish to proceed (Gert *et al.*, 1996). A few labs have tried such policies. Some put them into the consent documents that couples sign before testing, stating that the information will/will not be disclosed or that they have a choice of knowing/not knowing. The problem with placing these statements in consent documents is that the woman may wish to withdraw from testing, and by the time the couple both receive the forms (or the oral explanations) it may appear strange if she suddenly decides to withdraw. Accordingly, some geneticists have proposed telling her about possible accidental findings of nonpaternity in a special session, alone, before the informed consent, so that she can withdraw without prejudice. This approach may insult the vast majority of women who have nothing to fear and is totally impractical in countries where women are almost never seen alone in counseling. In sum, there is no simple solution.

Our 1995 question asked about some of these possible approaches. The question read as follows:

> A couple seeks genetic counseling because their child has a genetic disorder. You take the blood of both partners for testing. Genetic testing for the disorder shows that the mother's partner is probably not the biological father. Paternity testing subsequently excludes the possibility that he could be the father. The couple has not asked for this information. The man thinks he is the child's father. You tell the mother without her partner present, and she agrees that he is not the father.
>
> What are the geneticist's moral responsibilities in cases of non-paternity?
>
> 1. To tell the mother privately, in advance, that testing will disclose nonpaternity if present and give her the choice of not proceeding with testing
> 2. To tell the couple, as part of the informed consent before testing, that the tests will reveal non-paternity
> 3. To require that they make a pre-test agreement about who should know in cases of non-paternity
> 4. To tell the test results only to the mother
> 5. To urge her to tell her partner, with professional counseling support if requested
> 6. To urge her to tell the biological father
> 7. To urge her to tell her child at age 18
> 8. To tell the partner if the mother will not do so.

There was no consensus (Table 64). Warning the mother alone, before testing, was the most popular approach, followed by raising the topic in the consent document. Fewer (18%) supported the idea of a pre-test agreement about who should know, would urge the mother to tell her partner (42% in US, 31% elsewhere), would urge her to tell the biological father (34%, 43% in US), or would urge her to tell the child at age 18 (19%). As in 1985, very few (3%) would tell the spouse/partner if the woman would not do so. A slight majority (57%), however, thought that their responsibility went beyond disclosing the information only to the mother. In 1995, we asked respondents what they would do if a husband asked directly. A minority (36% in the US, 21% elsewhere) would tell him, usually after warning the woman first. Patients' views (mostly women's views) on this issue differed from those of geneticists; 74% thought the doctor should tell a husband who asks, on grounds that if he asks he probably already knows and has a right to the truth. Most said that the doctor should warn the woman first, before telling her partner. If the husband does not ask, 81% said the geneticist should tell only the mother.

TESTING CHILDREN

The increasing recognition of children's autonomy has led to a conflict with traditional beliefs about parental autonomy. One of the strongest arguments put forward by bioethicists against testing children for adult-onset disorders is that premature testing denies the future autonomy of the child (Brock, 1989; Buchanan and Brock, 1989). Persons who have already been tested as children cannot make their own decisions about testing once they become adults.

Although there are many potential benefits to testing asymptomatic children, the potential harms are such that many professional groups, including the American Medical Association (1995), American Society of Human Genetics (1995), National Society of Genetic Counselors (1995), and Institute of Medicine (1994), have urged caution. Although everyone agrees that children should be tested if there is a direct medical benefit in the near future, the various professional statements suggest that, in the absence of a direct medical benefit, testing for later-onset disorders or for carrier status is usually best postponed until adulthood.

We have summarized potential psychosocial harms elsewhere (Wertz *et al.*, 1994), as have our colleagues in the UK (Clarke and Flinter, 1996; Harper and Clarke, 1990). A child tested prematurely may become a family scapegoat for having a "bad" gene. Parents may favor the other children and withdraw family resources from the tested child. This is not entirely hypothetical. Several geneticists responding to the survey said that parents have requested testing of children for the Huntington mutation, in order to decide whether to save money for the child's college education. If the child had the mutation, the parents would not "waste" family resources. Children may also be stigmatized outside the family, by schools, by peers, by prospective partners, and ultimately by employers or insurers.

Even children who do not have the mutation in question may suffer from testing. They may feel "survivor guilt" if a sibling has or will develop a genetic disorder, while they remain healthy (Fanos, 1997; Fanos and Nickerson, 1991). Their parents may not trust the test results and may continue to believe that the child really has the mutation after all, a phenomenon known as the "vulnerable child syndrome" (Green and Solnit, 1964). Although "vulnerable children" are usually the survivors of serious illness, whose worried parents are forever looking for signs of a new bout of disease, a similar permanent worry could attach to children who have had genetic testing.

The various professional statements urge the health care provider to weigh the benefits and harms of testing in each case. Unfortunately, this may be impossible in the absence of research on the psychosocial effects of testing. To date, the only North American research on 'children' (actually adolescents age 15 and over) who have been tested reports a follow-up of Tay–Sachs carrier testing in high schools in Montreal, Canada. These studies suggest that learning that one is a carrier does not ordinarily damage self-image or lead to social stigmatization (Clow and Scriver, 1977a,b). One worrisome finding, however, suggests that some noncarriers (7%) would not mate with a carrier. A more recent high school study of cystic fibrosis carrier screening suggests a greater likelihood of peer stigmatization than did the Tay-Sachs studies (Page *et al.*, 1992).

In English-speaking nations other than the US (Australia, Canada, South Africa, UK) the percentages of geneticists who would test for Huntington, Alzheimer, and alcoholism were lower than in the US (Table 65). The low percentages who would test in the UK (2%, 2%, and 4%) are especially striking. This reluctance to test may stem in part from greater experience with requests for testing in the UK than anywhere else in the world except Switzerland and the Netherlands. In the UK, 76% had had requests to test children for adult-onset disorders, as compared with 38% in the US and 49% in western Europe. Percentages in Northern/Western Europe who would test are lower than in the US. In Southern Europe, Eastern Europe, the Near East, Asia, and Latin America, however, majorities would test children for all conditions listed. Willingness to test in these regions appears to stem from cultural beliefs about the rights of parents and their authority over their children.

Denmark and Switzerland stand out from other Northern/Western European nations as having higher percentages who would test minors for Huntington or predisposition to Alzheimer disease, though they were still in the minority. Russia, the Czech Republic, and Greece had the highest percentages of practitioners who would test for these disorders in Southern/Eastern Europe, while Italy had the lowest percentage. As in the US, fewer women than men geneticists would test for these conditions.

In the US, large majorities of geneticists, primary care physicians, and parents favored testing children for familial hypercholesterolemia and cancer syndromes. About half of geneticists and three-quarters of primary care physicians would test for genetic susceptibility to alcoholism. (Many US physicians believe that alcoholism may be prevented if susceptible people are identified early, but there is little evidence for this.) About one-quarter of US geneticists would test minors for Huntington disease or predisposition to Alzheimer disease. Fewer women than men would test for these conditions (Wertz, 1997a).

In giving reasons for their responses, 66% said they wished to avoid harm, through early treatment or prevention for familial hypercholesterolemia, cancer, or possible alcoholism, or to avoid harm from premature knowledge of Huntington disease or Alzheimer disease. 43% said the information was necessary to make decisions about medical management. Fewer mentioned parental autonomy (13%) or preparing for the future (9%) as reasons for testing. More (25%) mentioned protecting the child's autonomy as a reason not to test, and 13% said there was no need to know at this time. Few (3%) mentioned the possibility of stigmatization as a result of testing. A substantial minority (38%) had had requests to test children for adult-onset disorders. The most frequently reported condition for which parents sought testing was Huntington disease.

In contrast to geneticists, the majority of both primary care physicians and parents thought parents should be able to have their minor children tested for both Huntington and Alzheimer. Most members of the US public thought parents should be able to have their children tested for conditions that were "treatable if found early" (84%), or "preventable" (81%). A slight majority (53%) approved of testing for conditions that were "neither preventable or treatable." These results point to a dichotomy between geneticists and the rest of the US medical community, whose views closely paralleled those of their patients.

If a test is carried out, whether and when to tell the child the results becomes a question. On the one hand, it may be difficult for parents to conceal the test results entirely. Their demeanor toward the child may convey some change in perception of the child's health. On the other hand, there is at present no way of making certain that the child will have access to the results, if desired, upon reaching adulthood.

Substantial minorities of geneticists and primary care physicians would wait until a child reached the age of legal majority (18 in the US, 16 in the UK and Canada) before telling the child about a test result for familial hypercholesterolemia or a genetic cancer syndrome. These responses appear to be somewhat contradictory to the stated purpose of testing, which is to begin treatment or monitoring early. If not told the reason, the child may wonder about the purpose of medication, diet or medical testing. Even fewer parents than providers would tell the child about test results for these two conditions before the age of legal majority.

The majority of geneticists would not tell the child about susceptibility to alcoholism before legal majority, perhaps because they considered genetic susceptibility stigmatizing. Primary care physicians, however, would tell the child early, perhaps as part of an "early intervention" program to prevent alcoholism.

The majority of geneticists would not tell the results of tests for Huntington disease or Alzheimer disease before legal majority (Table 65). Primary care physicians were almost evenly divided, with a small majority favoring waiting until the age of legal majority. Among parents, 31% would tell the child the results of an Alzheimer test before majority, 55% would wait until majority, and 14% would never tell the child.

In the US, parents may approach testing laboratories directly (depending on the state). In a recent survey of testing children among 105 research laboratories in the Helix research network (an informational network of laboratories, most university-based) that also provided clinical services, 45% reported that they provided tests directly to consumers, without a physician intermediary (Wertz and Reilly, 1997). Of the 46% that had policies on testing asymptomatic children, most had only fragmentary or single-disease policies, such as refusal to test for Huntington disease. Yet 22% of those who performed Huntington tests had tested asymptomatic children under the age of 12. The laboratories surveyed were not commercial laboratories. There are no published studies of the private sector.

In a separate question, asked only in the US, Germany, Japan, and Hungary, 20% of US geneticists would urge a woman with a dominant breast cancer gene to have her 13-year-old daughter tested now; 71% of parents said it was moderately (22%), very (17%), or extremely (32%) likely that if they themselves had a breast cancer gene they would encourage their own 13-year-old daughters to be tested now (Table 65).

The Huntington disease exclusionary test, described in Table 66, is partly an issue of autonomy, but also analogous to questions about testing children for adult-onset disorders. If the fetus is carried to term after a test shows an elevated risk for Huntington disease, the resulting child will in effect have been tested. The Huntington gene was identified shortly after we began distributing questionnaires, making this question obsolete, or so we thought. However, a few requests for "non-disclosing" tests still occur. Geneticists found our question extremely difficult to answer and in many countries there was no consensus, though a slight majority would perform the test as requested (Table 66), using the bluntly stated "principle of autonomy" as the major rationale. In some countries, including Australia and the UK, half would require commitment, in advance, to abortion of an at-risk fetus, in order to avoid testing a child for an adult-onset condition. Most would provide prenatal diagnosis with no strings attached.

ADOPTEES' AND BIRTHPARENTS' RIGHTS

One of the best predictors of a person's future health, including susceptibilities to common diseases, is the biological status of the individual's parents as they age. There was worldwide consensus that adopted children should have the right to know, on reaching legal adulthood, (1) the complete medical history of their biological parents up to the time of the adoption (without the parents' names); (2) the health status of their biological parents post-adoption until the parents' deaths (without the parents' names); and (3) the causes of death for their biological parents, post-adoption (without the parents' names) (Table 67). To enable provision of such information would require centralized record-keeping and means of communication (while protecting identities) of an extent unknown in most nations today, except in Scandinavia. Nevertheless, many countries, states, and adoption agencies have established voluntary registries that the parties may update if they wish. There was far less support for disclosure of birthparents' names than for medical information, but in some countries large minorities thought adoptees had rights to know names (Table 67).

Information in cases of adoption can flow both ways. Sometimes a birthparent can benefit from information about the health of the child, especially if the birthparent intends to have more children. The case in column 4 of Table 67 described an adopted child who develops retinitis pigmentosa (RP), a disorder leading to blindness, at age 10. This particular form of RP is X-linked, meaning that it is carried by the mother alone and transmitted to male children. If the birthmother is still fertile, she could benefit from this information. Most respondents thought that the geneticist should tell the adoption agency and ask them to tell the biological mother about the diagnosis.

In sum, except for medical information related to adoption, there is no consensus about how to balance individual and family needs for information. There may be no acceptable "global ethic" in this area, given the varying emphases that cultures place on individuals and families. If such an ethic develops, it will require moral compromises and perhaps some redefinitions of privacy.

CHAPTER 7

PRIVACY, DISCRIMINATION, AND PUBLIC HEALTH

For many people in the US, privacy is the single most important ethical issue in genetics. Possible loss of insurance is their greatest fear, though this may reflect a general dislike of insurance companies rather than what is actually happening in genetics. In this chapter, we discuss access to genetic information (Tables 68 and 69), telling a school system a genetic diagnosis (Table 70), DNA fingerprinting (Table 71), genetic testing in the workplace (Tables 72–74), public safety (Table 75), and "genetic discrimination" (Tables 76–78). We have already described the meanings of "privacy" and "confidentiality" at the beginning of Chapter 6.

Geneticists worldwide distrusted insurance companies and employers (Tables 68 and 69). Close to 100% said that insurers and employers should have no access to genetic information without an individual's consent. These included one-third to one-half who thought there should be no access at all, even with consent, because insurers or employers could effectively coerce consent by refusing insurance or employment. There were similar responses for schools. However, most thought that public health departments should have access, on an anonymous basis, for epidemiological research (Table 68). Most thought that the individual adult should have the choice of knowing or not knowing his or her personal genetic information, but more thought that people should be told, even if they do not want to know, for conditions that could affect others, such as HIV (55%) or a balanced chromosomal translocation that could lead to Down syndrome in the children (42%). HIV was included in this question because at the time some people thought that laws on privacy of HIV might serve as models for laws on genetic privacy. Responses for HIV differed from responses for genetic conditions, with more in many countries favoring automatic access for spouses and blood relatives in cases of HIV, but did not differ with regard to access for institutional third parties.

Table 69 shows the percents who thought there should be access without consent to genetic information in a medical record. (We asked the same questions for forensic records and records in commercial DNA banks, with no significant differences.) Once again, responses show an overwhelming distrust of insurers and employers, and division about access for spouses and blood relatives. In the United States, 53% of patients, 20% of the public, 22% of geneticists, and 25% of primary care physicians thought spouses should have access without consent; 44% of patients, 9% of the public, 21% of geneticists, and 18% of primary care physicians thought blood relatives should have access without consent. The differences between the patients and the public may be partly attributable to demographics (patients were mostly women in the Northeast, while the public were from all regions), and partly to salience of family relationships (patients were mothers bringing children to genetics clinics, while the public may not have considered these issues before).

The majority, except in Russia, China, India, and Japan did not think that law enforcement agencies should have access without consent. In fact, these agencies can gain access with a search warrant or equivalent. Even in cases involving public safety, most geneticists thought employers should not have access to genetic information in medical records.

SCHOOLS AND GENETIC INFORMATION

Schools sometimes ask parents for genetic information, especially if a child is having problems. We described the following case:

> A couple with a mildly retarded six-year-old son comes to see you. They have received a questionnaire from the school system asking whether their son has ever had genetic testing and evaluation, and if so, what were the results.
>
> Past testing showed that the boy is XYY. The parents ask you whether they should disclose this test result to the school system. They are afraid that if they tell the school, their son will be stigmatized as a potential criminal. On the other hand, they are afraid that if they do not tell the school, their son will not receive the special early attention that could prevent later problems. How would you counsel?
>
> 1. Do not answer the school's questionnaire at all
> 2. Answer it, but do not mention the test results
> 3. Tell the school the symptoms but not the diagnosis
> 4. Tell the school the diagnosis and symptoms
> 5. This case could not occur in my country.

Patient questionnaires described "a genetic condition sometimes associated with antisocial behavior" instead of XYY. One of the most surprising findings of our survey was that 96% of patients would tell the school system about the genetic diagnosis. Most said that they thought this would help the child because the school could develop a better educational program. Some said that the school administrators, doctors, and parents should all sit down together and decide what was best for the child. Almost none mentioned the possibility of stigmatization or misuse of the information by the school system. In sharp contrast to their negative feelings about insurers, most people saw the school system as entirely benign and had no hesitation about providing information. One half of primary care physicians would also disclose the diagnosis.

In contrast, geneticists in most countries were much more cautious (Table 70), with many choosing to disclose only the symptoms, not the diagnosis, because the label XYY could stigmatize the child. The 30-year-old myth of XYY as associated with criminal behavior refuses to die. Although it has vanished from genetic textbooks, it appears in many social science texts as an example of scientists jumping to ill-considered conclusions and of blaming social deviance on genes. What college students often remember is that XYY is linked with crime, not that subsequent research showed that the link was erroneous. Today's geneticists know better, though they remain aware of the history and of the possibilities for stigmatization. Surprisingly, however, some in Europe and Latin America still thought that XYY caused crime, and even commented that the child might present a danger to other students.

When we asked this question, many people thought the situation would never become real. In the intervening years, we have heard an increasing number of reports of school systems asking very similar questions, at least in the US. Few respondents would not answer the school system's questionnaire at all. Not answering questions from a school system usually leads to more trouble than it is worth.

DNA FINGERPRINTING

In the United States, all states now have requirements for DNA fingerprinting in certain types of crimes (usually rape and murder). The UK is fingerprinting all persons who are arrested for all crimes, regardless of whether they are convicted, on the theory that most people who are arrested have committed, or will commit, another crime. Most survey respondents thought that DNA fingerprinting of criminals was, as we say in colloquial English, the greatest thing since sliced bread (Table 71). Most would require DNA fingerprinting not only for people convicted of "serious" crimes (including but not limited to sex crimes), but also for people charged with (but not convicted of) serious crimes. Most of those who thought DNA fingerprinting should be done also thought that once collected, the fingerprints should be kept on permanent file. There was strong support in most nations, including 80% of US geneticists, 73% of primary care physicians, and 86% of patients, for DNA fingerprinting of the armed forces, in order to identify casualties. There was considerably less support in most nations, including the US, for DNA fingerprinting of newborns to prevent mixups in hospitals, but 73% of patients favored this. Once DNA fingerprinting reached the level where it might affect the average respondent (passports for professionals, credit cards for US patients, who may not have passports), the level of support dropped drastically. Most people did not want DNA fingerprinting required, even if it could prevent identity fraud. (However, the majority of patients thought that welfare recipients should have DNA fingerprinting.)

WORK-RELATED SCREENING

Few geneticists – perhaps 5% worldwide- have ever been involved in occupational screening. In most countries this is done, if at all, by occupational physicians. For over 20 years, social critics have expressed concern over future widespread genetic testing of job applicants and employees for susceptibility to occupationally-related diseases, and consequent refusals of employment (Duster, 1990; Draper, 1991). The earliest and best-known example was the US Air Force Academy's rejection of candidates with sickle cell carrier status, on the unproved assumption that having one gene for sickle cell might cause a sickle cell "crisis" (extreme, painful reaction when sickled red blood cells lodge in the capillaries) under extreme conditions such as loss of cabin pressure. The Academy changed its policy in 1981, but sickle cell remained the major use of genetic testing by American companies, with no evidence that carrier status was related to any known occupational disease. Most companies never used genetic tests of any kind. A survey of 366 major companies in 1989 showed that fewer than 2% were using genetic testing, and 5% had used it in the past (mostly for sickle cell), (US, Office of Technology Assessment, 1990). More (16%) anticipated using new genetic tests as they became available. However, tests with predictive value and practical utility (cost-effectiveness) for most companies have not yet become available. Fear of genetic testing in the workplace remains as strong as it was 20 years ago, but the testing itself remains largely futuristic.

There are, however, some outrageous examples of misuse in unexpected places. The Lawrence Berkeley Laboratories, run by the University of California and the US. Department of Energy, did sickle cell testing on clerical employees until 1994, without the employees' knowledge or consent (Norman-Bloodsaw vs. Lawrence Berkeley Laboratory, 1998). There was no medical rationale for these tests. Although no one lost promotion or pay, the employees sued for invasion of privacy and social discrimination and won in Federal court. (Although Lawrence Berkeley was sequencing part of the human genome, laboratory personnel were not tested.) We do not know how much, if any, workplace testing occurs in other countries. The two cases we presented would have been largely hypothetical to most of our respondents.

We described the following two cases:

1. Assume that an accurate, simple, and reliable mass screening test has been developed for alpha-1-antitrypsin deficiency. This raises the possibility that factory workers who will be exposed to dust and smoke would be screened.
2. As part of an executive health promotion plan, a prominent private company wishes to test executives for predispositions to heart disease, cancer, and diabetes. Assume that there are highly accurate tests available.

There was worldwide consensus in both cases that the tests should be offered and should be voluntary, that there should be no financial incentives to take them or penalties for refusal, that costs should be reimbursed by public insurance, that there should be no access to results without the worker's consent for employers, government health departments, or labor unions (Tables 72 and 73). Only the first case is reported; there were no differences in regard to the second case, with the exception noted below. About one-quarter said there should be no access at all for employers, governments, or unions, even with consent. We asked a similar question in 1985. In all nations previously surveyed, except Sweden, more in 1995 thought screening should be voluntary than thought so in 1985, and fewer in all nations thought that the government health department should have access without consent. There were two notable exceptions to the overall consensus that screening in the workplace should be voluntary: only 51% of patients and 39% of the US public agreed. The patients – mostly working class – seemed to think that mandatory screening protected the worker. They took a different view of screening for executives, with most (81%) saying it should be voluntary. These answers probably reflect the respondents' social class.

In cases where screening shows heightened susceptibility and the workplace cannot be made safer, most respondents would let the worker decide about whether to stay at the same job or move to a less hazardous job within the company, at the same pay (Table 74). However, majorities in the Netherlands, Portugal, Spain, Latin America, China, and Thailand would not give the worker a choice and would simply move the worker if a less hazardous job were available. If there is no less hazardous job to which a worker can move within the company, most respondents thought that the worker should be allowed to stay on, after explanation of the risks (Table 74).

Responses to these questions illustrate overriding support for autonomy, and also, in some countries, economic necessity for the worker.

PUBLIC SAFETY

Sometimes a worker's genetic status poses a threat to others. The classic case described at genetics meetings involved the pilot of a 747 who has the mutation for Huntington disease; a critical error in judgment may be the first symptom. To our knowledge, no such pilot exists, but there have been air traffic controllers with the Huntington mutation and a tractor-trailer driver with actual symptoms of the disease. Many geneticists believe that individual confidentiality overrides safety concerns even in these cases.

We presented a less dramatic case:

> A public bus driver has familial hypercholesterolemia. His annual checkup shows a blood cholesterol of 400+ and high blood pressure. Last year he reported chest pains but now denies this. He could have a heart attack on the job, thus endangering his passengers. He has not told the transportation department because he fears that he will be fired or shifted to a position with lower pay or fewer benefits. He says he has only a few more years to go before he can retire comfortably. Your efforts to persuade him to retire on disability have failed.
>
> Would you (check only one)
>
> 1. Preserve the patient's confidentiality.
> 2. Tell the transportation department if the patient refuses to do so.
> 3. Warn the transportation department that passengers are at risk because of the medical status of a bus driver and ask them to have complete physical examinations for all their drivers, but do not tell them about your patient.
> 4. Warn the transportation department that passengers are at risk because of the medical status of a bus driver and ask them to have complete physical examinations for all their drivers. If the department will not do so, tell them about your patient.

For US patients and public, the case described a school bus driver (an occupation not found in some countries), which may account for their greater willingness to tell the transportation department about the driver's condition. Responses did not show worldwide consensus. More in the UK, US, Northern/Western Europe, and Argentina would preserve confidentiality than would do so elsewhere. Many Europeans wrote in that their disability laws would make it possible to move the driver to a desk job while maintaining the same pay, or would otherwise preserve both his present and future financial status and benefits. Others – also mostly Europeans – said that their laws governing physical examinations for transportation workers were so strict that required exams would have already identified this driver. Few anywhere would tell the transportation department directly, but close to one-third in all would choose the compromise alternative of warning the transportation department that passengers were at risk because of the medical status of one of the department's drivers, and ask the department to have complete physical examinations for all drivers. That few in English-speaking nations or Western/Southern Europe chose to add "if the department will not do so, tell them about your patient" suggests that in these nations people had faith in transportation departments. In Eastern Europe, Latin America, and Asia more chose this answer.

US patients and public favored telling the transportation department before passengers got hurt. Far fewer mentioned confidentiality.

GENETIC DISCRIMINATION

Respondents' reports about their countries' disability laws were highly varied (Table 78). In many cases, differences among respondents in the same country suggested that they were unfamiliar with their country's laws or how these laws could be interpreted for genetic conditions.

In the United States especially, there has been much talk about "genetic discrimination," a term that usually goes undefined. People are afraid that they will lose their health insurance if the company finds out about the results of a genetic test. Many are afraid that they will lose insurance even if their company finds out that they have participated in a genetic research project, without knowing the results (Lapham *et al.*, 1996). Willingness to participate in genetic research has declined, partly because of these fears. Meanwhile, there have been few reliable studies of the actual extent to which people are being denied insurance or employment on an unjustified basis. Most are collections of cases (Billings, 1992), somewhat along the lines of 19th century medical reports that listed the number of successful operations performed at particular hospitals. Others are surveys with extremely low response rates (Geller *et al.*). One recent survey has found, as we did, that there appear to be few actual cases where people were denied insurance on genetic grounds (Rich and Hall, 2000). Nevertheless, most people hate insurance companies, and many may be willing to attribute all denials of insurance, and all manner of diseases, disabling conditions, and cosmetic problems to genetics.

There is no uniform definition of "genetic discrimination." The word discrimination is usually associated with sex, race, or a history of oppression, and comes with social entanglements implying that insurers or employers have a purposeful plan to exclude people, when in fact many refusals may arise largely from insurance company ignorance of particular genetic conditions. Also, there are situations where exclusion of a particular condition or charging a higher premium may be actuarially fair, because treating the condition will be especially costly.

We wanted to look for cases where refusals of insurance or employment were clearly unfair and were on the basis of genotype. Accordingly, we did not use the word "discrimination." We limited questions to refusals on the basis of (1) carrier status alone (no symptoms), and (2) genetic predisposition, in the absence of symptoms. Our question read as follows:

> Have any of your patients or families:
>
> a) been refused life insurance because of carrier status?
> b) been refused life insurance because of a genetic predisposition in the absence of symptoms?
> c) been refused employment because of carrier status?
> d) been refused employment because of a genetic predisposition, in the absence of symptoms?

Providers were also asked to estimate how many persons had been refused in each category. Patients and public were asked about refusals for themselves or family members "because of genetic or inherited disability or disease," a statement that includes symptomatic as well as asymptomatic persons, and which allows respondents to create their own definitions of inherited disease.

Although the majority of reports of refusals of employment or insurance came from the United States (Table 76), there were also some reports from other English-speaking nations and Western Europe, suggesting that what is now mainly seen as a US problem may spread in future years.

In the US, 4% of geneticists reported that at least one patient had been refused employment on the basis of carrier status and 12% reported at least one refusal of life insurance on the basis of carrier status (We did not ask about health insurance, because most nations have national public health insurance.) (Table 76). Three percent reported at least one refusal of employment on the basis of a genetic predisposition, in the absence of symptoms, and14% reported at least one refusal of life insurance. In all, there were 321 reports of refusals of employment or insurance on the basis of carrier status or predisposition, and a total of 693 persons refused.

In the US, responses indicate that refusals solely on genetic grounds exist. Providers were asked to describe these cases, but many did not do so. Among those who did, it was evident that some refusals were based solely on company ignorance (e.g., for sickle cell carrier status, which has no health effects of actuarial interest). Others were not only unfair, but poignant for the families affected. In one case, a family "had to give up" a child with a costly condition because the father would lose the family's insurance if he changed jobs. (This situation has now been partially remedied by the federal Health Insurance Portability and Accountability Act of 1996.) Even if few people are affected by unfair refusals, it is important to protect them from financially disastrous outcomes.

However, the volume of reports in the US appears small in the context of overall numbers of providers and patients. These reports came from a total of 1,084 genetics services providers with a median of nine years' experience, who saw a median of six patients a week. Given their total patient volume, the numbers of reports of refusals are few, perhaps one-twentieth of one percent. It may be that, because genetics specialists often see people on a "one-time" basis, as a tertiary referral, they do not hear of such refusals.

The 499 primary care physicians, who are more likely to have ongoing contact with patients, reported even fewer refusals. There were 18 (5%) reports of refusals of health insurance on the basis of carrier status, with a total of 32 persons refused. There were four (1%) reports of refusals of employment on the basis of genetic predisposition, in the absence of symptoms. Twelve (3%) primary care physicians reported refusals of health insurance on the basis of genetic predisposition, with a total of 29 persons refused, and 20 (4%) reported refusals of life insurance on the basis of genetic predisposition, with a total of 31 persons involved. Primary care physicians had a total of 34 reports of refusals of insurance or employment, with a total of 92 persons refused. The 499 primary care physicians had a median of 14 years in practice and saw a median of 51–100 patients a week.

Among genetics patients, minorities reported that "because of a genetic or inherited disability or disease," they or a family member had been "denied or let go from a job" (2%), "refused health insurance" (3%), "refused health coverage of some services" (7%), "refused life insurance" (5%), or "refused school admission" (1%). When asked to give details of their refusals, almost all described situations that are characteristic of a broad general employment practice or general insurance practice. Examples of patient reports include: "I was refused a job as a firefighter because of chronic bronchitis;" "I was let go from a job as a proofreader because of cataracts due to Marfan syndrome;" " I hurt my back at work and was let go because I couldn't do the job;" "I was refused disability insurance because I'm overweight;" "I was refused additional life insurance because I just got out of the hospital after a heart operation, and the company wanted to wait until they saw how I was doing;" "I was refused cosmetic surgery on my child with Down syndrome, because the company doesn't pay for cosmetic surgery;" "I was refused payment for a third trimester abortion, because it had to be done out of state, and the company doesn't pay for out-of-state procedures;" "I was refused life insurance for my child with cystic fibrosis." (Some parents take out life insurance on their children with disabilities, perhaps hoping to recover their costs for the child's care.) None of the patients' reports pointed to specifically "genetic" discrimination. People were apparently objecting to what they perceived as unfair insurance practices in general, rather than practices specific to genetics. In this context, their characterization of just about all manner of diseases as "genetic" or "inherited" is interesting. Environment and lifestyle seem overshadowed by inheritance.

In the public survey, small numbers said that they or a member of their family had been denied or let go from a job (3%), refused health insurance (5%), or refused life insurance (4%) "because of a genetic disorder or a disability due to heredity." In the US, the proportion of individuals who were refused insurance approximately equals the total proportion of individuals who are refused private (medically-related) health insurance and life insurance (about 5%).

We conclude that:

1. "Genetic discrimination" exists. There is clear evidence in professionals' reports of refusals of employment or insurance on the basis of carrier status or genetic predisposition in the absence of symptoms.
2. "Genetic discrimination" is rare and is not a major reason for refusals of employment or insurance at present.
3. Most patient reports fell within general insurance or employment practices, not practice targeted at genetics.

Such reports may increase as more tests with greater predictive value appear. At least this is what many people fear. There is still no evidence that insurance companies plan large-scale genetic screening. Most get the information they need for actuarial purposes from the family history. Recently, however, discussion has arisen about whether family history should be put into the same category as genetic tests (that is, "genetic information") for purposes of law and policy, and whether insurance companies should be forbidden to ask about family history. This could lead to major changes in insurance practice.

Sometimes genetic testing places people at social risks of loss of insurance or employment. We posed the following case:

> A woman aged 35 expresses interest in being tested for autosomal dominant polycystic kidney disease, which leads to kidney failure. Her brother has recently been diagnosed with ADPKD and has urged her to be tested to see whether she could serve as a kidney donor for him in the future. You describe the medical aspects of the disorder to her and her chances of developing it. Would you tell her more? For *each* statement, indicate whether you agree or disagree that she should be told.
>
> a. I would tell her that a positive test may make it difficult to get life insurance.
> b. I would tell her that a positive test may make it difficult to obtain employment, if employers can get access to her medical records.

Many respondents said that denials of employment or insurance could not happen in their country, because laws prevented it or because privacy of medical records was so strict. This may explain why so many would not warn this woman. Some said that such undesirable outcomes could occur only in the United States, with its lack of legal protections. In the United States, as in most English-speaking nations, the Netherlands, and India, about four-fifths said they would warn the woman about possible loss of insurance and a majority would warn about risks to employment. It is questionable how many actually do so, however. A separate survey of Master's level genetic counselors suggests that many do not (Hoyle *et al.*, 1997).

PUBLIC HEALTH GENETICS

Several items in the survey fell under the general heading of public health genetics. These include newborn screening for treatable genetic disorders (Tables 79 and 80), carrier screening for cystic fibrosis (Table 81), and regulation of DNA banks (Table 82).

State-mandated newborn screening began with PKU, after it was discovered that a special, low phenylalanine diet could prevent mental retardation, but only if instituted shortly after birth. Now, governments in many countries require screening of newborns for various conditions, most commonly congenital hypothyroidism and PKU. The general rule has been that newborn screening should be mandatory only if early treatment would benefit the newborn and such treatment would be made available in a timely fashion on an affordable basis. With the advent of a new technology, tandem mass spectrometry, which automatically identifies a range of metabolic disorders, both treatable and not treatable, newborn screening is starting to move away from these parameters. Even if a condition is not successfully treatable, there may be a benefit to parents in knowing, in advance, that it will appear, so that they do not have to go from doctor to doctor seeking a diagnosis, and perhaps blaming themselves for their child's mysterious illness. Our survey did not deal with the rare conditions revealed by tandem mass spectrometry, but instead asked what conditions respondents would add to screening programs if DNA automation techniques were available. Early treatment of newborns makes no difference for any of the four conditions listed in Table 79. (Particular infants with cystic fibrosis have benefited, but there is no evidence that treatment benefits newborns with cystic fibrosis as a group.) There was widespread support for adding all four conditions to newborn screening programs. However, there was no consensus about whether the blood spots from newborn screening should be kept indefinitely (half in the US thought they should). Most respondents, however, thought that children should have access to their blood spots at the age of legal majority.

In a separate part of the questionnaire, we asked whether governments should require newborn screening for another untreatable condition, Duchenne muscular dystrophy, an X-linked condition that leads to loss of muscle control and to death in the teenage years. Respondents answered quite differently than for the four conditions in Table 79, with few supporting screening. It may be that this question, unlike the others, included the words "governments should require." Although "classical PKU screening programs" in the previous table are also government-mandated, the words were not used. The difference deserves further investigation. It may be that geneticists want to expand the list of newborn screens but are hesitant to make this a legal requirement.

Other evidence that newborn screening is moving beyond its classical goal of providing early treatment for newborns appears in responses to the question of whether "an important goal of newborn screening is to identify and counsel parental carriers before the next pregnancy." Three-quarters believed that it is (Table 80). Patients were asked a similar set of questions.

Their responses shed light on the interplay between their desire for information and their uneasiness about government intervention. Two major points emerge from the patient survey.

1. Most parents wanted to know if their newborn would develop a nontreatable disorder

The question read as follows:
"Some genetic conditions can be found at birth through a simple blood test. Sometimes there is no treatment for the child. In these cases, the main purposes of testing the newborn child are to find out if this child has a genetic condition and to let the parents know that they could have another child with the same condition. If you were a parent, would you want your newborn child tested right away so that you could find out if your next child might have a genetic condition?
71% said "yes", 11% said "no", and 18% said "don't know".
In write-in comments, parents said they had a right to know, that the information would help them relate to their child, and that they wanted the information so they could decide about having another child. These comments fit in with a strong belief in patient autonomy that ran throughout responses to the questionnaire.

2. Parents were uncertain whether the state should require newborn screening for nontreatable disorders. Although most wanted to know, many were hesitant about the state's requiring them to know. They would prefer to make the request on their own initiative

The question was a continuation of the previous question. It asked "Would you want your state to require such testing?"
27% said "yes", 50% said "no", and 24% said "don't know".

Mandatory screening is ordinarily followed by counseling, including information about the chances of future children having the same disorder. (Parents can refuse counseling but are unlikely to do so.) Mandatory screening for nontreatable disorders could therefore lead to what approximates mandatory genetic counseling, because of the state's involvement.

The questions on cystic fibrosis carrier screening were developed when the tests were new and the genetics profession in nations where most people were of European descent (cystic fibrosis is rare elsewhere) thought that most of the population would be interested in having tests. Since then, we have learned that few people are interested, except pregnant women whose obstetricians put forward the test as part of prenatal care. Most others regard the information as of little use. The questions presented in Table 81 are of historical interest but do not represent current practice. Population screening for cystic fibrosis carrier status has never occurred anywhere, and today most people having tests are pregnant women. Responses show that more geneticists favored preconception testing than testing during pregnancy or before marriage, and that few would test teenagers. There was no consensus about whether tests should be offered to everyone (an expensive proposition) or only to members of high-risk ethnic groups (whites).

The private DNA banks for personal storage of cells that were envisaged in Table 82 have not emerged. However, immense cell repositories now exist for research purposes, with an estimated 282 million samples in the US alone (National Bioethics Advisory Commission, 1999). Many samples in pathology laboratories predate informed consent. Traditionally, pieces were sliced off paraffin (wax) blocks containing the samples and shared with researchers without the "donor's" knowledge. Even if a blood or saliva sample were given specifically for research, with informed consent, the question arose as to whether the sample could be used in new research projects not foreseen at the time of the original donation, without going back to the original donor and getting a new informed consent for the new research, a costly and time-consuming process. A further question arose as to whether existing samples in collections could be used for research without consent if the samples were "anonymous" or rendered anonymous by removing all identifiers.

Anonymity is never total, because any DNA sample can be traced to its individual source, given enough technological and human resources, though such efforts would have to be highly motivated and would be extremely rare. Sometimes the sociodemographic and disease information accompanying anonymous samples also makes it possible to trace them to individuals. The issues are so complex that the US National Bioethics Advisory Commission (NBAC) devoted several years of discussion, resulting in a comprehensive document with recommendations, entitled Uses of Human Biological Materials: Ethical Issues and Policy Guidance (1999).

Cell repositories are adding samples from all over the world for research on asthma, diabetes, cancers, and cardiovascular disease. Issues of informed consent, new research uses, and ownership will likely be decided by the developed nations where commercial companies are located. (The Human Genome Diversity Project, a proposal to collect samples from diverse groups for research on population genetics and migration, is entirely separate from these large-scale disease-related cell collections. This project has encountered opposition from some indigenous groups and has not been funded.)

When the survey began, most respondents were unfamiliar with these issues. There was no consensus about whether existing collections could be shared with other researchers if the samples were anonymous (NBAC has since recommended that they can). Majorities in the survey, including two-thirds in the US, thought that a new permission should *not* be required for uses of a sample in new research, and three-quarters thought that donors should *not* have a right to share in profits from commercialization of research, with an especially strong rejection of this idea in Northern/Western Europe (Table 82). As with earlier questions on family access (a question that might arise if a donor died in such a way as to leave no other DNA sample), there was no consensus about access for blood relatives.

CONCLUSIONS

In sum, responses to questions on privacy were mixed. Dislike of insurance companies and distrust of employers influenced many responses toward absolute protection of privacy from intrusions by these institutions. On the other hand, faith in schools and belief that disclosure of information will benefit their children led most patients to say they would disclose a genetic diagnosis to a school system. Responses to public safety questions suggest that confidentiality and public safety vie for dominance, depending partly on a country's laws protecting a worker's financial status. Patient and professional views on this issue differed in the US, France, and Germany. Most respondents were enthusiastic about DNA fingerprinting for forensic purposes but not for passports or credit cards.

The survey documented the existence of refusals of insurance or employment solely on the basis of genotype, in the absence of symptoms. Some of these refusals could appropriately be labeled "genetic discrimination." Such refusals were rare in proportion to the numbers of patients seen, and most were part of insurance practice generally rather than inequities based specifically on genetics.

CHAPTER 8

THE GEOGRAPHY OF ETHICAL REASONING

People may arrive at the same conclusions on the basis of different ethical values or courses of reasoning. They may also start from the same premises and arrive at different conclusions. The moral bases of reasoning may be as important to international discussion as answers to specific cases.

RESPONDENTS' REASONING

We asked respondents to explain why they had chosen their answers (Table 83). These explanations were in respondents' own words, on several blank lines provided after most questions. Although some left these lines blank, geneticists gave a total of 44,178 qualitative responses in their own words. We assigned numeric codes to these, using an expanded version of a set of categories originally developed for our 1985 survey. The Codebook appears in the Appendix. The codes were summed across responses to all questions, using the "Multresponse" procedure in the SPSS (Statistical Package for the Social Sciences) program. Readers should remember that the unit of analysis in the first part of this chapter is responses, not respondents.

We coded up to two moral reasons for each comment. In developing categories, the list of principles – autonomy, nonmaleficence, beneficence, and justice – found in standard Western bioethics texts appeared to be the simplest approach and also best reflected the general structure of respondents' thinking, at least as expressed in writing and translated into English. We realize that categories might have appeared differently if based on Asian approaches. However, many Asian respondents used equivalents of nonmaleficence, beneficence, justice, and respect for persons.

We also coded responses for the "bases of reasoning." Did respondents mention moral principles as such ("patient autonomy," "avoidance of harm," "rights") without tying these to the situation at hand? Did they base their actions on experience, for example, by describing a similar case in their practice? Bioethicist Albert Jonsen describes these two viewpoints in his essay on "Balloons and Bicycles" (Jonsen, 1991). The balloonist sees the overall picture, while the bicyclist has the best view of the details of a particular situation. For example, imagine a multi-car accident on a major highway. The balloonist sees the extent of the accident, the traffic jams extending for miles, and the difficulties of getting rescue vehicles to the scene. The bicyclist sees the injuries of several motorists at close hand and their particular medical needs.

Principles-based ethics is analogous to the balloonist's orientation, and situation-based ethics is analogous to the perspective of the bicyclist. Both have something to contribute to the moral appraisal of the whole situation.

Most respondents were neither balloonists nor bicyclists (Table 84). They preferred an intermediate approach that connects principles directly to situations, without describing the principles in abstract, universal terms. A short-hand version of the principle is given (e.g., "the patient has a right to know the test result," instead of "respect for autonomy"). Bioethicists Beauchamp and Childress (1994) call these short-hand versions of principles "action-guides," which are prescriptions derived from principles to guide decisions and actions. Few geneticists or physicians think in terms of abstract ethical principles. Action-guides are the major bases of medical professionals' ethical decision making in the United States. Apparently professionals around the world used the same type of shorthand; 47% of qualitative responses outside the US were based on action-guides, as compared with 9% on principles alone and 19% on experience. (33% of responses had no discernible bases, or simply repeated what the respondent would do.) The low percents who referred to experience may reflect the small size of the space allowed for comments. A survey using interviews might have elicited far more reliance on experience as a basis for actions.

We coded responses for their reference to consequences of an action, including both openly mentioned consequences and implied consequences ("the family needs this information" – implying that they will use it but making no specific mention of its use). A minority of comments pointed to consequences – 27% in the US and 33% outside the US, with little variation among countries. An exception was China, with 59% of responses directly mentioning consequences. Usually these consequences were the helping or hindering of Chinese geneticists' stated goal of "reducing population quantity and improving population quality." Every Chinese respondent described this goal several times.

We also looked for references to relationships in the comments, including both expressed and implied relationships (Table 85). An ethics of relationships is an alternative rarely used in Western cultures, though some feminists have supported it as more humane and more reality-oriented than a principles-based ethics (Noddings, 1984). The US National Society of Genetic Counselors (NSGC) has a relationships-based Code of Ethics (1992). The Code explains how counselors can base their actions on examination of the relationships involved in a case, including relationships with the patient, with the patient's family, with society, and with oneself (personal and professional integrity), and also on the patient's own relationships with spouse/partner, family, and others. We had expected to see some of these relationships mentioned or implied in the comments. Accordingly, we coded each comment for up to two relationships, using the types of relationships described in the NSGC Code as a guide. Few comments – 16% both in the US and worldwide – mentioned or even implied relationships of any kind. Especially few mentioned relationships with society; China was an exception, with 10% mentioning benefits or burdens to society. This does not mean that relationships did not exist or were not important. The small space provided for comments may have discouraged their mention. It appears, however, that relationships were not the first considerations that came to mind when people gave reasons for their actions.

To the standard four bioethical principles, autonomy (respect for persons and their choices), nonmaleficence (avoidance of harm), beneficence (actively doing good for others), and justice (as equity and fairness), we added a fifth, deserts. Deserts are the basis of affirmative action programs for racial minorities and women in the US, for Scheduled Castes (former "Untouchables") in India, and special education and employment preferences for people with disabilities in many countries. Adding deserts as a special "principle" allowed us to code comments referring to patient characteristics or disabilities. The category also includes "undeserts" in the form of negative and pessimistic comments on people with disabilities, such as came from China, India, Russia, and many other developing or Eastern European nations.

More comments (38%) fell under the heading of "autonomy" than any other principle (Table 86). Israel led the way with 54%, followed by Belgium (48%), Finland and the United States (45%). China (20%) and India (25%) had the fewest autonomy-related comments.

Nonmaleficence ranked second, appearing in 24% of comments, ranging from 20% in the US to 34% in Greece. Deserts came third, with 13%, ranging from 8% in Finland and Thailand to 24% in Hungary. As in the 1985 survey, comparatively few mentioned reasons related to beneficence (9%) or justice (8%). China was an exception, with 27% of comments related to beneficence, as compared with 4% in Canada, the UK, and the US, and 4–7% in Western and Southern Europe. Clearly, Western medical ethics is dominated by respect for individuals and their autonomy, combined with a desire to prevent harm to these individuals and families. Promoting the common good, deserts of individuals and groups with special characteristics, and justice (equitable access) play lesser roles in clinicians' thinking.

PROFESSION, GENDER, AND GEOGRAPHY

Professional Cultures: Geneticists, Genetic Counselors and Primary Care Physicians in the US

In the United States, three distinct groups of professionals provide genetic services. (For details, see Chapter 10, Organization of Genetic Services.) Their differences in ethical views are instructive, particularly in a field that is entering mainstream medicine. Over half of US respondents (555) were specially trained genetic counselors with Master's degrees. The rest were M.D.s or Ph.D.s (529). Counselors differed from doctoral-level geneticists ($p < 0.05$) on 295 (60%) of the 491 sub-parts of ethical questions and on 25 (96%) of 26 questions about having had experience with the cases described. Some of the major differences are described in Tables 86–89.

Although most (94%) Master's-level counselors are women, comparisons between counselors and women doctoral-level geneticists suggest that most differences between counselors and geneticists are associated with profession rather than gender. Counselors differed from women M.D.s on 46% of issues and from men M.D.s on 60%. Genetic counselors appear to have a distinct ethic that places greater emphasis on (1) client autonomy, including the autonomy of children; (2) nondirectiveness in counseling after prenatal diagnosis; (3) support for people's decisions even if the counselor disagrees; (4) individual privacy, rather than duties to warn relatives at risk or preserve public safety. In most cases where individual rights or privacy conflicted with the potential welfare of the family, counselors placed the individual's rights first.

The exception was accidental findings of nonpaternity, where counselors, like women physicians, placed the welfare of the family above the man's right to know. More counselors would help couples obtain prenatal diagnosis for sex selection by offering referrals. In most cases where there was a conflict between parental autonomy and the health of a potential child, counselors placed autonomy first. Counselors were more reluctant to use abortion themselves in 20 of the 26 situations described; and more would present giving up a child for adoption as an alternative to abortion after prenatal diagnosis. They were also somewhat more optimistic about disability than either women or men M.D.s. More counselors thought that people with severe disabilities enriched society or that disabilities could be overcome with social support.

US counselors were at the opposite end of the spectrum from women physicians in other nations in regard to many issues, including directiveness, disability, confidentiality, and conflicts between individual and family or individual and society. Counselors saw their primary relationship as with their individual client.

Genetic counselors' distinctive ethics is not only appropriate for a nation in which individual autonomy is the primary social value, but may also serve as a survival mechanism for an emerging profession that must cooperate with the established field of clinical genetics. By being nondirective, counselors can avoid conflict with physicians who may hold different opinions. Counselors cannot be faulted by physicians for presenting information and letting clients make their own decisions. By offering referrals instead of refusing requests for sex selection, they can fulfill a duty to the client without being directive.

The third professional group, primary care physicians, differed from US genetics professionals as a whole in some important respects. Primary care physicians were (1) less likely to agree that patients were entitled to whatever services they can pay for; (2) more directive in counseling, though their direction in some cases was to urge women to carry a child to term; (3) more ready to hold parents responsible (blameworthy) for knowingly having children with genetic disorders, if the birth could have been avoided; and (4) less likely to believe that national public health insurance should pay for genetic services. When genetic counselors and geneticists are considered separately (Table 90). However, it appears that many of the differences between primary care physicians and genetics professionals are accounted for by differences between M.D.s and counselors, with their special ethic of nondirectiveness and support for patient decisions. Nevertheless, primary care physicians (especially obstetricians) are more directive and generally more "paternalistic." Their counseling after PND – whether optimistic or pessimistic – is a very different world from that of genetic specialists. They see far more patients (over 120 a week) than geneticists or counselors (6) and live in a different clinical world. Their write-in comments were concerned about the evils of teenage pregnancy, sexually transmitted diseases, patient irresponsibility, welfare fraud, and socialized medicine. However, they were aware of the importance of genetics; 75% had referred at least one patient to a genetic specialist during the preceding year.

In US genetics, it is not clear whose ethic will prevail. Much depends on how, or whether, patients express their own views, which are not always in accord with the views of professionals. The general trend in medicine is away from paternalism and toward increased (sometimes exaggerated) respect for autonomy, as patients become "consumers."

Most nations have no separate profession for genetic counselors. The differences between primary care physicians and genetics services providers (usually M.D.s) may not be so stark, though the Clinical Genetics Society in the United Kingdom found dramatic differences with regard to testing children, with far more general practitioners willing to test children for adult-onset disorders (Clinical Genetics Society, 1993). More research is needed on the views of ordinary physicians, and more discussion is needed among medical specialties about developing uniform approaches to ethical issues.

The Role of Gender

Worldwide Differences

A comparison between women and men MD geneticists shows important worldwide differences. Women were (1) more open to requests for prenatal diagnosis for sex selection; (2) more likely to override individual confidentiality in order to warn relatives about their own genetic risks; (3) More likely to think that spouses and blood relatives should have access to an individual's genetic information without consent (even though only a minority thought so in the US); (4) Less likely to think that parents should be able to have their children tested for mutations for adult-onset disorders such as Alzheimer disease, Huntington disease, predisposition to alcoholism or breast cancer. Women also had greater respect for the right of a 16-year-old to refuse testing. Although fewer women than men were married or had children, these characteristics were unrelated to attitudes about testing children.

Women physicians were more likely than men to regard genetic information as belonging to the whole genetic family rather than as the property of one individual. In four of five disorders listed (Huntington disease, hemophilia A, familial hypercholesterolemia, and Li–Fraumeni syndrome), fewer women physicians worldwide would maintain confidentiality and more would tell relatives if they asked or would refer the matter to the family physician, who presumably has a closer relationship with the family. Outside the United States, women would also tell relatives about a fifth disorder, schizophrenia. Women's responses to this question point to respect for the family as a network of relationships.

US Men and Women Geneticist–Physicians

In the US, profession appears to play a greater role in ethical views than gender. Nevertheless, gender is important. Women comprised 45% of the 338 US geneticists with M.D.s.

US men and women geneticist–physicians reported similar levels of experience with most of the cases described; men had more experience with accidental findings of nonpaternity and more direct questions from husbands who wanted to learn if they were the father of a child. Women were as likely as men to tell a man who asked directly whether he was the father of a child. Women physicians were similar to men in directiveness in 22 of 26 counseling situations after prenatal diagnosis.

Women geneticist–physicians differed from men on 13% of 491 sub-parts of ethical questions. In addition to the worldwide differences discussed above, these differences fell into eight major areas (Tables 86–88). (1) More women would perform prenatal diagnosis for sex selection or for a deaf couple wanting a deaf child, or would offer referrals. (2) In counseling after prenatal diagnosis, more women would be nondirective about trisomy 21 and more men would counsel pessimistically; more women would discuss carrying a fetus with Klinefelter syndrome to term and placing the child for adoption. More men would be unbiased if the mother's life were in danger, and more women would urge termination (Table 87). (3) More women than men would themselves abort for three of the 24 fetal conditions (obesity, cleft lip/palate, and achondroplasia), all affecting appearance rather than intelligence. (4) In cases where genetic testing, done for other purposes, may also accidentally reveal nonpaternity, fewer women physicians than men physicians thought it necessary to warn the woman alone before testing but more would require a pre-test agreement from the couple about whether the information should be disclosed. (5) More women would warn a patient's relatives about their genetic risks, against the patient's wishes. (6) Fewer women would test children for mutations for adult-onset disorders. (7) Fewer women would discuss some reproductive options, especially sterilization, IVF with a donated egg, or surrogacy, unless the client asked. (8) More women physicians than men agreed with the autonomy-oriented statements that "patients have a right to whatever services they can pay for out of pocket", and that "withholding any requested service is paternalistic."

Women Geneticists Outside the United States

Women constituted 50% of the doctoral-level medical geneticists in the 35 other countries in the survey. They differed from men on 43% of 491 subparts of the ethical questions, but many of their responses ran counter to those of US women geneticists. In virtually every preconception case, women were more directive than men, in the interests of preventing genetic harm to the future child. Sometimes this directiveness on behalf of the potential child included legal sanctions. More women would suggest sterilization for a single blind woman on welfare and would suggest involuntary hospitalization for a pregnant woman with phenylketonuria (PKU) who would not stay on the prescribed diet.

In every case where there was potential conflict between parental autonomy and the health of the potential child, women placed the child or "child-to-be" first. Those who would perform sex selection often justified this on the basis of the child's quality of life, stating that children had a right to be born into families where they were wanted and would be happy.

Women also took a more pessimistic view of disability. More women thought that it was unfair to the child, to siblings, and to society to have a child with a disability if the birth could have been prevented. More women thought that people at high risk for having children with genetic disorders should not have children at all. More women would support parental decisions not to perform lifesaving operations on handicapped newborns. More thought that reducing the number of "bad" genes in the population was an important goal of genetics.

More women than men would themselves abort in the first trimester for 17 of 24 fetal conditions and in the second trimester for 13 conditions, including Huntington disease, neurofibromatosis, severe obesity in the absence of a known genetic syndrome, rubella, PKU, toxoplasmosis in the fetus, achondroplasia, predisposition to schizophrenia, predisposition to alcoholism, cleft lip/palate in a girl, and predisposition to Alzheimer disease. Personal attitudes were significantly related to directive counseling practice.

Women's relatively pessimistic views about disability outside North America may reflect a widespread belief that most of the care (and in some cultures also blame) will fall on the mother. That more women than men were willing themselves to abort for many conditions listed may reflect a social reality that makes the lives of families affected by disability very difficult in some countries, especially developing nations where there are few support services. However, this pessimistic view of disability also applied in Western Europe, where there are advanced support services for people with disabilities. Potential people with disabilities were seen in terms of their effects on others and on society rather than solely in terms of their individual rights.

Women physicians outside the United States rejected individual privacy to a greater extent than men and verged on a belief in what some might call "voluntary eugenics." Although most geneticists outside North America regard "prevention" of the births of children with disabilities as a major goal of medical genetics, women seemed to accept this with greater enthusiasm than men.

Women also placed greater responsibility on the patient. More women than men thought that people had a responsibility to know genetic information that might affect their own or their children's health, and fewer women than men thought that people had a right not to know test results. Women were more likely than men to honor some requests from women clients, such as prenatal paternity testing for a woman involved with two men, and prenatal diagnosis solely to resolve a woman's anxiety, in the absence of advanced age or family history of genetic disorder.

Women physicians outside the United States were more likely than men to say that patients had the right to make their own decisions or, at least, the right to referral for procedures that the provider considered immoral, and that children should be born only if they were wanted, by which they meant that children of an unwanted sex might be better off not being born at all than living with a family who wanted a child of the other sex. The latter comment is especially pertinent in societies that desire male children and where a geneticist in pediatrics may have seen mistreated girls.

Feminist Theories

In psychological studies of moral decision making, Gilligan (1982) found that women tended to use a relationships-based ethics that saw individuals as embedded in a network of relationships with family, colleagues, friends, and, perhaps, society. Sometimes this is called an "ethics of care." According to the Gilligan hypothesis, women faced with the dilemma of preserving patient confidentiality or telling an employer that a patient's medical condition threatens public safety (e.g., a school bus driver at high risk of heart attack or stroke) would tend to look at the overall situation, and might say, "Everybody should sit down together and work out some solution that will protect the driver's livelihood and also protect the children." Men, according to this model, would tend to argue on the basis of principles of medical confidentiality or a duty to protect public safety. In our study, women patients, but not women professionals, appeared to fit this model.

According to feminist theory, women's viewpoints are molded by their experiences in family life, which differ from men's. Philosopher Susan Sherwin claims that women's choices, whatever these choices may be, are embedded in a world view that sees human beings in terms of relationships. As Sherwin points out, "Ethical models based on the image of historical, self-sufficient, atom-like individuals are simply not credible to most women. Because women are usually charged with … caring for children, the elderly, and the ill as well as … physically and emotionally nurturing men both at work and at home, most women experience the world as a complex web of interdependent relationships, where responsible caring for others is implicit in their moral lives." (Sherwin, 1982, p. 74.)

The views of women physicians in other countries fit this model, which is suggestive of an older medical ethic stating that it is a doctor's duty to provide advice, direction and guidance. According to this ethic, nondirectiveness is abandonment of the patient. Directiveness may be interpreted as the "maternal thinking" that feminist Sara Ruddick (1989) believes can benefit society. According to this model, women see the individual in a context of relationships and place this context above the atomized individual. In cases where there was conflict between an individual's wish for confidentiality and the presumed benefit of the extended family, women outside North America placed the family first. The health of the child-to-be, the marital relationship, the extended family, the safety of others, and sometimes the welfare of society came before the "rights" or "privacy" of individuals (words that do not have the same force in all nations).

The survey results suggest that there is no universal "women's ethics." In the United States, women and men physicians responded similarly to most questions, although their areas of difference-sex selection, telling relatives against a patient's wishes, not testing children-were suggestive. Women physicians in the United States were more similar to men physicians than to women genetic counselors. Outside the United States, however, women differed from men on about half the questions and answered in diametrically opposite directions from women genetic counselors in the United States. In the United States, ethical views were affected by professional locus more than by gender.

Ethics of care vary according to cultures and professions. Honoring a request with which the professional disagrees may be one form of an ethics of care, because it respects the person; denying the same request, on the basis that it would harm family, child, society, or the moral order, may be another. These contradictions point to a basic weakness in an ethic of relationships; it provides little specific guidance and can be used to justify contradictory actions.

Perhaps the more interesting question is how women's ethical views can instruct practice. Although US men and women physicians agreed on most ethical questions, the areas of difference are worth some attention. Women appear to bring to medicine a sharper awareness of family interrelationships and needs. They appear more willing to place themselves in the position of the other and to listen to requests with which they disagree. These are qualities essential to the communication that is the basis of good genetics services.

Patient Culture

The 476 US patients had some very different ideas from the three groups of professionals (Tables 91 and 92). Patients (1) Placed greater emphasis on autonomy; (2) Felt entitled to requested services; (3) Supported use of PND and genetic testing to greater extent than professionals but were more reluctant than professionals to abort; (4) Took a familial view of genetic information that included spouses and blood relatives; (5) Regarded disclosure of information to schools as benign; (6) Were somewhat more optimistic than professionals about disability. On some questions, patients' views were closer to the two groups of M.D.s (geneticists and primary care physicians) than they were to the genetic counselors. These included knowing carrier status before marriage; people at high risk should use PND; it is not fair to the child to have a child with a genetic disorder; directive counseling and sterilization for a blind woman on welfare. These patients, however, represent only one segment of American society, white working-class mothers, mostly from the Northeast and Midwest. A much broader survey would be needed to include our rich variety of minorities, religions, and cultures. Meanwhile, professionals need to become aware that what the professional thinks is most ethical is not always what people want (or need). Many patients held views of marriage and family that would give spouses and relatives automatic access to information. Many also believed that responsible parents should use medical technology and were somewhat judgmental about those who did not.

INTERNATIONAL COMPARISONS OF PATIENTS

Our colleagues surveyed French patients at five clinics and German patients at two clinics using some questions (in translation) from the US patient questionnaire. In all, 593 (65%) in Germany and 394 (51%) in France responded. Comparisons between American, French, and German patients suggest that on some points they were remarkably similar (Table 93). Most (72–85%) thought doctors should tell pregnant women that a blood test may lead to a decision about abortion; that parents should be told fetal sex if they ask (86–92%); that women at high risk should have prenatal diagnosis (80–93%); that counseling should precede prenatal diagnosis (78–94%); that a birthmother who gave up a child for adoption should be informed if the child develops a genetic condition (93–96%); that people should tell partners their genetic information (78–89%); that DNA fingerprinting should be performed, and the fingerprints kept on permanent file, for convicted sex offenders (93–97%), persons convicted of serious crimes (83–95%), persons charged with sex offenses (77–82%), and persons charged with other serious crimes (72–79%); that doctors should respect parents' decisions not to have a lifesaving operation on a newborn (68–90%); that parents have a right to know fetal sex (86–92%); that patients are owed referrals (86–88%). Most in the US (96%) and France (92%) would tell a school system a diagnosis of XYY (there were no data for Germany). Minorities would maintain confidentiality of a patient who refuses to warn relatives at risk (12–25%); thought carriers at high risk should not marry each other (4–7%); would maintain confidentiality of a bus driver at risk for a work-related accident (6–20%); thought parents had a right to select sex (4–13%) thought genetic tests should be included in pre-employment physical exams (1–7%). Similar percents in all three nations thought that people at high risk should not have children (40–46%), that there should be automatic access to an individual's DNA for blood relatives (32–44%), family doctor (43–59%), or spouse/partner (42–55%) and that patients have a right "not to know" (35–40%) genetic information after taking a test. Many (45–60%) said patients had a right to any service they could pay for out of pocket, or that parents should be able to have children tested for predisposition to Alzheimer disease (34–61%).

There were some important international differences. American patients were more autonomy-oriented; more believed that withholding any service was a denial of patients' rights, that referrals should be offered outside the country, if necessary, that preconception sex selection should be available with no restrictions and that parents should be able to have their children tested for adult-onset disorders. French patients tended to see genetics in a "preventive" light. More believed it was unfair to both child and society to have a child with a genetic condition if the birth could have been avoided. The French favored directive counseling in order to prevent births in certain situations, such as for the couple who both carry a cystic fibrosis mutation, and thought women should abort after a positive prenatal diagnosis. German patients were considerably less likely than Americans or French to say that parents had the right to choose the quality of their children, or to support government-required premarital carrier testing. More German patients than American patients also favored a "right not to know" results after taking a test. The Germans, probably in the light of history, were understandably apprehensive both of unlimited parental choices and of government intervention of any kind. Yet, about 30% more German patients than American patients, on average, would abort for most fetal conditions (there were no data for France).

Patients' views deserve considerably more investigation than they have received. Future projects should attempt to survey patients on a worldwide basis. Patients and families (sometimes called consumers, users, counselees, or simply "people") are the endpoint of genetic services. It is time that they take their rightful place in developing an ethics for the future.

REFERENCES FOR CHAPTERS 1–8

Adams, M. (ed.). *The Wellborn Science: Eugenics in Germany, France, Brazil, and Russia*. Oxford: Oxford University Press, 1990.

Agelli, Maria, Porter, Kathryn S., McQuillan, Geraldine M., and Kington, Raymond S. NHANES: Sociodemographic Factors Associated with Consent to Future Research and to Genetic Research Among Adult Participants. Paper presented at A Decade of ELSI Research. Washington, DC, Jan. 16, 2001, 1999.

Allan Guttmacher Institute. *Clandestine Abortion: A Latin American Reality*. New York and Washington: Allan Guttmacher Institute, 1994.

Allen, Garland E. Eugenics and American social history. *Genome* 1989; 31: 885–889.

American Medical Association Council on Ethical and Judicial Affairs. *Testing Children for Genetic Status. Code of Medical Ethics, Report 66*. Chicago: AMA, 1995.

American Society of Human Genetics Social Issues Subcommittee on Familial Disclosures. Professional Disclosure of Familial Genetic Information. *American Journal of Human Genetics* 1995; 62: 474–483.

American Society of Human Genetics Board of Directors and American College of Medical Genetics Board of Directors. Points to Consider: Ethical, Legal and Psychosocial Implications of Genetic Testing in Children and Adolescents. *American Journal of Human Genetics* 1995; 57: 1233–1236.

American Society of Human Genetics Board of Directors. Eugenics and the Misuse of Genetic Information to Redirect Reproductive Freedom. *American Journal of Human Genetics* 1999; 64(2): 335–338.

American Society of Human Genetics Social Issues Committee and the American College of Medical Genetics Social, Ethical and Legal Issues Committee. Genetic Testing in Adoption. *American Journal of Human Genetics* 2000; 66(3): 761–767.

Anderson, Betsy, quoted in Wertz, D.C., Gregg, R., *et al.* Optimizing Genetic Services in a Social, Ethical and Policy Context. Suggestions from Consumers and Providers in the New England Regional Genetics Group. *The Genetic Resource* 1996; 10(2): 13.

Angastiniotis, M.A., Kyriakidou, S., and Hadjminas, M. How thalassemia was controlled in Cyprus. *World Health Forum* 1986; 7: 291–297.

Baird, Dugold. Sociological considerations of maternal and infant capabilities. In: Kretchmer, Norman and Hasselmeyer, Eileen G. (eds), *Horizons in Perinatal Research*. New York, 1970, pp. 11.

Baker, Bellis. *Human Sperm Competition*. Chapman and Hall, 1995.

Beauchamp, Tom L. and Childress, James F. *Principles of Biomedical Ethics*, Fourth Edition. New York: Oxford University Press, 1994, p. 28.

Billings, Paul R. *et al.* Discrimination as a consequence of genetic testing. *American Journal of Human Genetics* 1992; 50: 476–482.

Brock, D.W. Children's competence for health care decision-making. In: Kopelman, L.M. and Moskop, J.C. (eds), *Children and Health Care: Moral and Social Issues*. Dordrecht, The Netherlands: Kluwer Academic Publishers, 1989, pp. 181–212.

Buchanan, A.E. and Brock, D.W. *Deciding for Others: The Ethics of Surrogate Decision Making*. Cambridge: Cambridge University Press, 1989, pp. 215–266.

Burleigh, Michael. *Death and Deliverance: Euthanasia in Germany c. 1900–1945*. Cambridge: Cambridge University Press, 1994.

Cerda-Flores, R.M., Barton, S.A., Marty-Gonzalez, L.F., Rivas, F., and Chakraborty, R. Estimation of nonpaternity in the Mexican population of Nueva Leon. A validation study with blood group markers. *American Journal of Physical Anthropology* 1999; 109: 281–293.

Chan, C.K. Eugenics on the rise: a report from Singapore. *International Journal of Health Services* 1985; 15: 707–712.

China. Law of the People's Republic of China on Maternal and Infant Health Care. Order of the President of the People's Republic of China No. 33, October 27, 1994.

Clarke, Angus (ed.). *Genetic Counseling: Practice and Principles*. London: Routledge, 1994.

Clarke, Angus and Flinter, F. The genetic testing of children: a clinical perspective. In: Marteau, Theresa and Richards, Martin (eds), *The Troubled Helix: Social and Psychological Implications of the New Human Genetics*. Cambridge: Cambridge University Press, 1996, pp. 64–176.

Clinical Genetics Society. The Genetic Testing of Children: Report of a Working Party of the Clinical Genetics Society, Angus Clarke, Chair. Edgbaston and Birmingham, UK, CGS, 1993.

Clow, C.L. and Scriver, C.R. Knowledge about and attitudes toward genetic testing among high-school students: the Tay–Sachs experience. *Pediatrics* 1977a; 59: 86–91.

Clow, C.L. and Scriver, C.R. The adolescent copes with genetic screening: a study of Tay–Sachs screening among high school students. *Progress in Clinical Biology Research* 1977b; 18: 381–393.

Cohen, Pamela and Dorothy C. Wertz., Genetic counseling practices in Germany: a comparison between East German and West German geneticists. *Journal of Genetic Counseling* 1997; 6: 61–80.

Draper, Eleanor Elaine. *Risky Business: Genetic Testing and Exclusionary Practices in the Hazardous Workplace*. Cambridge: Cambridge University Press, 1991.

Duster, Troy. *Backdoor to Eugenics*. New York: Routledge, 1990.

Dworkin, Gerald. *The Theory and Practice of Autonomy*. New York: Cambridge University Press, 1988.

Fanos, Joanna H. and Nickerson, B.G. Long-term effects of sibling death during adolescence. *Journal of Adolescent Research* 1991; 6: 70–82.

Fletcher, John C. *Coping with Genetic Disorders: A Guide for Clergy and Parents*. San Francisco: Harper & Row, 1982.

Fraser, Clarke F. Genetic counseling. *American Journal of Human Genetics* 1974; 26: 636–661.

Galton, Francis. *Inquiries into Human Faculty and Its Development*. London: Macmillan, 1883.

Geller, Lisa N., Alper, Joe, Natowicz, Marvin, and Beckwith, Jonathan. Individual family, and societal dimensions of genetic discrimination: a case study analysis. *Science and Engineering Ethics* 1996; 2: 71–88.

Gert, Bernard, Berger, Edward M., Cahill, George F., Clouser, Danner K., Culver, Charles M., Moeshler, John B., and Singer, George H.S. *Morality and the New Genetics*. Boston: Jones and Bartlett, 1996.

Gilligan, Carol. *In a Different Voice: Psychological Theory and Women's Development*. Cambridge, Mass.: Harvard University Press, 1982.

Gouldner, Alvin W. *Enter Plato; Classical Greece and the Origins of Social Theory*. New York: Basic Books, 1965.

Green, M. and Solnit, A.J. Reactions to the threatened loss of a child: a vulnerable child syndrome. *Pediatrics* 1964; 34: 58–77.

Hall, Mark A. and Rich, Stephen S. Laws restricting health insurers' use of genetic information: impact on genetic discrimination. *American Journal of Human Genetics* 2000; 66: 293–308.

Harper, Peter S. and Clarke, Angus. Should we test children for "adult" genetic diseases? *Lancet* 1990; 335: 1205.

Harris, Rodney and Reid, Margaret. Medical genetic services in 31 countries: an overview. *European Journal of Human Genetics*, Genetic Services in Europe 1997; 5(Suppl 2): 3–21.
Hoyle, Cynthia L., Warren, N.S., and Wertz, D.C. Discussion of genetic discrimination issues by genetic counselors and their clients. *Journal of Genetic Counseling* 1997; 6(4): 455–456.
Institute of Medicine. *Assessing Genetic Risks*. Washington, DC: National Academy Press, 1994.
Jonsen, Albert R. Of Balloons and Bicycles or the Relationship Between Ethical Theory and Practical Judgment. *Hastings Center Report* 1991; 21: 14–16.
Juengst, Eric T. The ethics of prediction: genetic risk and the physician–patient relationship. *Genome Science and Technology* 1995; 1(1): 21–30.
Kessler, Seymour. Psychological aspects of genetic counseling. VII. Thoughts on directiveness. *Journal of Genetic Counseling* 1992; 1(1): 9–18.
Kitcher, Philip. *The Lives to Come*. New York: Simon and Schuster, 1996.
Lapham, Virginia E. *et al.* Genetic discrimination: perspectives of consumers. *Science* 1996; 274: 271.
La Fleur, William R. *Liquid Life: Abortion and Buddhism in Japan*. Princeton: Princeton University Press, 1992.
Locke, John. *Two Treatises*. Reprinted in *The Locke Reader*, Yolton, John W. (ed.), 1977, p. 289.
MacIntyre, Sally and Soomans, A. Non-paternity and prenatal genetic screening. *Lancet* 1991; 338(8771): 869–871.
McLaren, Angus. *Our Own Master Race: Eugenics in Canada 1885–1945*. Toronto: McClelland and Stewart, 1990.
Mao, Xin and Wertz, Dorothy C. China's genetic services providers' attitudes towards several ethical issues: a cross-cultural survey. *Clinical Genetics* 1997; 52: 100–109.
Murray, Christopher J.L. and Lopez, Alan D. (eds). *The Global Burden of Disease: A Comprehensive Assessment of Mortality and Disability from Disease, Injuries and Risk Factors in 1990 and Projected to 2020*. Boston, MA: The Harvard School of Public Health on behalf of The World Health Organization and The World Bank, 1996, Summary.
Mill, John Stuart. *On Liberty and Other Essays*. Oxford: Oxford University Press, 1855 [1991].
National Society of Genetic Counselors. Code of ethics. *Journal of Genetic Counseling* 1992; 1(1): 41–44.
National Society of Genetic Counselors. Prenatal and childhood testing for adult-onset disorders. *Perspectives in Genetic Counseling* 1995; 17: 5.
Noddings, Nel. *Caring: A Feminine Approach to Ethics and Moral Education*. Berkeley: University of California Press, 1984.
Norman-Bloodsaw v. Lawrence Berkeley Laboratory (9th Circuit, February 3, 1998).
Page A. *et al.* Attitudes of high school students toward carrier screening for cystic fibrosis. *American Journal of Human Genetics* 1992; 51(Suppl): Abstract 17.
Planned Parenthood of Southeastern PA v. Casey, 505 US, 833 (1992).
Penchaszadeh, Victor B. Delivery of genetic services in developing countries. In: Khoury, Muin J., Burke, Wylie, and Thomson, Elizabeth J. (eds), *Genetics and Public Health in the 21st Century*. Oxford: Oxford University Press, 2000, pp. 321–328.
Rapp, Rayna. *Testing Women, Testing the Fetus: The Social Impact of Amniocentesis in America*. New York: Routledge, 2000.
Reed, Sheldon. Heredity counseling. *Eugenics Quarterly* 1954; 1: 47–51.
Reed, Sheldon. A short history of genetic counseling. *Social Biology* 1994; 21: 332–339.
Reilly, Philip R. *The Surgical Solution: A History of Involuntary Sterilization in the United States*. Baltimore: Johns Hopkins University Press, 1991.
Ruddick, Sara. *Maternal Thinking: Toward a Politics of Peace*. Boston: Beacon Press, 1989.
Sasse, G., Muller, H., Chakraborty, R., and Ott, J. Estimating the frequency of nonpaternity in Switzerland. *Human Heredity* 1994; 44: 337–343.
Schama, Simon. *The Embarrassment of Riches. An Interpretation of Dutch Culture in the Golden Age*. New York: Knopf, 1987.

Sen, Amartya. More than 100 Million Women are Missing. *New York Review of Books*, 20 December 1990, pp. 61–66.

Sen, Amartya. East and West: the Reach of Reason. *New York Review of Books*, 20 July 2000, pp. 33–38.

Sherwin, Susan. *No Longer Patient: Feminist Ethics and Health Care*. Philadelphia, PA: Temple University Press, 1982.

Sing, C.F. *et al*. Studies on genetic selection in a completely ascertained Caucasian population II. Family analysis of 11 blood group systems. *American Journal of Human Genetics* 1971; 23(2): 164–198.

Sivard, Ruth. *World Military and Social Expenditures 1996*. World Priorities, Box 25140, Washington, DC 20007, pp. 48–51.

Sorenson, James R., Swazey, Judith P., and Scotch, Norman A. *Reproductive Pasts, Reproductive Futures: Genetic Counselling and Its Effectiveness*. New York: Alan R. Liss, 1981.

Stepan, Nancy L. *The Hour of Eugenics: Race, Gender, and Nation in Latin America*. Ithaca, NY: Cornell University Press, 1991.

Teicher, J., Tsipis, J., Wertz, D.C., and Lerner, B. Disability awareness training in the graduate genetic counseling training program: a survey of recent graduates. National Society of Genetic Counselors Annual Meetings, Denver, CO, Oct. 24–27, 1998, p. 164.

Toqueville, Alexis de. Bradley Phillips (ed.), In, *Democracy in America*, vol. I. New York: Alfred A. Knopf, 1951.

UNICEF. *The State of the World's Children, 1998*. Oxford and New York: Oxford University Press, 1998.

United Nations Population Fund. *The State of World Population*. New York: UNFPA, 1998.

US Congress, Office of Technology Assessment. Genetic Monitoring and Screening in the Workplace (OTA-BA-1455). Washington: US Government Printing Office, 1990.

US Congress, Office of Technology Assessment. Genetic Testing and Monitoring in the Workplace. (OTA-BP-BA 67) Washington: US Government Printing Office, 1991.

United States, National Bioethics Advisory Commission. *Research Involving Human Biological Materials. Ethical Issues and Policy Guidance*. Rockville, MD: NBAC, August 1999.

United States President's Commission on Ethical Issues in Medicine and Biomedical and Behavioral Research. *Screening and Counseling for Genetic Conditions*. Washington, DC: US Government Printing Office, 1983.

Wertz, Dorothy C. Prenatal Diagnosis and Society. Royal Commission on New Reproductive Technologies (Canada), *New Reproductive Technologies: Ethical Aspects*. Ottawa: Royal Commission, 1993, pp. 191–332.

Wertz, Dorothy C. Is there a "Women's Ethic" in genetics? A 37-nation survey of providers, *Journal of the American Medical Women's Association* 1997a; 52: 33–38.

Wertz, Dorothy C. Society and the not-so-new genetics: what are we afraid of? *Journal of Contemporary Health Law and Policy* 1997b; 13: 299–346.

Wertz, Dorothy C. International perspectives on privacy and access to genetic information. *Microbial and Comparative Genomics* 1997c; 2: 53–61.

Wertz, Dorothy C. Ethical and social issues in prenatal sex selection. *Social Science and Medicine* 1998; 46: 255–273.

Wertz, Dorothy C. The difficulties of recruiting minorities to studies of ethics and values in genetics. *Community Genetics* 1998; 1(3): 175–179.

Wertz, Dorothy C. and Fletcher, John C. *Ethics and Human Genetics: A Cross-cultural Perspective*. Heidelberg: Springer-Verlag, 1989.

Wertz, Dorothy C. and Fletcher, John C. A critique of some feminist challenges to prenatal diagnosis. *Journal of Women's Health* 1993; 2: 173–188.

Wertz, Dorothy C. and Gregg, Robin. Genetic services in a social, ethical, and policy context: a collaboration between consumers and providers. *Journal of Medical Ethics* 2000; 26: 1–4.

Wertz, Dorothy C. and Knoppers, Bartha M. "Serious" genetic disorders: can or should they be defined? *American Journal of Medical Genetics* 2002; 108: 29–35.
Wertz, Dorothy C., Fanos, Joanna H., and Reilly, Philip R. Genetic testing for children and adolescents: who decides? *JAMA* 1994; 272: 857.
Wertz, Dorothy C. and Reilly, Philip R. Laboratory policies and practices for the genetic testing of children: a survey of the Helix network. *American Journal of Human Genetics* 1997; 61: 1163–1168.
Wertz, Dorothy C., Rosenfield, Janet M., Janes, Sally R., and Erbe, Richard W. Attitudes toward abortion among parents of children with cystic fibrosis. *American Journal of Public Health* 1991; 81(8): 992–996.
White, Mary Terrell. "Respect for autonomy" in genetic counseling: an analysis and a proposal. *Journal of Genetic Counseling* 1997; 6(3): 297–314.
Whitner v. State, 492 S.W. 2d 777 (SC 1997).
Wilcke, J.T., Seersholm, N., Kok-Jensen, A., and Dirksen, A. Attitudes toward an unsolicited approach in relation to status of genetic disease: exemplified by α1-antitrypsin deficiency. *American Journal Medical Genetics* 2000; 94: 207–213. *Medline abstract.*

PART II

REPORTS FROM OUR COLLEAGUES: GENETICS AND ETHICS IN INDIVIDUAL NATIONS

INTRODUCTION TO PART II

The ten chapters in this section were written by our international colleagues. In the first three, we have compiled materials from all nations, except for the six (China, Germany, Hungary, India, Japan, and Peru) that have individual chapters around a theme, so that readers who desire to make comparisons, for example, of genetic services may do so without having to read about a nation in its entirety. In each of these chapters, nations are organized according to the geographical and culture areas used in the tables: English-speaking nations, Northern/Western Europe, Southern Europe, Eastern Europe, Near East, Latin America, and Asia. With the exception of the sections on the US, all material in these chapters has been provided by our international colleagues, who are credited after each piece.

Chapter 9, "Geneticists Speak Out," includes passages that we found especially poignant or telling. They include descriptions of past political oppressions (Argentina), pleas for women's lives (Peru), the role of traditional healing (South Africa), failure of adoption programs (Poland), views on disability (Greece), devastating effects of poverty and illiteracy (India), and family social structure (Thailand).

Chapter 10, "Organization of Genetic Services" describes numbers of genetics centers, geneticists, professional organizations (and sometimes their ethical guidelines), methods of reimbursement, including the place of genetics in national health insurance structures, services for minority groups and people with disabilities, inequities in distribution, numbers and types of services provided and percents of populations served. These descriptions vary in emphasis and comprehensiveness, but most include the authors' general evaluations of current services. They provide a useful resource for readers desiring a quick overview of a particular nation or wishing to make comparisons between particular nations.

Chapter 11, "Laws Affecting Genetic Services," includes what our colleagues perceived as the most important laws actually affecting practice. Most of these are abortion laws, but some colleagues included laws on gamete or embryo donation, IVF, or genetic privacy. This chapter does not pretend to be a comprehensive compilation of all laws related to genetics. This would be a massive effort requiring a book of its own. This also is not a chapter on laws affecting genetic research (though some authors included them), since we asked our colleagues to concentrate on practice.

Six chapters (12–14, 16–18) submitted by our colleagues were so interesting that we decided to present them in their entirety. Five are from countries with which most readers will probably be unfamiliar, at least regarding genetics: China, India, Japan, and Peru. There are two chapters from China, one by China's senior bioethicist (Chapter 12) and one by a younger genetics researcher working in England (Chapter 13). We chose to include both chapters because they present different and valuable material, although we saw no ideological differences between the authors. Chapter 14 is a comprehensive view of genetics in a Germany that has been reunited since our previous survey in 1985. Results of the German patient survey, which were originally included in tables for this chapter, have been transferred to the general tables for the entire volume, in order to make easier comparison with US patients, who turn out to be remarkably similar to German patients in many respects.

By way of encouraging similar efforts to involve the public, we have added a brief report (Chapter 15) on reviewer responses to a TV series in Hungary that used some questions from our international survey.

CHAPTER 9

GENETICISTS SPEAK OUT: CULTURE, ETHICS, AND HISTORY

ENGLISH-SPEAKING NATIONS

Australia

Australian clinical geneticists work very much within the traditional Western framework of medical ethics, in that their main focus is on individual welfare, rather than social goals.

The dominant cultural influences on Australian society are British and American. The culture is, therefore, strongly individualistic, and this is reflected in the importance given to individual autonomy in medical decision making. However, over the last few decades Australia has received increasing numbers of immigrants from a large number of races and cultures, and has increasingly recognized the culture of the original inhabitants of Australia. Some newer cultures within Australia place great emphasis on family considerations. These can create problems in the current practice of genetics, as clinical geneticists try to be sensitive to the traditions of minority groups, while maintaining the dominant individualistic ethos, which demands that the autonomy and benefit of the individual should take priority over family considerations.

The increasing ethnic diversity of the Australian population will undoubtedly bring about more ethical issues and concerns for geneticists. Patients from some Asian countries expect direction from doctors regarding decision making, such as with prenatal testing, whereas most Australian clinical geneticists adhere to non-directive counseling. Regarding the testing of a child for an adult-onset condition, Australian geneticists, like their northern European and American colleagues, will respect the autonomy of the child and delay testing until adulthood when the individual may be able to make his or her own decision. However, some cultures see the child as part of a collective family in which the parents have the power to direct their children's lives and thus the ability to decide about genetic tests. This cultural diversity will also have an impact on genetic counseling regarding autopsy examinations on fetuses and stillbirths, which is increasingly recognized by geneticists as contributing significantly to accurate diagnosis of birth defects. Australian geneticists are becoming more skilled in the use of interpreters, but there are still ethical concerns about the equity of services, as interpreter counseling interviews may take longer time and there is little or no guarantee that the meaning conveyed is the same as that received.

Alison Colley and Keith Joseph

South Africa

While about 90% of pregnant women are black, only about 5% of prenatal diagnoses are performed on women from this community. The traditional world-view and system of thought in the black population appears to favour collective rather than individual thinking (Hammond-Tooke, 1989) and this system may be accompanied by an external locus of control for the individual. Decision-making processes are more complicated, because with self-directing behavior is limited and consultation with significant others is often required. The mother is often the key figure and source of wisdom, and some women may not consent to amniocentesis until they have consulted their mothers (Mofokeng, 1990). The second philosophy relevant to genetic services is fatalism. This philosophy may interfere with decisions regarding termination of pregnancy for genetic reasons or uptake of prenatal diagnosis. If an individual believes supernatural powers are in complete control of the universe, he or she will be immobilized and unable to alter the course of events and make use of freedom of choice.

Beliefs regarding disease causation should be considered when offering genetic counseling. Albinism, for example, may be attributed to maternal impression during pregnancy, or to the ire of the ancestors, to witchcraft, or the breaking of strict taboos (such as having intercourse too soon after the birth of an infant). Cultural practices may determine mate selection. The preferred marriage partner in several groups was a first cousin; this leads to an increase in recessive disorders such as albinism (Kromberg and Jenkins, 1982). Parents' beliefs about the origin of life are also important. Many individuals believe that the father gives the genes and the mother is only an incubator for the developing fetus, so that explanations of recessive inheritance may be discarded. Furthermore, several myths surround genetic disorders; for example, the belief that albinos do not die naturally but simply "disappear". Albinos are therefore stigmatized and not fully accepted as normal people (Kromberg *et al.*, 1984). These myths need to be recognized, exposed, and discussed, for genetic counseling to be appropriate and effective.

Attitudes toward reproduction and the value of children also affect genetic service delivery. Family planning is an unpopular service in some communities where a woman's status is determined by the number of children she bears. However, there is an increasing demand among women for family planning services, education, and recognition of their rights and, as a result, the number of births per woman is falling. The birth rate was about 3.3 per woman in 1991 and 2.8 in 1995 (Sidiropoulos *et al.*, 1996). The simultaneous use of both Western medical and traditional healer services is widespread.

Jennifer G.R. Kromberg and Trefor Jenkins

References

Hammond-Tooke, D. *Rituals and Medicines. Indigenous healing in South Africa.* Johannesburg: AD Donker, 1989.

Kromberg, J.G.R. and Jenkins, T. Albinism in South Africa: some genetic counseling issues. *Journal of Biosocial Science* 1984; 16: 99–108.

Mofokeng, Z. Attitudes of black women to mental retardation and termination of pregnancy. *BSc (Hons) Research Report*. University of the Witwatersrand, Johannesburg, 1990.

Sidiropoulos, E., Jeffery, A., MacKay, S., Gallocher, R., Forgey, H. and Chipps, C. *South Africa Survey 1995/96*. Johannesburg: South African Institute of Race Relations, 1996.

United Kingdom

In a market system there will be more concern about inequity of access to health care, while in a socialized system there may be concern to ensure that clinical standards are uniformly high. The issues raised by insurance in relation to genetic assessment or genetic testing will be relevant to all health care access in a market system, but in Britain this is much less important – access to mortgages for house purchase, for which mortgage lenders require life insurance, being the major concern rather than access to health care. The potential misuse of genetic information by the insurance industry provides another strong argument – as if one were needed – for a socialized system of health care, whether on the British model or through universal insurance as in much of Europe.

The profile of genetic issues in the media is high, and reporting tends to be polarised – it alternates between wild enthusiasm and strident alarmism. There was strong public interest in the cloning of Dolly the sheep in 1997, but the reaction against it was not as strong as the Yuk! reaction in the US; there is a greater willingness to consider the potential uses of genetic manipulation for medical purposes. There is probably rather more public concern about, and opposition to, the use of genetic engineering in agriculture than in health care.

Some feminists have criticized the medicalisation of pregnancy and the inappropriate technical domination that accompanies widespread prenatal screening. Disability rights groups share many perspectives with these particular elements within feminism, although there remains a tension between the feminist demand for complete control over reproduction and the disability movement's concern for the rights of people with disability to exist, to be respected, and not to suffer stigmatization or discrimination.

British society has so far only inadequately considered these wider implications of prenatal screening programmes. These programmes were introduced by enthusiasts and often promoted by those with vested professional or commercial interests in their adoption, often on the ethically questionable grounds that screening and the termination of affected pregnancies would cost less than the medical and social care of affected individuals.

An overemphasis on client autonomy can be perceived by the client as abandonment in the face of a bewildering mass of information that is too complex for a lay client to assimilate – especially if the client is under pressure to make an important, personal decision in a limited time. There may well be a place for clinical geneticists to recommend courses of action to their clients/patients in specific contexts, such as (i) where genetic testing may target surveillance measures or preventive treatments onto those at risk of a genetic disease who may benefit directly from the intervention, and (ii) in relation to the disclosure of genetic information to other family members if it may have implications for their health or the health of their children in the future. The decisions will still rest with the individual, but the professional may legitimately recommend a course of action and explain why (s)he does so.

Angus Clarke

SOUTHERN EUROPE

Greece

In view of the importance given for centuries by Greeks to health, it is not strange that pessimism is a characteristic in genetic counseling by Greek geneticists.

A handicap, physical or mental, is considered a "stigma" in the graditional Greek family. In fact, the carrier state for thalassemia or sickle-cell anemia is called a "stigma". In a country where Spartans used the Kaiadas chasm to dispose of malformed infants, even today families try to conceal genetic diseases, as a barrier for marriage for the other, healthy children in the family.

Therefore, it is no surprise that handicapped or mentally retarded individuals do not appear outdoors in Greece as much as in other countries.

Christos Bartsocas and Alamitsi-Puchner

Italy

The Italian media are greatly emphasizing the wonderful prospects of the human genome project, in terms of disease therapy and prevention, of personalization of drug therapy, and so on. However, they seldom provide correct scientific information and are prone to point out the implications of a genetic discovery in outrageous terms, rather than providing the basis for a thorough discussion. People's scientific knowledge has not significantly improved, but their expectations about the magic omnipotence of science and their illusions of control over hereditary and common disorders have greatly increased.

On the Web, commercial labs are advertising molecular analyses for all sorts of disorders and for paternity assessment. Doctors are requiring more and more genetic tests, even though testing is not always appropriate. They often do tests without any sort of counseling, let alone assisting the clients' autonomous decision making. Social pressure to perform genetic tests also contributes to the increasing demand for genetic testing. Amniocentesis, for instance, has become a "must" for middle and high-class pregnant women, even when there is no indication to perform it and its cost is not covered by the National Health Service (NHS).

As an result of the all factors above, the characteristics of people asking for genetic services nowadays are different from those of a few years ago. Whether or not they are completely or correctly informed, they "feel" more knowledgeable. Their attitude is not that of passive receivers of mysterious notions. On the contrary, they will actively compare the information provided by the doctor or the counsellor to a preconceived, internal set of beliefs. People's attitudes result from the growth of the illusion of control and the praise that our culture, as all western cultures, attributes to medical/biological information. The belief is that it is "better to know". Meanwhile, there has been minimal change in the traditional doctor–client relationship. The patient certainly has a less reverential attitude towards the medical authority. However, the paternalistic attitudes of doctors and their usual directiveness remain mostly unchanged. Recognizing that the individual is the only one who can decide what value should be attributed to the information written in his/her genome implies a major turn in the traditional doctor-client relationship. Some changes are visible, and a more open attitude toward these new values is spreading among doctors who have had training as medical geneticists (see also Chapter 10). However, even among these specialists there is a strong conflict between the desire to behave according to these new ideals of nondirectiveness and a paternalistic attitude. Paternalism leads them to provide directive advice, in accordance with their sense of professional responsibility toward the client and their deep belief that they know what is best for their clients and for society.

Marina Frontali and Gioia Jacopini

EASTERN EUROPE

Czech Republic

The medical profession is highly valued in this society, and the physician's decisions are very often accepted by the patient without any questions. The physician's paternalistic position is also strengthened by the system of financial evaluation of physicians' work. In the socialist era, the physician was compared to the humanist who worked only for the benefit of the patient and without any financial profit for him or herself. Physicians can still express their opinions much more openly than in other countries without fear of legal repercussions. Legal action against physicians is very rare in the Czech Republic.

Jana Zidovská and Lenka Foretova

Hungary

In Hungary, medical genetics has been burdened both by the tragic consequences of Nazi fascist eugenics before and during the Second World War and by the devastating effect of Soviet Lysenkoism during the 1950s. Genetics, including human and medical genetics, had a new start in the late 1960s; however, a certain part of our society considered medical genetics as a "guilty" or at least suspect professional activity.

Andrew Czeizel and Laszlo Timár

Suggested Reading

Andrew, E. Czeizel. *The Right to be Born Healthy. The Ethical Problems of Human Genetics in Hungary*. New York: Alan R. Liss, 1988.

Poland

A number of meetings of human geneticists with members of antiabortion groups have been organized. The main argument used by the geneticists in these discussions is the pro-life or pronatalistic character of prenatal diagnosis. We have been able to prove that prenatal diagnosis reduces the number of induced abortions. Data from the Warsaw program of prenatal diagnosis shows that terminating one pregnancy, from which a severely affected child would be born, protects 2–8 normal pregnancies which would otherwise have been aborted due to maternal anxiety (Zaremba *et al.*, 1986).

One genetics department in Warsaw got in touch with an American organization – Genesis of Pittsburgh – which seeks to begin an adoption process for special needs children born in Poland. Owing to that, a Polish woman with prenatal test results indicating a disorder may be offered the option of having the affected child adopted by a United States citizen if she chooses to carry it to term. So far no patient has chosen this option, but at least the possibility exists. One Catholic pro-life organization in Warsaw – Pro Familia, could not manage to find a single family in Poland that would be willing to adopt an affected child.

Jack Zaremba and Elzbieta Fidzianska

Reference

Zaremba, J., Wiceniewski, L. and Mazurczak, T. Results of prenatal diagnosis programme – medical part. In: Zaremba, J. (ed.), *Programme of Prenatal Diagnosis for the City and Province of Warsaw*. Warsaw: Institute of Psychiatry and Neurology, 1986, pp. 13–67.

Russia

Due to difficult economic problems, a large number of geneticists have left the genetic service. At the end of 1989, there were about 400 members of the USSR Society of Medical Genetics, but by the time of the survey we could identify and locate only 102 genetic professionals.

Those who insist on total prohibition of abortions, even for sound medical reasons, have an ethical obligation to suggest and ensure practical long term help for the prospective mother and her ill, handicapped or invalid child, irrespective of the severity of its condition.

Vladimir Ivanov and Vera Leonidovna Izhevskaya

NEAR EAST

Turkey

In the classical form of the family, the male spouse and his family predominate. Therefore prenatal diagnosis can be forced by the husband, nonpaternity cannot be kept from the husband, and the "unwanted sex" idea comes from the low position of women in society. This traditional view sometimes interferes with counseling so as to prevent women from making decisions but gives more rights to men.

The portion of the national budget spent on health is less than 3%.

Isik Bökesoy

LATIN AMERICA

Argentina

Argentina is a country battered by war until the advent of democracy in the last decade. Only lately there began a stage of national reconstruction, which is highly promising, but will take time, a time during which Argentines in general, and medical genetic professionals in particular, will learn to exercise their freedom of mind and speech, denied to all of us practically since birth.

In Argentina, professionals can be easily found among those who endured the military dictatorship, but not among those who generated it. The consequences of this persecution are still being felt today by those who should have developed freedom of thought and speech. Ethical issues related to medicine in general and medical genetics in particular have begun to be openly discussed at an institutional level only very recently. This delay resulted from Argentina's recent history. The age of those surveyed reveals that most have endured, directly or indirectly, the effects of the last military dictatorship. The repression compelled people to keep silence and was intended to stifle freedom of thought. Therefore, there is an opinion gap due to this lack of reasoning and thinking. This is currently being reversed. We believe that if this survey were to be conducted in a couple of years it would reveal more sound and mature responses, due to greater institutional and personal development.

Religious beliefs, so deeply rooted in a great part of the Argentine population, hinder freedom of opinion about such conflicting matters as abortion. Some Argentine geneticists believes that pre-natal diagnosis is not fundamentally different from post-natal diagnosis. This view arises from the assumption that the embryo is a human being since its conception and therefore the interruption of a pregnancy is not a viable alternative after the diagnosis of a disease.

Therefore, there could possibly be a confrontation of values between the physician's personal religious beliefs and the responsibility to respect the patient's freedom of choice when receiving genetic counseling. The institutional framework that exists to discuss ethical issues in the medical community is very weak. Ethics committees are still rare, and the expertise of their members is not always appropriate. There is a need to express individual opinions, but these are not always forthcoming. Once again, recent Argentine history could justify fears of speaking openly on some conflicting issues. This is the legacy of a time during which speaking frankly was considered extremely dangerous.

Elba M.P. de Giorgiutti

Brazil

The Brazilian senator Eva Blay (1994) presented impressive statistics about the health of Brazilian women. Despite the fact that 95% of the births here are assisted by trained health personnel, the index of maternal mortality is as high as in Botswana or Honduras (200 women dying per 100 thousand liveborn children, while in developed countries this number does not exceed 3–5). Every three days a pregnant woman dies through lack of medical assistance, while waiting to be admitted to a hospital. Data relative to 1986 indicated that of each 100 maternal deaths 11 are due to abortions performed without medical assistance. The World Health Organization estimates that around 3 million illegal abortions occur each year in Brazil.

Brazilian medical geneticists resemble their colleagues from other nations in many aspects. Some differences, however, were noted, and this is not necessarily bad. After all, *diversity* is the key characteristic that has enabled our species to evolve, and to develop culture and a sense of morality.

Francisco M. Salzano and Lavinia Schüler-Faccini

Reference

Blay, E. Uma questão de ética: direito à maternidade voluntária. *Bioética* 1994; 2: 96–99.

Chile

Vital statistics show that Chile has a relatively low level of mortality from abortion. During the last twenty years (1974–1993), women's mortality rate from abortion has decreased from 0.5 to 0.1 per 100,000 population (from 111 down to 39 cases per year). Hospitalizations for spontaneous abortion have decreased from 44,500 cases to 43,000, while the population has increased from 10 to 14 million. The number of illegal criminal cases of clandestine abortion is unknown.

The general trend of cultural beliefs of the Chilean people is to follow the "culture of life", maintaining the survival of the fetus and carrying to term.

Ricardo Cruz-Coke

Mexico

Mexican geneticists are among the least prone to provide non-directive counseling for most conditions, particularly if compared to Europe or English-speaking countries. There is probably no easy explanation, but an important factor could be the perception in Mexico that one key goal of medical geneticists is the prevention of disease, *not* for eugenic reasons, but in order to avoid suffering for families with affected children.

There is a tendency to answer general questions in one way, and then change one's opinion when the question refers to a specific situation. A good example is that although there is a general consensus among respondents that genetic counseling should be neutral, fewer than 30% would be unbiased in counseling in 26 different situations. There is a clear paternalistic attitude, probably stemming from desires to: (a) prevent disease; (b) start treatment on time; (c) make the best possible use of scarce resources; and (d) protect the patients/interests as well as those of their families and third parties. Many respondents counsel according to their personal value system, with little regard for the patient's autonomy.

Ruben Lisker, Salvador Armendares,
Alessandra Carnevale and Osvaldo Mutchinick

Peru

Women's status in Peru, as in other developing countries, is unequal and unfair to women. Their human rights are not respected, not only in education and work, but also at home and in social recognition. There is still a long way to go to solve this problem and to put women in the equal place where they belong. This situation is a handicap in promoting public education to encourage couples, and especially women, to have freedom of choice about parenthood. This freedom of choice is essential for the development of genetics.

Teresa Perez de Gianella

ASIA

India

Most striking is the per capita income, which is 79 times less than that in the US, and 116 times less than that in Japan. Adult literacy is also almost 1/2 of that observed in USA/Japan. It is thus easy to see that the majority of the people are struggling for survival, and have limited resources for their basic needs like food, housing, education and health. For them "survival is more important than ethics" (Siva Subramanian, 1986). This position is not well appreciated in the West.

In the West the major objection to directive counseling comes from advising abortion in the presence of a fetus with a defect. This is not an issue in India, where the presence of a fetus affected with some disease is generally considered an indication for abortion by the majority of the people.

The burden of genetic disease is high in developing countries, due to the limited facilities for rehabilitation and the socio-economic cost of looking after these children. The quality of life is more important than sanctity of life. Therefore prevention, prenatal screening and prenatal diagnosis of genetic disease are important for the people.

I.C. Verma and Kusum Verma

Reference

Siva Subramanian. In India, Nepal, and Sri Lanka quality of life weighs heavily. *Hasting Center Report*, August 1986, pp. 20–22.

Thailand

The trend of counseling reflects "eugenic principles", namely a belief that prevention is better than having burdens to society. Since Thailand is a developing country, the budget plan is targeted toward prevention rather than supporting the disabled. Social status in Thailand depends on income rather than religion or ethnicity. The low income group does not have much knowledge about genetic diseases and prenatal diagnosis because of under-education. This group has other essential things in life to worry about, such as food and money to support their families. Besides, they cannot afford the high price of genetic tests, since most Thai people do not have health insurance and have to pay from their own pockets. Funds to support the poor have become less, due to the country's financial problems. The moderate and high income groups seek more information about treatment and prevention of genetic disorders when their family members have suffered from genetic diseases.

Ethics in Thailand is based mostly on religion (Buddhism) and culture. The basic precepts of Buddhism are taught early in life to cultivate good morality in children. These early lessons influence the Thai way of life and decision making.

Most Thai people are opposed to abortion, except for medical reasons. They believe that life begins at the stage of the embryo *in utero*. To abort the embryo or fetus is equal to killing, which is prohibited in Buddhist conduct. But in cases of pregnancy with an abnormal fetus, people will allow termination of pregnancy, because they believe that if the baby were to be born, it would suffer more than if not born. Disability in families is believed to be a sin and disgrace, so people are pessimistic about abnormal or handicapped family members and try to keep the condition secret.

The main purpose of prenatal diagnosis in Thailand is to terminate the pregnancy, provided that the fetus is abnormal, mentally retarded, or has a genetic disorder. Most people independent of a social status, hold a pessimistic view about disability. Some old-fashioned and conservative families will not accept prenatal diagnosis because they believe that everything has its destiny and no one can change it. They do not believe that some genetic disorders can be prevented.

In Thai culture, most couples still live with their parents after the marriage, preferably with the husband's parents. The advantage of living in the same house is saving on household expenses. The grandparents can take care of the children while the young couple are working. The grandparents are happy that they are not left alone and have someone around to take care of them when they are ill. This has advantages on both sides, but also involves compromises. In the new generation, Thai couples prefer to live on their own, provided that they have enough income. Still, they are not totally individually autonomous. Most wives have to ask their husbands' opinions when making a decision. When they are dealing with major issues such as prenatal testing and termination of pregnancy, couples will seek advice from their parents, relatives, and friends. They do not want to make a decision by themselves, because they are afraid of making a wrong one. If the outcome turns out to be bad, they will be blamed. They usually ask others for their opinions, so that they will have someone to share the blame and they will not feel guilty. This pattern may stem from lack of self-confidence and caring too much for other people's points of view. People's values depend on the acceptance of society, families, and friends.

Most Thai people believe in monogamy. If marital infidelity is disclosed, it is disgraceful and embarrassing for the entire family. Marital infidelity is a taboo for Thai women. Thai wives are supposed to be loyal and faithful to their husbands. If a husband finds out that his wife has an affair, the marriage will end up in divorce and cause shame and disgrace to the woman's family. The woman herself will be stigmatized with infidelity for the rest of her life. Even though infidelity in women is considered sinful and dishonorable, it is acceptable for Thai men to have affairs. Most Thai men believe that having many wives is the symbol of wealth and power. These wives are not legally married, because Thai law permits only one legal wife. In Thai culture men are domineering and Thai women are submissive. Most women who suffer from their husband's marital infidelity will not reveal this to anyone because it causes humiliation and disgrace to the woman herself and the family. If false paternity is disclosed to the husband, it is sure to be a disaster to marital life. In general, Thai couples will not ask for paternity testing, as this is asking for trouble. They would prefer to keep it secret and live in harmony and peace.

The doctor–patient relationship in Thailand is not paternalism for all. The doctor gives information to all patients, but the way information is provided depends on the patient's level of education. Mostly, educated patients can make decisions on their own (non-directive counseling). Uneducated patients are likely to ask for doctors' opinions. The most popular question is that if the doctor were a patient, what he or she would do. Sometimes directive counseling is necessary if patients do not know what to do at all. They should still receive full information. Most Thai people, independent of social status, have a pessimistic view of disability.

Valairat Dhamcharee, Apiurat Mutirangura, and Orasri Romayanan

CHAPTER 10

ORGANIZATION OF GENETIC SERVICES

ENGLISH-SPEAKING NATIONS

Australia

Genetics services in each of the six states of Australia are provided by the State Government Departments of Health, through Clinical Genetics Units, usually located in major teaching hospitals. Tasmania (the smallest state) receives services from Victoria, the North Territory receives services from South Australia, and the Australian Capital Territory from New South Wales (NSW).

Australian geneticists uniformly are medical specialists with postgraduate qualifications in pediatrics, or less commonly in obstetrics or adult medicine. The Australian training program has been under the auspices of the Human Genetics Society of Australasia, but as of 1995 training has been through the Royal Australasian College of Physicians. The majority of Australian geneticists (approximately 80%) have overseas training in either the United Kingdom, the United States, or Canada.

Clinical geneticists in Australia mainly work in large metropolitan cities and their suburbs in hospital-based, publicly funded clinical genetics units; approximately 50% supervise trainees. Australia has compulsory national health insurance and there is only very limited private, fee-for-service genetics co-existing in capital cities.

Rural outreach services, from city clinical genetics units to main country towns, have been set up and will undoubtedly expand in the larger states, but there will continue to be inequities in access to clinical genetics services because of geography. Australia is increasingly a multicultural society, particularly in the state capital cities, and counseling frequently involves language interpreters, who are readily available to inner-city genetics units. A telephone interpreter service is available for rarely encountered languages and for rural clinics.

There are state-based newborn screening programs for conditions where screening has been shown to be cost-effective: hypothyroidism, phenylketonuria, and galactosemia. Cystic fibrosis screening was done in NSW by using immunoreactive trypsin starting in 1981, and by using delta F508 since 1993. South Australia added cystic fibrosis to their newborn screening in 1990 and Victoria added it in 1991. Congenital adrenal hyperplasia was added in NSW in 1995.

Disability services are on a state-wide basis. In New South Wales, where 17% of the population has a disability, the legal basis for disability policies is the *Disability*

Services Act of 1993, which requires government agencies to develop and report on plans to improve the provision of services. There has been a trend away from institutions for people with mental disabilities, and the current emphasis is on creating and promoting opportunities and services to allow people with disabilities and those who care for them, to participate in society and have a better quality of life. Children with learning difficulties and physical handicaps are encouraged to attend regular schools when this is possible with appropriate help.

In 1992, 20.4% of all perinatal deaths in Australia and 27% of infant deaths were due to congenital malformations. This has not changed remarkably over the last decade. Major congenital malformations, diagnosed at birth or within 28 days of life, affected 1.5% of infants born from 1981 to 1992. Some apparent increases, such as for ventricular septal defects, may reflect an improvement in the quality and quantity of perinatal ultrasound scanning. Neural tube defects declined in occurrence between 1982 and 1992. The decrease was most pronounced for anencephaly. Induced abortions after prenatal diagnosis of neural tube defects increased, which probably accounts for the decrease. However, only South Australia and the Northern Territory have mandatory reporting of induced abortions, which adds to a general under-reporting of fetal abnormalities. The most inclusive reporting is that from genetics laboratories. Maternal serum screening to detect at-risk pregnancies for Down syndrome occurs in an *ad hoc*, inequitable way between and within states and territories, with the exception of South Australia, which has had a centralized maternal serum screening program for neural tube defects since 1983, and has offered screening for Down syndrome since 1990. In 1994, approximately 85% of all pregnant women requested neural tube defect screening and 80% requested Down syndrome screening in South Australia. From 1986 to 1991, screening detected over four-fifths of neural tube defects (Chan *et al.*, 1993). In NSW, virtually all women over age 40 and almost half of those age 35–40 received PND (Table 1). In the other states, screening is available through the public health system, and is publicly funded, but the Royal Australasian Colleges and State Health Departments have not issued policy statements or guidelines.

There is a general trend away from restricting testing to parents who would terminate an affected pregnancy, due to the desire to respect parental autonomy. The increase in demand for prenatal testing has preceded an increase in perinatologist positions and feto-maternal units in major teaching hospitals.

Pre-implantation diagnosis is in the state of research and development, with limited access and strict protocols. However, it is reasonable to expect its implementation into clinical practice in the next decade. The concern of Australian clinical geneticists lies with the possible consequential step to germline gene therapy; they have less concern regarding somatic cell gene therapy.

Table 1. NSW Prenatal Diagnostic Tests by Maternal Age Groups, Compared with Confinements in 1994

Maternal age	Amniocentesis	Chorionic villus sampling	Fetal blood	PND as % of confinements
<35 years	554	232	19	1.1
35–40 years	2434	1209	10	45.7
>40 years	573	319	4	97.5

At present, Australia does not have codified ethical standards for clinical geneticists or genetic counselors. This concept of codified standards is, however, being embraced by the Human Genetics Society of Australasia, along with broader guidelines for a national approach to genetics services to ensure quality in all aspects of genetics. The Society has been discussing ethical, social and legal issues along with the optimal provision of services.

Alison Colley and Keith Joseph

References

Chan, A., Robertson, E., Haan, E., Keane, R., Ranieri, E., and Carney, A. Prevalence of neural tube defects in South Australia, 1996–91: effectiveness and impact of prenatal diagnosis. *BMJ* 1993; 307: 703–706.

The Privacy Implications of Genetic Testing. Information paper number 5. Privacy Commissioner, The Human Rights and Equal Opportunity Commission. Commonwealth of Australia. Genetic Testing and Life Insurance (position paper) Life, Investment and Superannuation Association of Australia, Jan. 1996.

Canada

Canada's population has continued to increase slowly since 1985 and passed the 30,000,000 mark early in 1997.[1] The trend toward urban versus rural living also continues, and, overall, 77% of the population now live in urban centers, while 33% are concentrated in the three major conurbations of Toronto, Montreal, and Vancouver. There are, however, marked provincial differences, with the eastern Atlantic provinces and the Northwest Territories having more than 40% rural dwellers. The birth rate remains low, at 12.8/1000 population, and there has been a striking decline, especially in those more rural provinces that previously had the highest fertility. The higher proportion of women who are now delaying childbearing until their thirties has an impact on the provision of maternal age-related genetic services such as prenatal diagnosis.[2] Fortunately, Canada continues to be one of the world's healthiest countries, as measured by its low infant mortality rate of 6.3/1000 births and, as of 1997, a further increase in life expectancy at birth to 80.89 years for women and 74.55 for men.[1]

Over the past decade and a half, population increase has occurred almost equally through natural expansion and immigration. Policies regarding immigration fluctuate, but in recent years over 200,000 new immigrants have arrived in Canada annually.[1] The origin of these newcomers has also been changing, with the proportion from Europe diminished to under 20%, and the majority now coming from Africa, the West Indies, South and Central America and, especially, Asia.

In 1986, the Science Council of Canada launched a project to review the present and potential future role of genetic knowledge and technology in the country's health care, with a view to exploring and stimulating attention to related policy issues. The resultant report, Genetics in Canadian Health Care, 1991,[3] addresses the difficulties of measuring the occurrence of genetic disease in the population. A few provinces, such as Alberta, British Columbia and Nova Scotia, have been able to maintain excellent registers of congenital anomalies, but the federal Congenital Anomalies Surveillance

System has disappeared. Its function, like several other aspects of vital statistics collection at the federal level, was taken over in 1993 by the newly created nonprofit independent national agency, the Canadian Institute for Health Information.[4]

In theory, everyone has equal access to services, since they are provided under the national health insurance system. In practice, however, access is influenced by many social and demographic factors. Also, since the mandate for health matters is carried by the provinces, there is variation from place to place. Some provinces have attempted to improve accessibility and efficiency by coordination of clinical genetic and community health services through the creation of provincial genetic programs. The most comprehensive of these has been in Alberta.[5]

Establishment of standards for the delivery of clinical and laboratory genetic services, as well as for the training and qualifications of those providing them, led to the formation of the Canadian College of Medical Geneticists (CCMG) in 1975, launched with a group of 31 fellows. Entrance examinations have been held since 1977; by 1987, there were 100 fellows, and by 1993, there were 155.

While requirement for a common core of genetic knowledge persists, the CCMG qualifying exam is now offered in clinical genetics for those with an M.D., medical genetics for those with a Ph.D., and cytogenetics, molecular or biochemical genetics for those with either an M.D. or Ph.D. In the 1998 CCMG list of fellows and excluding honorary, emeritus, or retired fellows, 49% were in clinical genetics, 5% in medical genetics, 24% in cytogenetics, 11% in molecular genetics, and 12% in biochemical genetics; a few were doubly qualified. Over the years, the proportion of CCMG fellows with M.D.s or M.D./Ph.D.s has steadily increased in relation to those with a Ph.D. only – from just over 50% in 1977 to 70% in 1998.

The training and certification of medical specialists in Canada, other than family physicians, is in the hands of the Royal College of Physicians and Surgeons (RCPS). It was not until 1989 that the Council of RCPS finally recognized medical genetics as a free-standing specialty and instituted the appropriate committee structure to evaluate the training of candidates, accredit training centers, and conduct examinations. The first RCPS certification exams in medical genetics were held in 1992 and yearly thereafter. The CCMG has agreed to reciprocity with RCPS for MD training in clinical genetics, and those candidates who are successful in the RCPS examination are automatically accepted as CCMG fellows if they wish.

The allied health profession of genetic counseling has developed somewhat separately from the CCMG. Until the eighties, the majority of individuals working in this capacity gained their education on an inservice basis. While they came from a variety of backgrounds, the majority were nurses. The introduction of individuals with a Master's degree in genetic counseling has been a comparatively recent innovation, although there is now an accelerating trend. This pattern, compared to the US, is explained by the slow development of Canadian training opportunities; for several years only McGill University in Montreal offered a Master's in genetic counseling. McGill was joined in 1997 by the University of British Columbia and in 1998 by the University of Toronto. All these programs accept very small numbers.

In order to provide a forum for discussion and a route to action on shared professional concerns, those working in a genetic counselor capacity in Canadian clinical genetic centers incorporated in 1990 as the Canadian Association of Genetic Counselors (CAGC). Membership has been on the basis of job description, but a major activity of CAGC has been the development of educational modules and to a system of certification. Currently, membership in CAGC is approximately 150. The first certification examination was held in late 1998.

Reimbursement for molecular genetic diagnostic testing, whether through provincially based laboratories or external purchase, is a provincial responsibility. To date there has been almost no development of private commercial genetic laboratories for DNA testing in Canada. While there is no direct charge to individuals for genetic testing, controls are exerted over what is often a costly and complex process. Justification for testing must be provided along with related counseling. There is no licensing system in place in Canada as yet.

Indications and guidelines for prenatal diagnosis were developed in 1974–1983 by a collaborative group representing the Society of Obstetricians and Gynecologists of Canada (SOGC), the Canadian Pediatric Society (CPS) and the CCMG. A revised version was prepared by SOGC and CCMG and published in 1991,[6] and a further update in 1998. While they were not intended to be binding, these guidelines have been closely followed by obstetricians and geneticists and are recognized as providing a consistent standard across the country. Recognition of risks, such as maternal age, reproductive or family history and transmission of information to patients about the availability of prenatal testing are understood to be clinical responsibilities by both the courts and the medical profession. Provincial medical care insurance programs cover the cost of prenatal diagnosis for pregnant women who meet the guidelines and wish to make use of such testing. A recent survey of prenatal diagnosis in Canada[7] documented a fourfold difference between the lowest ranking and highest ranking provinces in the proportion of women eligible for prenatal diagnosis on the basis of advanced maternal age who made use of the procedure. Uptake of prenatal diagnosis for adult onset conditions has also been much lower generally than anticipated.

The CCMG, through its Ethics and Public Policy Committee, has prepared statements on predictive testing for Huntington disease,[8] population screening for cystic fibrosis[9] and the genetic testing of children.[10] In regard to Huntington disease, one of the first Mendelian disorders for which predictive testing was available, the particularly close-knit relationship among Canadian genetic centers permitted a collaborative trial of predictive testing,[11] the results of which contributed to guidelines that are applicable to many adult onset genetic diseases.

The CCMG was one of the first such organizations to draw up and agree to a set of Guidelines for Professional and Ethical Behaviour.[12] There was resistance from the CCMG membership to calling this a "Code" to which rigid adherence would be expected, since there was widespread recognition of rapid change and of the multiplicity of different individual situations that might arise. The guidelines[12] were based on the existing Code of Ethics of the Canadian Medical Association and dealt with responsibilities to patients, society and the profession. Nondirectiveness in genetic counseling was specifically advocated, as was the need to disclose information and to consult. A revision of the guidelines was undertaken and accepted by the CCMG in 1996. Changes were minor, but further emphasized non-discrimination in the provision of services, dealt with the patient's right not to receive results and the occasional appropriateness of breaking confidentiality to contact relatives at risk, as well as further strengthening the statement relating to the non-eugenic objectives of genetic services. In the light of current debate about the obligation to recontact when new information becomes available, these guidelines place the responsibility on the patient to maintain contact with the genetics clinic.

Elizabeth J. Ives

References

1. Canadian Source Book, Section 5. 33rd Annual Edition, Southam Inc., 1998.
2. The Health of Canada's Children: A CICH Profile. Second Edition, Canadian Institute of Child Health, 1994.
3. Scriver, C. (ed.) Genetics in Canadian Health Care. *Science Council of Canada, Report 42*. Ottawa: The Council, 1991, pp. 75–78.
4. The Health Report. *Maclean's Magazine* 1998; p. 15.
5. Lowry, R.B. and Bowen, P. The Alberta Hereditary Diseases Program: a regional model for delivery of genetic services. *Canadian Medical Association Journal* 1990; 142(3): 228–232.
6. Canadian guidelines for prenatal diagnosis of genetic disorders: an update. *Journal of SOGC* 1991; 13(6): 14–31.
7. Stranc, L.C., Evans, J.A., and Hamerton, J.L. Prenatal diagnosis in Canada – 1990: a review. *Prenatal Diagnosis* 1994; 14(13): 1253–1265.
8. Guidelines for the Molecular Genetics predictive test in Huntington disease. *Journal of Medical Genetics* 1994; 31: 555–559.
9. Statement of the American Society of Human Genetics on Cystic Fibrosis Carrier Screening. *American Journal of Human Genetics* 1992; 51: 1443–1444.
10. Working Party of the Clinical Genetics Society: Report on the Genetic Testing of Children. *Journal of Medical Genetics* 1994; 31: 785–797.
11. Benjamin, C.M., *et al.* Proceed with care: direct predictive testing for Huntington disease. *American Journal of Human Genetics* 1994; 55: 606–617.
12. CCMG. *Professional and Ethical Guidelines*. Ottawa, 1986; Revised 1996.

South Africa

The total population of South Africa is approximately 41 million (Central Statistical Service, 1995) and the number of births per annum approximately half a million (SA Statistics, 1993). The birth rate is falling slowly, as the majority Black population increases its use of family planning clinics, as these clinics become more accessible, and the education levels of women rise. The infant mortality rate is also falling. For the Black population in urban areas (where it is 26 per 1000, in comparison with 12.3 for the White population), it has almost reached that of some developed countries. In the rural areas, however, it is still high (about 100 per 1000) and comparable to that found in other developing countries (de Beer *et al.*, 1988). Including all populations, the infant mortality rate was approximately 46–50 per 1000 in 1995 (Sidiropoulos *et al.*, 1996). South Africa has an annual birth rate of less than 35 per 1000 people (SA Bureau of Census, 1992, quoted by Cooper *et al.*, 1993), while the world average is 26 per 1000 people and for the rest of Africa the average is 45 per 1000 people (Cooper *et al.*, 1993). The birth rate varies with population group; in Johannesburg it is 10.0 for Whites, 19.5 for Asians, 25.4 for so-called Coloureds (mixed race) and 29.9 for Blacks (Medical Officer of Health Report 1988/89), giving an average urban birth rate of 16.9 per 1000. A national literacy study undertaken by Harvard University and the University of Cape Town and published in 1995 showed that 80% of Black Africans and 40% of Whites in South Africa had less than 7 years of schooling (Sidiropoulos *et al.*, 1996). However, life expectancy is improving; the average for 1991 was 63.4 years, while in 1995 it was 65.1 years (Sidiropoulos *et al.*, 1996).

Most childhood mortality is still due to infectious diseases. However, as primary health care is becoming more accessible, and with free treatment for all children under six years and all pregnant women since 1995 (Wilkinson *et al.*, 1997), congenital malformations will become more apparent as causes of child mortality.

Organization of Clinical Genetic Services

The Genetic Services Division of the Department of Health initiated a National Birth Defects surveillance system in collaboration with the Department of Community Health, University of Cape Town (Genetic Services. Progress report 1993 and 1994) some years ago, with the cooperation of a small sample of hospitals and clinics across the country.

Genetic services, however, have been available in Johannesburg since the 1950s, when a small cytogenetic laboratory was established (Jenkins, 1990). More formal services began in 1973, when the first weekly genetic counseling clinic was opened at the Children's Hospital in Johannesburg (Jenkins *et al.*, 1973). Chairs of Human Genetics were created at the Universities of Cape Town in 1972 and Witwatersrand in 1975. A third chair was created at the University of Pretoria in 1989, and there are smaller genetic services at a further four universities: the University of Stellenbosch, Orange Free State in Bloemfontein, Natal in Durban, and the Medical University of South Africa near Pretoria. Some of the staff of these universities provide outreach programs to rural areas around the country, and neighbouring countries (without their own service) use the services. Those offering genetic counseling include both medical personnel (about 72%) and genetic counselors.

The services offered depend on the size and type of department providing them. The major universities offer complete services, which include genetic counseling clinics, laboratory services in cytogenetics, molecular genetics, and biochemical genetics, as well as community genetic programs. The smaller centers may only offer either genetic counseling, with or without cytogenetic back-up, or a laboratory service without counseling.

The number of chromosome studies undertaken on blood samples varies by center, but in the cytogenetic laboratory of the University of the Witwatersrand/South African Institute for Medical Research (Wits/SAIMR), in Johannesburg, for example, approximately 1500 samples are analyzed per year. This number increased annually until 1990, when private pathology laboratories began offering the service. Their staffs now perform about 50% of the tests. This development has also affected the Wits/SAIMR's prenatal diagnosis service, which has decreased dramatically from approximately 1000 cases in 1988 to about 450 in 1996. Maternal serum triple screening is being offered routinely by many obstetricians in private practice, in conjunction with private pathologists, but such screening is not routine in public hospitals.

The South African government has been slow to recognize community needs, and although a Genetic Services subdirectorate of the Department of Health was established in the 1970s, its budget was very small. Altogether about 16 full-time genetic nurses were appointed and trained in short courses (2 weeks) to provide community genetic services across the country. This service has been reduced and decentralized to structures within each of the nine provinces. A policy on genetic services is being developed by the central government with input from all stakeholders and providers, including academic departments of medical/human genetics. The relatively new democratically elected government places a strong emphasis on equity and primary health care.

The training of clinical geneticists and genetic counselors is in the early stages. Registration of medical practitioners as specialist clinical geneticists has been formalized with the SA Medical and Dental Council (SAMDC), and guide-lines for training are being considered by the South African Society of Human Genetics. Medical geneticists receive only in-service training at present, and few have taken the opportunity of obtaining further training overseas. An M.Sc. level program in genetic counseling was established at Wits/SAIMR in 1989. The reluctance of government departments and hospitals to create suitable posts has stifled further development. However, registration of genetic counselors has been finalized with the SAMDC, and 12 people are registered presently. Genetic counseling is provided by both medical geneticists and genetic counselors or associates. About half of the latter group have Ph.D. degrees. As the majority of the population gain access to health services and start to demand a more extensive genetic service, more medical geneticists and particularly more non-medical genetic counselors will have to be trained.

Prenatal Diagnosis

Prenatal diagnosis services are provided in the major centers, and the smaller towns either send the patient's samples or the patient to the nearest center. Country-wide data are not presently available to assess utilization. Most patients receive some counseling prior to the prenatal diagnosis procedure, although the quality depends on the provider, the minimum being given by the private obstetricians and the maximum by the genetic counselors in the academic centers. The geneticists believe that the counseling should be provided in an unbiased manner, using non-directive methods (Kromberg and Jenkins, 1997).

The cut-off age for prenatal diagnosis for advanced maternal age is 35 years in most centers, although the provincial hospitals in the Cape have specified that their minimum age is 38 years due to limited funding. Parental anxiety is also accepted as an indication, and when the patient cannot be reassured, prenatal diagnosis is performed. Nearly 2% of tests are done for this reason. As in any other developing community, the demand for services comes first from the parents of affected children and only later from women of advanced maternal age. In the local Black community, parents of affected children present for counseling, and if appropriate undergo prenatal testing, but older pregnant women are not yet fully using the prenatal diagnosis service.

Genetic services are generally focused in the major cities, and the country-wide distribution is inequitable. This situation is partly a result of history, since the services developed around the major academic centers and the state Department of Health has played only a minor role. Consequently, the users of services were initially from the higher socio-economic, better-educated groups, generally from the White population who had medical insurance and lived in the industrialized part of the country (Kromberg and Berkowitz, 1986). The number of Black users is increasing dramatically.

The lower socio-economic groups are not receiving equitable services and often have neither easy access to services nor referral channels. Even when they have medical insurance they may be treated by well-meaning doctors, who, aware of the cultural background and limitations of the social environment, treat their patients as stereotypes and not as individuals, resulting in a poorer quality of care. In a sample of 92 Black mothers of children with Down syndrome (Kromberg *et al.*, 1992), the majority (68%) had attended a private practitioner in the first trimester, and 55% were older than 35 years. However, none had been offered prenatal diagnosis, although 73% would have accepted it and 52% would have requested termination of pregnancy for an abnormal result. The value of health, however, may be seen differently in a community where infant deaths are common, the life span is shorter, violence is rampant, and regard for human life is reduced. The conservative religious views among some sections of South African society have led to problems in the liberalization of the abortion laws and in introducing widespread screening programs (e.g., for Tay–Sachs disease in the orthodox Jewish community) and prenatal programs. The uptake of prenatal diagnosis among women (mostly from the White population) of advanced maternal age, for instance, was only 35% in 1980 (Schreier *et al.*, 1986).

Cost–Benefit of Early Diagnosis

Prenatal diagnosis followed by selective abortion has been widely acknowledged as a cost-effective procedure. The majority of couples who had fetal abnormalities detected in the decade 1976–1985 (Kromberg *et al.*, 1989) requested termination of pregnancy (e.g., in 97% of trisomy 21 cases, 100% of trisomy 13 and 18, 66% of XXX, and 80% of XXY). Similarly, the majority of parents with a fetus diagnosed *in utero* with a neural tube defect requested termination. The saving, for example, per cystic fibrosis birth avoided was estimated at R2,500,000, and per thalassemia birth avoided was R1,900,000 in 1994 (Kromberg and Jenkins, 1994). This program therefore saved the state millions with regard to the provision of special education facilities, therapy, hospital and medical specialists services, and saved families a great deal of trauma and burden. It is well recognized that prenatal diagnosis for the prevention of Down syndrome and neural tube defects is cost-effective in developing countries (WHO, 1989). In the policy document being prepared for the development of genetics services in South Africa, the services for preventing these disorders have been made a priority.

Presently, there is no newborn screening generally available locally, so that cases of phenylketonuria (PKU) and hypothyroidism are not being detected and treated early. Findings from previous studies suggested that PKU was not as common in the Black population as it is in European and North American populations. There is also some evidence that PKU may be rarer in South African Whites than it is in Europe, possibly as a result of founder effect (Jenkins, 1974). Hypothyroidism, however, occurs at significant frequencies in Black and White South Africans, and there is probably a need for neonatal screening. Galactosemia has an incidence of about 1 in 16,000 Black births in South Africa, 4–5 times higher than in European populations, and it could therefore be included in a screening program (Manga *et al.*, 1997).

Need for Staff

The suggested model for genetics services is that of a hub and spokes, where the academic center of excellence is the hub with spokes reaching out, using staff employed by the Department of Health and supervised by the academics, right into the remotest parts of the country. For this model to be successful, training of new staff is essential. The WHO suggests that one clinical geneticist is required for every million people and four genetic counselors or nurses are required for every clinical geneticist. For the South African population of about 40 million, 40 geneticists, and 160 genetic counselors, are required. The country presently has 12 clinical geneticists, 12 genetic counselors, and about 16 genetics nurses, so that much development has to occur to allow the service to expand. The short-term goal would be to train more genetic associates and nurses on two levels: M.Sc. (post-graduates) in genetic counseling (a course is available at Wits/SAIMR) and perhaps post-basic nurses could be offered a 6-month training course in genetics for a diploma that would be recognized for advancement by the Nursing Council (this course still has to be developed). The aim could be to get at least two such people per province trained in the next 5 years. A longer term goal would be to train further clinical geneticists, either with a pediatric, pathology, obstetrics, or internal medicine background or with medical genetics as a recognized specialization.

Services for Persons with Disabilities

Services for persons with disabilities are generally poor, inadequate and underdeveloped. There are limited facilities for remedial teaching, and children with special needs are frequently in the mainstream with no support services (Kromberg *et al.*, 1997). There is also an inequitable distribution of services; those who can pay can access services as good as any being offered in developed countries, while the others receive inferior and inadequate services or none at all.

Genetic Counseling

The findings of the international survey suggest that the doctor–patient relationship is still somewhat authoritarian and that patients tend to seek advice rather than counseling from their doctor. This latter attitude affects genetic counseling, since the patient then demands an opinion from the counselor, who, although most favor unbiased and non-directive counseling, may be tempted to describe an approved course of action, or what he or she would do in the same situation. This may lead to counselors who are pro-abortion having a higher rate of patients requesting abortion for genetic disorders. These attitudes may, however, be changing. A recent local study (Girson, 1996) on chorionic villus sampling shows that a number of patients whose fetuses were diagnosed with cystic fibrosis decided against terminating the pregnancy. Extenuating factors included having had previous abortions and having a young mildly affected first child. However, counseling may also be becoming more non-directive as the counselors' experience and confidence with the technique increases.

Jennifer G.R. Kromberg and Trefor Jenkins

References

Central Statistical Service. Demographic statistics 1995. Pretoria: Central Statistical Service, 1995.

Cooper, C., Hamilton, R., Mashabela, H., MacKay, S., Sideropoulos E., Gordon Brown, C., Murphy, S., and Markham, C. *Race Relations Survey 1992/93*. Johannesburg: SA Institute for Race Relations, 1993.

De Beer, C, Buch, E., and Mavrondonis, J. Health Care in South Africa: how far from health care for all? In: *A National Health Service for South Africa*. Center for the Study of Health Policy, Department of Community Health, University of the Witwatersrand, Johannesburg, 1988: pp. 77–82.

Genetic Services: Progress report 1993 and 1994. *Genetic Services, Department of Health*. Pretoria: Government Printer, 1995.

Girson, L. A follow-up study of women who have had chorionic villus sampling for prenatal diagnosis. *MSc Research Report*, University of the Witwatersrand, Johannesburg, 1996.

Harris, R., Emery, A.E.H., Johnston, A.W., Pembrey, M.E., Winter, R., and Insley, J. Role and training of clinical geneticists. *Report of the Clinical Genetics Society Workshop Party*. London: Eugenics Society, 1983.

Jenkins, T. Prevention of hereditary disease. *South African Medical Journal* 1974; 48: 1500–1506.

Jenkins, T. The role of screening in the prevention of inherited disease in South Africa. *South African Medical Journal* 1977; 51: 832.

Jenkins, T. Medical genetics in South Africa. *Journal of Medical Genetics* 1990; 27: 760–779.

Kromberg, J.G.R. and Berkowitz, D.E. Managing genetic disorders. *SA Family Practice* 1986; 7: 279–285.

Kromberg, J.G.R., Bernstein, R., Jacobson, M.J., Rosendorff, J., and Jenkins T. A decade of mid-trimester amniocentesis in Johannesburg: prenatal diagnosis, problems and counseling. *South African Medical Journal* 1989; 76: 344–349.

Kromberg, J.G.R., Christianson, A.L., Duthie-Nurse, G., Zwane, E., and Jenkins, T. Down syndrome in the Black population. *South African Medical Journal* 1992; 81: 337.

Kromberg, J.G.R. and Jenkins, T. Insurance cover for common congenital abnormalities. Report for Bensure Management Services, 1994.

Kromberg, J.G.R. and Jenkins, T. Ethics and South African geneticists. Southern African Society of Human Genetics Congress, Pilanesberg, 18–21 May 1997, Abstract Book: 51.

Kromberg, J.G.R. and Jenkins, T. Genetic services in South Africa and common ethical issues related to genetic information. *Specialist Medicine* 1997; XIX: 54–60.

Kromberg, J.G.R., Christianson, A.L., Manga, P., Zwane, M.E., Rosen, E., Venter, A., and Homer, S. Intellectual disability in rural Black children. *South African Journal of Child and Adolescent Mental Health* 1997; 9: 2–11.

Manga, N., Jenkins, T., and Lane, A.B. The molecular basis of transferase galactosaemia in the South African Black population. SA Society of Human Genetics Congress, Pilanesberg, 18–21 May 1997, Abstract Book: 43.

Schreier, A., Kromberg, J., and Hofmeyr, J. Prenatal risks in the older woman. *Proceedings of Conference on Priorities in perinatal care in South Africa*. Rustenberg, 25–28 February, 1986: pp. 151–154.

Wilkinson, D., Sach, M.E., and Karim, S.S.A. Examination of attendance patterns before and after introduction of South Africa's policy of free health care for children aged under 6 years and pregnant women. *BMJ* 1997; 314: 940–941.

World Health Organization. Community approaches to the control of hereditary diseases. Report of WHO Advisory group, Geneva, 1985 (document HMG/WG/85.10).

World Health Organization. Report of a WHO meeting on Advances in Diagnosis, Treatment and Prevention of Hereditary Diseases. Geneva, 20–22 November, 1989 (document WHO/HDP/ADV/89.3).

World Health Organization Hereditary Diseases Programme. *Guidelines on Ethical Issues in Medical Genetics and the Provision of Genetic Services*. Prepared by Wertz, D.C., Fletcher, J.C., Berg, K., and Boulyjenkov, V. Geneva: WHO, 1995.

United Kingdom

Clinical genetic services are made available in the United Kingdom largely through the National Health Service, and there is little private activity in this field. There is at least one Genetics Center in each health region in the UK, the largest serving a population of approximately 5.5 million. Although there may be some social class bias in referral to these services, reflected in Tudor Hart's "Inverse Care Law", there is no reason to think it is greater for clinical genetics than for any other aspect of health care. Each region offers a network of clinics held in district hospitals in addition to clinics in the Regional Center, and this helps to promote access to the service for those living in more remote areas.

The staffing levels for clinical genetics services recommended by the Royal College of Physicians – 2 consultants and 4 genetic nurses or associates per million population – have not been achieved in most regions, and the need for further staff is clear, given the recent rapid increase in referrals relating to familial cancers.

Clinical genetics services are organized as a tertiary service and are distinct from routine antenatal screening services made available through hospital obstetric and community midwifery services. They are also distinct from the haemoglobinopathy screening services that are organized on a more local basis, especially in areas with a substantial proportion of ethnic minority groups (Black, Mediterranean, south and east Asian). Newborn screening is made available universally for phenylketonuria and congenital hypothyroidism. In some areas, screening is provided for cystic fibrosis. Screening for sickle cell disease is also made available where the carrier frequency of sickle cell disease is considered to justify it.

Clinical Genetics and Genetic Screening

Clinical genetics in the UK is seen as a service provided to individuals and families, while prenatal screening programs are seen rather as a public health activity. The goals of clinical genetics are therefore stated in terms of providing information, options and support to families and individuals and helping them to make the most appropriate decisions about reproduction or genetic testing, given their circumstances and beliefs. Prenatal screening programs can either be understood as having similar goals, providing information and enabling women or couples to make informed reproductive decisions, or as public health programs whose goals can be framed in terms of reducing the birth incidence of individuals with particular genetic conditions or congenital malformations (especially Down syndrome and neural tube defects). The underlying purpose of such a "public health" approach is the sparing of resources. There is a clear tension between these "informed choice" and "public health" models of prenatal screening, and these programs can be portrayed as a form of eugenics introduced through the "backdoor," in Troy Duster's phrase (Duster, 1990).

Although prenatal screening programs are delivered by maternity services, clinical genetics units are involved in such programs; they provide the necessary cytogenetic laboratory services, they have a role in prenatal genetic counseling and they are expected to monitor the population frequency of genetic disorders and congenital malformations. With a growing emphasis in Britain on the thorough evaluation of medical screening services, however, and the willingness of health services researchers to employ a wide range of relevant outcomes, including social and psychological consequences as well as resource implications, a serious re-evaluation of prenatal screening programs may become feasible. Prenatal screening may then move away from the "public health" paradigm towards that of "informed choice".

Prenatal Diagnosis

Antenatal screening for Down syndrome and for structural anomalies is made available to most pregnant women in the UK, and is accepted by the large majority. National figures are difficult to ascertain. Of those with an increased chance of a fetus affected by Down syndrome, about 80% accept the offer of amniocentesis. Of those women who have had an amniocentesis performed, the vast majority proceed to have a termination in the event of an abnormal result. When the amniocentesis result is unanticipated (e.g., neither Down syndrome nor "normal"), or when a structural anomaly of uncertain significance is found on fetal ultrasound, then patients' decisions are much more variable and depend heavily upon how they are given the relevant information.

The birth incidence of infants with neural tube defects is falling for secular reasons and because of the widespread promotion of folic acid for women planning to conceive, but part of the decline in incidence is attributable to the termination of pregnancies with an affected fetus detected on scan. The reduction of births of children with Down syndrome is also in part attributable to terminations of pregnancy, but policy about the offer of screening – and what type of screening is made available – is too variable geographically to define the effect of any one policy. In any case, many clinical geneticists in Britain would not accept such crude outcomes as a useful measure of their work and would wish to distance themselves from the implicit devaluation of affected individuals with these causes of disability.

The uptake of prenatal diagnosis for specific diseases common in particular ethnic minority groups is very dependent upon the access of members of these communities to genetic counseling services. Where culturally appropriate access is provided, as in the North London Greek Cypriot community, the birth rate of individuals with beta-thalassemia is greatly reduced.

Ethics of Counseling

Whether or not there is advice to be dispensed, the ethos of genetic counseling in Britain is usually first to concentrate on listening to clients – to understand their concerns – and only then to focus on the medical issues (e.g., diagnosis, prognosis, risk of recurrence, risk of complications) that may need to be confronted, even if they cannot always be resolved. While most clinical geneticists have no formal training in (non-genetic) counseling or psychotherapy, there is a growing recognition that more attention needs to be paid to the processes of communication within genetic counseling. The ethos of genetic testing in routine antenatal screening programs is very different. Empirical studies have shown that inadequate information is often presented to pregnant women and the offer of screening in practice often becomes the routinisation of testing. Maternal serum screening and fetal ultrasound anomaly scanning, for example, may be "offered" without the woman experiencing any real sense of choice.

Blame is sometimes attributed to women or couples who have infants affected by Down syndrome, on the grounds that prenatal screening should allow the condition to be avoided if the woman complies with the screening program. To respond to such attributions of blame by merely explaining that screening will not detect all cases is an inadequate response because, in effect, it is colluding with the implicit view that women should comply with screening and that Down syndrome (and similarly "abnormal") conceptions should be terminated.

Angus Clarke

Reports (see also Laws Affecting Genetics Services)

Nuffield Council on Bioethics. Genetic Screening. Ethical Issues. London: Nuffield Council on Bioethics, 1993.

Nuffield Council on Bioethics. Mental disorders and genetics: the ethical context. London: Nuffield Council on Bioethics, 1998.

Royal College of Physicians. Prenatal Diagnosis and Screening. Community and service implications. London: Royal College of Physicians of London, 1989.

Royal College of Physicians. Ethical Issues in Clinical Genetics. London: Royal College of Physicians of London, 1991.

Other Relevant Books

Abramsky, L. and Chapple, J (eds). *Prenatal Diagnosis: The Human Side*. London: Chapman & Hall, 1994.

Clarke, A. (ed.). *Culture, Kinship and Genes*. London: Macmillan, 1997.

Clarke, A. (ed.). *The Genetic Testing of Children*. Oxford: Bios Scientific Publications, 1998.

Duster, T. *Backdoor to Eugenics*. New York and London: Routledge, 1990.

Harper, P.S. and Clarke, A. *Genetics, Society and Clinical Practice*. Oxford: Bios Scientific Publications, 1997.

Marteau, T.M. and Richards, M.P.M. (eds). *The Troubled Helix: Social and Psychological Implications of the New Human Genetics*. Cambridge: Cambridge University Press, 1996.

United States

Structure of the Profession

The United States has approximately 2847 board-certified clinical and medical geneticists at the doctoral level and 1410 genetic counselors, mostly with Master's degrees. Clinical genetics certificates are M.D.s who have passed a special examination given every three years ($n = 1006$). Ph.D. "Medical Genetics" is the equivalent certification for those with Ph.D.s. ($n = 150$). There are also certifications in Clinical Cytogenetics ($n = 522$), Molecular Genetics ($n = 284$), Clinical Biochemical Genetics ($n = 180$), and Clinical Biochemical Molecular Genetics ($n = 49$). Although some people in these specialties are primarily researchers, many clinicians hold dual or triple certificates. A total of 2191 certificates have been granted plus an additional 1410 for genetic counselors.

Three years' experience in genetics is required before taking the examinations. Many people from other countries receive American Board certification and practice and reside outside the United States; hence the "approximate" figures. Until 1993, the American Board of Medical Genetics (ABMG) certified M.D.s, Ph.D.s, and Master's level genetic counselors in the specialty of Genetic Counseling, with the same examinations for all. A total of 631 persons were certified in Genetic Counseling under this board. In 1993, a separate American Board of Genetic Counseling (ABGC) was established for Master's-level counselors, under the auspices of the National Society of Genetic Counselors. This was necessary because when genetics came under the umbrella of American Medical Association (AMA) recognized specialties in order to qualify for professionally-set standards of insurance reimbursement, the ABMG could not permit certification of Master's-level counselors under its aegis. Hence the need for two Boards. A total of 779 counselors have been certified by ABGC, for a total of 1410 counselors altogether.

The American Society of Human Genetics (ASHG), with over 6700 members, includes clinicians, researchers, counselors, and others. Anyone may join. Annual conferences now draw over 5000 registrants and have over 2000 poster presentations, making them the largest genetics conferences in the world. The much smaller American College of Medical Genetics (ACMG), founded in 1991, oversees ABMG certification and represents the specialty to the AMA. Under AMA rules, 51% of ACMG full members must be physicians, and 51% of these must be members of the AMA. Until genetics came under AMA aegis, few geneticists were AMA members; some joined AMA in order to make the College possible. Full membership in ACMG is open only to those who have practiced full-time in genetics (including Master's-level counselors) for at least three years. ACMG meetings attract about 600 persons a year, mostly clinicians rather than researchers. (A special organization for genetic researchers, The Human Genome Organization, known as HUGO, is internationally-based and draws about 750 conference attendants.)

The National Society of Genetic Counselors (NSGC), founded in 1979, has over 1800 members, including many awaiting certification by the ABGC. Some have Ph.D.s, M.D.s, or R.N.s. NSGC has its own national conference, with over 900 attending. There are 33 genetic counseling programs worldwide, including 25 in the United States, three in Canada, three in Australia, and one each in South Africa, and the United Kingdom. In the United States, a college degree, but not a science major or science courses, is required for entry. Counseling programs typically last two years, with intensive training in biology, genetics, and the psychosocial aspects of counseling. Counselors-in-training observe counseling sessions and conduct counseling under supervision. Some are required to design and complete their own research projects. Some programs require students to follow the daily activities of families affected by disabilities on a regular basis for a specified period, in order to better acquaint them with the realities of disabilities and help them appreciate these families' satisfaction with their lives. Many recent graduates of counseling programs, however, feel that they did not receive enough experiential training about disability (Teicher *et al.*, 1998). Most genetics programs graduate between four and seven students a year, who find jobs almost immediately in genetics centers.

Nurses also play a role in genetics, but a small one in comparison to their numbers. There are 2,162,000 registered nurses in the United States. Fewer than 200 belong to the International Society of Nurses in Genetics (ISONG). For many years, it was hoped that nurses would supply a growing need for genetic counseling and support. So far, efforts to educate nurses about genetics have produced only small numbers of graduates. There are three Clinical Nurse Specialist in Genetics programs, but most have graduated fewer than 10 students. Most nurses are afraid of a heavy commitment to science.

The ASHG and ACMG keep no sociodemographic data on their members or on Board certificates. According to 1990 census data on US physicians as a whole (21%)

of all physicians are women, 80% are White, 4% African-American, 11% Hispanic, 0.1% Native American (American Indian), and 11% Asian-American. A look at those attending genetics meetings suggests that the demographics of genetics professionals may differ from those of physicians generally. Most geneticists appear to be White, with some Asians. Until recently, there were fewer than ten African-American geneticists. There are almost no Native Americans in genetics. Representation of women among M.D.–Ph.D. geneticists appears to be greater than among physicians generally, however.

The NSGC keeps demographic data. Its members are 96% women, 93% White, 1% African-American, 1% Hispanic, 4% Asian-American and 1% other (Schneider and Kalkbrenner, 1998). In all, 34% are age 21–30, 40% age 31–40, and 19% age 41–50. Overall, this is a homogeneous group, despite efforts to increase diversity.

Professional Codes of Ethics The NSGC has a Code of Ethics (1992) based on the counselor's relationships with patient, patient's family, profession, society, and self. This Code assumes that moral actions will flow from good relationships. Relationships come first. If a relationship is open, honest, and caring, the best course of action will emerge. Special sessions at annual NSGC meetings discuss how to use this Code in specific cases.

The ASHG and ACMG do not have overall codes of ethics, and there are no plans to develop such codes. The preferred approach is to issue statements on specific topics, such as genetic testing of children, genetic information and adoption, behavioral genetics, informed consent for genetic research, cystic fibrosis carrier screening, genetic testing and insurance, patenting, freedom of choice, eugenics and the misuse of genetic information to restrict reproductive freedom, and professional disclosure of familial genetic information. These statements are not binding on members and do not constitute a legal "standard of care" that can be used in lawsuits. There are now 25 "Points to Consider" statements, available on the ASHG website (http://www.faseb.org/genetics/ashg/pol-00.htm). There are no statements on some of the cases in the survey that respondents found most difficult – sex selection and the deaf couple, and parenthood for persons with disabilities.

Since the foundation of ACMG, the ethical and social issues committees of both organizations have coordinated their efforts, and both boards of directors have approved policy statements. The effects of these statements on practice is unknown.

The Council of Regional Networks of Genetic Services (CORN), a coordinating body for the country's ten regional networks (voluntary organizations that included geneticists, counselors, and patients), also produced a statement supporting the principles of autonomy and justice (Baumiller *et al.*, 1996). Except for its rejection of sex selection, it contained few specific points about the practice of genetics. The federal government (DHHS, Bureau of Maternal and Child Health) has since discontinued funding for CORN and the regional networks, and the Ethics Committee of CORN has been disbanded.

Distribution of Services There are no national data in the US on how many people receive genetic counseling, have genetic tests, have prenatal diagnosis or abort following prenatal diagnosis. CORN attempted for many years to gather such data, but had to depend on voluntary and usually unreimbursed cooperation of hospitals.

Data were never complete. During the years that CORN gathered such data, it appeared that the majority of patients were under the age of 14, with more boys than girls (because X-linked conditions affect boys). There were no data about race or methods of payment for services.

Some states keep data about state-funded services. California keeps data on use of its maternal-serum alphafetoprotein program, which offers screening (through a blood test) to all pregnant women at minimal costs, with the cost including ultrasound and amniocentesis if necessary. According to California records, 75% of pregnant women have the screening and 25% refuse, usually on moral grounds (Cunningham and Tompkinson, 1999). The US Centers for Disease Control (CDC) also tries to extrapolate trends by examining epidemiological statistics, for example, births of children with Down syndrome in metropolitan areas where figures may be available. A few CDC publications have reported a decrease in births of children with Down syndrome to women over 35, but there has been no decrease among younger women, who bear most of the children (CDC, 1994). A much larger decrease in births of children with neural tube defects began before prenatal diagnosis became available and may be attributable to dietary changes increasing intake of folic acid during pregnancy (Cragan *et al.*, 1995).

Since there are no national registries of genetic conditions, support groups (voluntary organizations of patients and families) are often the best sources of information. Not everyone with a genetic condition joins a support group, but membership figures provide a general estimate of numbers. On the basis of support group memberships, we can estimate that there are 30,000 people with cystic fibrosis and 30,000 people with the mutation for Huntington disease, including those who will develop symptoms in the future.

Payment for Genetic Services Almost alone among developed nations, the US has no universal national health insurance. In 1996, the latest year for which statistics are available, 42 million people – 15.4% of the population – had no health insurance of any kind, including Medicaid, the federal government's program for poor people (US Census Bureau, Statistical Abstract of the US, 1998, Table 18, p. 125). The percents uninsured are even higher in the age groups having children: 29% for ages 18–24, and 22% for ages 35–44. Overall, Hispanics (34%) and African-Americans (22%) are more likely to be uninsured than Whites (14%). These inequities affect prenatal care and birth outcomes. Twice as many African-Americans (7.4%) and Hispanics (6.8%) as Whites (3.4%) receive prenatal care only in the third trimester or not at all. Over twice as many African-American babies (13%) as White babies (6.3%) are low birthweight. Over twice as many African-American babies (9.2 per 1000 live births) as White babies (3.9 per 1000) die in the first month of life. Over twice as many African-American babies (14.2 per 1000 live births) die in their first year of life as White babies (6 per 1000) (Census Bureau, Table 134, p. 98). Congenital anomalies are the single largest cause of death in the first 12 months of life, accounting for about 22% of infant deaths (Census Bureau, Table 136, p. 98). Over four times as many African-American women (18.5 per 100,000 live births) die in pregnancy and childbirth as White women (4.2 per 100,000 live births). These inequalities have existed ever since the United States began keeping birth registrations in 1913. Social and health care inequalities demonstrated in infant and maternal mortality statistics exist throughout health care, including genetics.

At present, there is no popular movement for health care reform. Most people are insured and seem to care little about those who are not. Exclusion from insurance coverage on genetic grounds is not responsible for the large numbers of uninsured persons. Most people receive their insurance through "employer packages" covering all employees, without any questions about an individual's medical status or any tests or examinations. Others are covered under Medicare, a government program for people over 65, or Medicaid, a government program for poor people administered by the states. Only about 8% of insured people apply individually and must therefore fill out applications about their medical status and history, or supply medical records. Requirements for genetic testing are very rare and are applied mainly to help those who would otherwise be refused insurance. For example, someone at 50% risk for Huntington disease could be tested presymptomatically, and if found not to have the Huntington mutation, could be covered by insurance. If a person did not want to have the test, or the test showed the Huntington mutation, the person could usually receive health or life insurance with an exclusion for Huntington disease. Most people who are denied insurance by one company eventually receive it from another (Hall and Rich, 2000).

However, there is a major loophole in insurance regulations that could permit widespread exclusion of people with genetic conditions. Insurance is regulated by the 50 states, not the federal government, except for government programs such as Medicare. One kind of insurance, however, is not regulated at all – employer self-insurance. Employers who act as their own insurers are exempt from all regulations. The US Supreme Court (*McGann* v. *H&H Music Company*) has ruled that employers may exclude or severely limit coverage for AIDS and other conditions. No laws would prevent exclusion of sickle cell anemia, Down syndrome, or cystic fibrosis, though no employer has tried to do so. Employees are not usually aware that their employer is self-insured, because the insurance is administered by an insurance company or health maintenance organization. About 55–60% of all persons insured are covered by self-insuring (unregulated) employers.

Patient Organizations Usually called support groups, these organizations provide information, referrals, community and public education, and above all, psychosocial support for individuals and families. Many people find them the most up-to-date sources of information about tests, treatments, and research. Sometimes support groups have collected funds or located families to donate cell samples to initiate research on rare diseases, such as ataxia telangactasia or Canavan disease.

There are two national umbrella organizations for support groups, the Genetic Alliance (Washington, DC), founded in 1986, which includes over 287 organizations for genetic or partially genetic conditions, and the National Organization for Rare Disorders (NORD). Support groups may also have regional and state chapters. Many publish their own informational brochures and have exhibits at ASHG, ACMG, and NSGC meetings. Some apply for research grants or testify before national commissions. They represent a growing and important force in national discussions.

Studies of Genetic Counseling The largest study of genetic counseling to date, of 1369 counseling sessions, was conducted by James R. Sorenson in 1977–79 at 44 clinics (Sorenson *et al.,* 1981). Most geneticists in that study were physicians, because few Master's-level counselors existed at the time. Prenatal diagnosis was not yet widespread, and was not included in the questionnaires. Patients completed questionnaires before and after counseling; professionals completed them only after counseling. Questions focused on session content; there were no questions on ethics. The study showed that neither doctor nor patient was aware, after a 45–60 min counseling session, of what the other party had wanted to discuss (Wertz *et al.*, 1988a). Yet geneticists were satisfied with 95% of the sessions; their satisfaction was unrelated to whether patients had learned their risk or diagnosis (Wertz *et al.*, 1988b). The Sorenson study also showed that couples frequently differed in their perceptions of a situation, both before and after counseling, and that psychosocial issues such as effects on the marriage or family life were not often discussed in counseling (Sorenson and Wertz, 1986). Although about half of those counseled reported at six months that counseling had influenced their reproductive plans, it appeared that counseling reinforced, rather than changed, existing plans (Wertz and Sorenson, 1987).

Our own survey of 476 counseling sessions was the second major study of counseling, but does not approach the Sorenson study in size or geographical distribution. As in the Sorenson study, we asked questions of patients both before and after a counseling session. Although our main focus was ethics, we asked questions about session content and about perceptions of counselor directiveness. We also asked counselors about session content. Counselors included M.D.'s as well as those with Master's degrees. This time it appeared that patients and counselors were aware, in most cases, of what the other most wanted to discuss (usually etiology or risk). Most patients thought that counselors were not urging them to take any particular actions, though many thought that the counselor's own preferences were for prenatal diagnosis and carrier testing. The psychosocial issues not discussed in 1977–79 were still not being discussed, nor would most counselors offer referrals to social workers or psychologists to discuss them.

Yet both counselors and patients were satisfied. It may be that patients go to genetic counseling primarily to get genetic information and do not expect to discuss these issues. One may wonder, however, why there was so little discussion of what a child with a genetic condition might be like as an adult, which is surely medical information that parents would find useful.

Studies of Minority Groups The US population is 72% White (non-Hispanic), 12% African-American (non-Hispanic), 11% Hispanic, 4% Asian-American, and 1% Native American. There are no large-scale studies of responses of racial and ethnic minority groups to genetics, although there have been attempts to develop culturally sensitive counseling (Fisher, 1996). African-Americans are more distrustful than Whites of medical research generally; in a public survey of 500 African-Americans and 500 Whites, conducted by Stephen B. Thomas of Emory University, 36% of African-Americans thought they were "very likely to be used as guinea pigs" in research, as compared with 16% of Whites (Thomas, 1998). This distrust results from a long history of such use, notably the Tuskegee syphilis study (Jones, 1981), which left Black men untreated for years after the discovery of penicillin.

It is difficult to locate and recruit minorities to surveys; a survey of African-Americans may cost four times much as a survey of the same number of Whites (Wertz, 1998). There were not enough minorities of any kind in our patient survey for comparative analysis. Comparison of the 11% in our public survey who were African-Americans with the remainder, however, suggests that African-Americans may approach ethical issues differently. African-Americans were more concerned with getting care for their families than with the privacy issues that have attracted so much attention in the United States. Fewer would abort for almost all of the 12 conditions in the public survey (Wertz, 1998). Most geneticists agree that many minorities are underserved and that more research is necessary on their ethics and values.

Dorothy C. Wertz

References

Baumiller, Robert C., Cunningham, George, Fisher, Nancy, Fox, Lynda, Henderson, Merrill, Lebel, Robert, McGrath, Geraldine, Pelias, Mary Z., Porter, Ian, Seydel, Frank, Willson, and Nancy Roper. Code of ethical principles for genetics professionals: an explication. *American Journal of Medical Genetics* 1996; 65: 179–183.

Centers for Disease Control. Down syndrome prevalence at birth – United States 1983–1990. *Morbidity and Mortality Weekly Report* Vol. 43, No. 33, August 26, 1994, 617–623.

Cragan, J.D. *et al.* Surveillance for anencephaly and spina bifida and the impact of prenatal diagnosis, 1955–1994. *Morbidity and Mortality Weekly Report* Vol. 44, No. SS-4, August 25, 1995, 1–12.

Cunningham, George and Thompkinson, D. Gwynne. Cost and Effectiveness of the California Triple Marker Prenatal Screening Program. *Genetics in Medicine* July–August 1999; 1(5): 199–206.

Fisher, Nancy (ed.). *Cultural and Ethnic Diversity. A Guide for Genetics Professionals.* Baltimore: Johns Hopkins University Press, 1996.

Hall, Mark A. and Rich, Stephen S. Laws restricting health insurers' use of genetic information: impact on genetic discrimination. *American Journal of Human Genetics* 2000; 66: 293–308.

Jones, H. *Bad Blood: The Tuskegee Syphilis Experiment – A Tragedy of Race and Medicine.* New York: Free Press, 1981.

National Society of Genetic Counselors Code of Ethics. *Journal of Genetic Counseling* 1992; 1(1): 41–44.

Schneider, K.A. and Kalkbrenner, K.J. Professional Status Survey, 1998. National Society of Genetic Counselors. *Perspectives on Genetic Counseling* 1998; 20: S1–S8.

Sorenson, James R., Swazey, Judith B., and Scotch, Norman A. *Reproductive Pasts, Reproductive Futures: Genetic Counseling and Its Effectiveness.* New York: Alan R. Liss, 1981.

Sorenson, J.R. and Wertz, D.C. Couple agreement before and after genetic counseling. *American Journal of Medical Genetics* 1986; 25(3): 549–555.

Teicher, J., Tsipis, J., Wertz, D., and Lerner, B. Disability awareness training in the graduate genetic counseling training program: a survey of recent graduates. National Society of Genetic Counselors Annual Meetings. Proceedings, Denver, Co, Oct. 24–27, 1998, p. 164.

Thomas, S.B. *Assessing Differences in Factors Associated with Participation in Medical and Public Health Research: A Random Household Survey of Black and White Americans.* Atlanta: Emory University, 1997.

Wertz, D.C. The difficulties of recruiting minorities to studies of ethics and values in genetics. *Community Genetics* 1998; 1(3): 175–179.

Wertz, D.C. and Sorenson, J.R. Client reactions to genetic counseling: self-reports of influence. *Clinical Genetics* 1987; 30: 494–502.

Wertz, D.C., Sorenson, J.R., and Heeren, T.C. Communication in health professional-lay encounters: how often does each party know what the other wants to discuss? In: Ruben, B.D.

(ed.), *Information and Behavior*, Vol. 2. New Brunswick, NJ: Transition Books, 1988a, pp. 329–342.
Wertz, D.C., Sorenson, J.R., and Heeren, T.C. 'Can't get no (dis)satisfaction': professional satisfaction with professional-client encounters. *Work and Occupations* 1988b; 12(1): 36–54.

NORTHERN/WESTERN EUROPE

Denmark

Demographic Information

Denmark is a small North European country with a population of 5.2 million people. The majority of the population lives in urban areas, with 25% in the capital, Copenhagen. The country is strongly secularized, though almost 90% formally belong to the Lutheran state church. The population is very homogeneous, although since the 1960s a growing segment of the population consists of immigrants and their descendants. Denmark has a literacy rate close to 100% and a high standard of living.

Eighty-five percent of health services are publicly financed, and the public health insurance system covers everybody regardless of employment. The system gives access to free health services, except for dental services. The expenditure on health care is 6.5% of the GNP. Although the National Board of Health plays a role in setting health policy guidelines, the financing and actual policy decisions are decentralized to regional governments.

The number of newborns approximates 70,000 per year. This figure has increased since the early 1980s; 1983 was an all-time low, with only 50,000 newborns. Maternal age has also increased, and the mean age for primipari is now 29. The fertility rate is 1.7 per woman, and the infant mortality rate 7.5 per thousand.

Testing and Counseling

Genetic testing and counseling are offered in two different institutional settings, i.e., in gynecological departments and clinical genetic centers. Prenatal diagnosis was first performed in 1970. The number of tests gradually increased until 1993, during which year 8,646 tests were performed. Since then the total number of prenatal tests has decreased, and in 1995 the figure was 7,776. The latest national figures are from 1995. In that year 4,695 amniocenteses and 3,081 chorionic villus samplings were performed. Even in relative terms the figure has decreased. After an all-time high of 12.8% in 1993, only 11.1% of all pregnancies that were carried to term were tested in 1995. The figure includes amniocenteses and chorionic villus samplings. In order to reduce the number of invasive tests, preliminary serum screening ('triple test') for abnormal levels of certain proteins is gaining importance.

The majority of fetal tests are routine chromosome analyses offered to pregnant women over 35 years of age. A minor number of tests are performed to detect fetuses with rare genetic diseases in at-risk pregnancies. Programs have thus been developed to test for the inheritance of cystic fibrosis, Duchenne muscular dystrophy, the hemophilias, Huntington disease and a large number of other diseases in at-risk families. It is estimated that in this area approximately 1,200 counseling sessions are performed annually by trained geneticists. In gynecological departments the counseling of pregnant women is sometimes offered by a gynecologist, but more often by a midwife, who performs the majority of routine counseling of pregnant women over 35 years.

In recent years the percentage of women with an age-related risk having prenatal diagnosis has decreased quite substantially, from about 80% to 60% of all pregnant women over 35 years. There is no clear explanation of this development. It is interesting, however, that although the number of prenatal tests performed has decreased since 1992, a higher percentage of the pregnancies are terminated due to a pathological finding (3.6% in 1995, 2% in 1993).

All newborns are tested for phenylketonuria (PKU) and hypothyroidism. The tests are non-compulsory but very few refuse. Undoubtedly a significant number of mothers do not realize that they have the option of refusing the PKU test of their newborn since the procedure does not require a written consent.

To a small but growing extent, adults seek genetic tests in order to plan their reproductive lives and for career-related purposes.

Denmark has six specialized clinical genetic centers doing laboratory work and genetic counseling. Some of these are associated with the universities, others belong in independent institutions. Altogether around 16 full-time medical staff are involved in genetic counseling in these departments. In addition, a similarly sized staff of part time or temporarily employed people are engaged in counseling and testing. The genetic centers offer counseling to people with genetic disease both in connection with pregnancy and otherwise. Only in 1996 was clinical genetics recognized as a medical specialty in Denmark.

Recently, the departments of clinical biochemistry in the larger hospitals have also begun to offer genetic tests to be used in various clinical settings. A certain competition between genetic centers, gynecological departments and departments of clinical biochemistry is becoming visible as the demand for genetic tests outside the traditional domain of clinical genetics is growing. This may lead to a separation of testing and counseling facilities. It is too early, however, to evaluate the consequences of this development for the quality of genetic services in Denmark.

Access to Services

The centers for clinical genetics as well as the gynecological departments are situated in the larger regional hospitals in the urban centers of Denmark. As Denmark is a small country, nobody needs more than a few hours to reach such facilities. It is difficult to estimate the extent of social and ethnic inequalities in access to genetic services. A growing segment of the population has immigrated from southern Europe or Asia. In population terms, Denmark is no longer the homogeneous country it once was. Greenland and the Faroe Islands are also part of the Danish Kingdom. In these sparsely populated and less accessible regions local genetic services do not exist.

In Denmark itself the use of testing and counseling facilities is not quite evenly distributed. Although approximately 12% of all pregnant women are tested, the regional variation is substantial. The wealthy and urban region of Frederiksborg Amt close to Copenhagen has a higher uptake rate (15%) than the more traditional rural region of Ringkøbing Amt in western Jutland (8%). There is no consensus as to whether this difference is an issue of social inequality or an expression of a different, though socially determined, ethical culture.

Lene Koch and Søren Nørby

France

Traditionally, in France genetics has been the prerogative of pediatricians, since genetic counseling has usually been organized on the premises of pediatric hospitals. Genetic counseling is always carried out by physicians, and no genetic counselors with other qualifications are recognized. Since obstetricians are now becoming involved in genetics, some centers specializing in prenatal diagnosis have been established.

Besides these traditional genetic practitioners, some physicians are gradually becoming involved in adult genetics. These were not included in the French survey, since they did not yet have any definite status in the health care system. These practitioners have different profiles from those of their pediatrician colleagues; those working in cancer genetics, for example, often have a background in medical research or clinical oncology. New network configurations are likely to result from innovative collaboration between biological research laboratories, industrial firms, and clinical units.[1,2] For the moment, the health system does not cover the cost of genetic testing for adults. Hospitals have to finance this activity themselves and cannot bill their patients directly. Providing genetic testing is in competition with other types of health care provided by the hospitals. When hospitals cannot pay for these tests, it is necessary to resort to research funds collected by the medical teams, but patients are not charged.

Since the National Health System provides everybody with the same health insurance contract, whatever their risk status, let us hope that no selective procedure will be applied taking risk as the criterion to determine people's access to health care. If, in the years to come, the present system no longer holds and comes to be partly or completely replaced by a private health insurance system, the question of equal access to health care could be problematic. Our country would then have to deal with the difficulties faced by other liberal health care systems without a comprehensive system for payment.[3,4]

The international survey presented French geneticists as physicians who place great importance on their patients' rights and duties. It refers to a strong tradition of clinical medicine where the use of genetics for public health purposes is not sustained. The legal framework adopted in France for the social diffusion of genetics is in agreement with the opinion of the majority of geneticists involved in the survey. In this context, the freedom of decision of the persons involved should be easier to ensure, since no legal barriers, such as those raised by the health insurance companies or by the job context, are at present applicable in France. The French government's approach consists of taking anticipatory steps to protect individuals from a discipline assumed to expose them to a panel of potential dangers. Even if the worldwide expansion of industrial networks and the deregulation of commercial markets may potentially destabilize national legislative attempts to protect individuals, it seems reasonable to assume that as long as the French National Health System is able to maintain a health insurance organization such as the present "Sécurité sociale", based on a single contract for all, regardless of people's genetic risk status, the diffusion of genetic risk information will probably be highly controlled in our country. This restrictive aspect has the disadvantage of limiting the diffusion of genetic services, but has the advantage of keeping social discrimination reasonably under control. In a scientific context where any new discoveries have rapid dynamics, it will also allow more time for new genetic knowledge to be tested and approved before being implemented in the real world of applied medicine.

Claire Julian-Reynier

References

1. Kenen, R.H. The at-risk health status and technology: a diagnostic invitation and the 'gift' of knowing. *Social Science & Medicine* 1996; 42(11): 1545–1553.
2. McCarthy, M. US lab starts breast-cancer-gene screening. *Lancet* 1996; 347: 1033.
3. Holtzman, N. *Proceed with Caution*. Baltimore: The Johns Hopkins University Press, 1989, p. 303.
4. Faden, R.R. and Kass, N.E. Genetic screening technology: ethical issues in access to test by employers and health insurance companies. *Journal of Social Issues* 1993; 49(2): 75–88.

The Netherlands

The Netherlands has a population of 15.5 million. With over 450 persons per square kilometer, it is one of the most densely populated countries in the world. Half of the population lives in the central western part of the country, where the main cities (Amsterdam, Rotterdam, Utrecht, The Hague) are located. Of the total population, nine percent were born outside The Netherlands and seven percent were born outside European Union countries. The main countries of birth outside the European Union are Surinam, Indonesia, Turkey, Morocco, and the Netherlands Antilles. Twenty-two percent of the population is under 18 years of age, while 13% is aged 65 years and over.

There are two types of medical expenses in The Netherlands: expenses associated with routine medical care, and exceptional costs associated with long-term care or advanced treatment that is too costly to be borne by individuals or to be adequately covered by private insurance. These exceptional costs are covered by a compulsory national health insurance scheme under the Exceptional Medical Expenses Act. Routine medical expenses are paid for by various insurance arrangements. Persons whose annual income is below a given threshold (59,7000 guilders or US$ 29,000 in 1997) and all recipients of social security are insured under the Health Insurance Act, which applies to about 63% of the population. About 5% of the population is covered by health insurance schemes for public servants, and 32% have private medical insurance. Although for some years the costs of clinical genetics services were met under the Exceptional Medical Expenses Act, they are now reimbursed under insurance plans that cover routine medical expenses.

In 1996, almost 58 billion guilders (slightly over US$ 28 billion) were spent on health care. The health care sector provides over 10% of all employment and about 10% of all investment in buildings in The Netherlands. Thirty-five percent of health care expenditures are financed under the terms of the Health Insurance Act, 33% via the Exceptional Medical Expenses Act, 15% through private insurance, 6% by civil servants' insurance, and 10% directly by the government. A yearly increase in costs of 1.3% is allowed. Since 1996, the costs of clinical genetics services have been constrained to a maximum budget (81 million guilders or US$ 39,326,000 in 1996).

Clinical Genetics Services

Clinical genetics as practiced today in The Netherlands has evolved from departments of human genetics and pediatrics of the universities and academic hospitals. In the mid 1970s, a committee of the Health Council advised funding a limited number of clinical genetics centers. This was followed by a decision in 1979 by the Association of Sick Funds and the Sick Funds Council to reimburse cytogenetic diagnosis, metabolic screening, and amniocentesis, and to subsidize genetic counseling and clinical enzymology. This decision was supported by the Ministry of Public Health and private insurance companies. Provision for reimbursement was through the creation of a separate foundation for clinical genetics, which then led to the establishment of seven foundations, one in each city with an academic hospital: Amsterdam, Groningen, Leyden, Maastricht, Nijmegen, Utrecht and Rotterdam. As Amsterdam has two academic hospitals, the actual number of clinical genetics centers is eight (approximately one per 2 million people).

The full spectrum of services within clinical genetics centers includes genetic counseling, metabolic screening, enzymology, cytogenetics, molecular genetics, fetal sampling, and advanced fetal ultrasound. Genetic counseling includes not only the communication process itself but also family history taking, inspection of medical records, physical examination, ordering diagnostic tests, and coordination of work-ups by other specialists, as well as contributions to follow-up care, such as referrals for prenatal diagnosis or periodic screening for malignancies. It includes psycho-social support by people trained in psychology or social work. Laboratories have agreed upon dividing the diagnostic work involved in cases of rare disorders.

There is agreement between the clinical genetics centers and health care insurers on a list of medical indications for access to diagnostic studies and complex genetic counseling. For instance, pregnant women who are 36 years of age and over, in the 18th week of gestation, are always accepted for prenatal cytogenetic diagnosis. For situations not described in the lists, no reimbursement will be forthcoming unless the insurer gives special consent. The primary care physician acts as the principal gatekeeper to services. Notwithstanding the small size of the country and the well-spread geographic distribution of the centers, several centers also have "out-reach" clinics in non-university hospitals.

Close to 50 physicians are trained in genetics, and another 20 are in training. There are about 30 cytogeneticists and 30 molecular geneticists.

Clinical Geneticists

Clinical genetics was recognized as a separate medical specialty in 1987. There are about 45 clinical geneticists. Certification as a clinical geneticist needs renewal every 5 years. Requirements for renewal include time spent in clinical genetics work and regular attendance at accredited conferences and courses.

Training in clinical genetics can be offered only in centers accredited by a committee of the Royal Dutch Medical Society after a site visit and peer review. Requirements for accreditation include the presence of training programs for the great majority of medical specialties; the presence of, or formalized cooperation with, a department with fundamental research in human genetics; minimum yearly numbers of activities in the field of genetic counseling (300), cytogenetics (500), and molecular genetics (300), and presence of the leading journals in the fields of human genetics and dysmorphology.

Table 2. Activities in Dutch Clinical Genetic Centers, 1996 (approximate numbers)

Activity	Numbers
Genetic counseling	7,600
Postnatal molecular diagnosis	12,500
Postnatal cytogenetic diagnosis	12,500
Postnatal enzyme studies	2,900
Postnatal metabolic screening	13,500
Prenatal molecular diagnosis	300
Prenatal cytogenetic diagnosis	13,000
Prenatal biochemical diagnosis	110
Amniocentesis	9,000
Chorionic villus biopsy	4,000
Advanced fetal ultrasound	8,000

Training takes four years and involves supervised counseling of at least 300 couples or individuals, training in cytogenetics and molecular genetics (at least 4 months each), courses in statistical genetics, communication skills, and attendance at other educational meetings at local and national levels.

Clinical geneticists and trainees meet bi-monthly to discuss clinical cases and obtain information about new developments. Ethical issues are also discussed at these meetings. The Dutch Society of Clinical Genetics and the Dutch Society of Human Genetics have taken initiatives towards implementation of quality control and development of standard protocols.

In principle, there is equal access to genetics services in the Netherlands, within the framework of the agreed upon medical indications. In practice, there may be some obstacles to equal access. For instance, it is well known that people from several ethnic minorities make less use of genetics services than autocthonous people (those of the majority culture). Other possible hindrances are lack of knowledge about risks and options among the public and primary care physicians, religious hesitations, and financial considerations. The latter applies especially to those people whose insurance only reimburses costs above a certain sum in exchange for a lower contribution.

Table 2 shows the numbers of activities performed in clinical genetics centers in 1996.

Leo P. ten Kate

Sweden

Sweden is the third largest country in Europe (excluding the former USSR). It covers 450,000 km^2 and stretches 1,650 km in a north–south direction from above the Arctic Circle. The population is just below 9 million (20 inhabitants per square kilometer), of which 10.5% are first generation immigrants, mainly from other Scandinavian countries, the Mediterranean region, and the Middle East. The population is concentrated in the south and middle regions and the coastal part of the north.

Good health care with equity is the goal of the Swedish health care system. A fundamental feature of the system is its overwhelmingly public character. Responsibility for health care and medical services rests primarily with the county councils. Clinical genetics services are a part of, and completely integrated into, the national health care system. For health care purposes the counties co-operate in six regions. Hospitals are subdivided into three hierarchical categories: regional, district, and local. The regional hospital is also a district and local hospital for the surrounding areas. Medical faculties are associated with the six regional hospitals, which provide top clinical care. The district hospitals provide a comprehensive range of specialist care, while the local hospitals offer basic specialist care. Primary care centers with general practitioners, district nurses, and nursing homes are found in every local administrative district.

Swedish health care is predominantly hospital-oriented, in spite of government encouragement of primary health care centers. The county councils have the ultimate responsibility for the hospitals and also direct the hospitals by way of a politically elected board; private alternatives are found only in a few large cities, and their service is often not reimbursed. The majority of the health care centers are directed by the local government, but there is a trend toward privatization; very recently some hospitals have been sold to private companies. Clinical genetics services are a part of the Swedish health care system. The Health and Medical Services Act of 1982, which states that equal access to good quality services must be available to all citizens, regulates access to health care. Access to hospitals and specialists in most counties is direct, and thus no referral from a general practitioner is required.

Good health care with equity is the goal of the Swedish health care system, and thus all who require genetic counseling should have access to it. However, it is rather clear that geographic and local political circumstances may interfere with these goals. For example, patients have been refused cancer genetic counseling and molecular investigations due to the costs. To avoid inequalities the Southern Health Care Region has changed the funding for genetic services from the usual fee-for-service based system to a system based on purchasing the service from the genetics center. In the genetics center in Lund, more patients are referred from the urban counties than from the more distant rural counties. This may reflect doctors' attitudes to referral, or reflect a different attitude about genetic disorders and thus a lack of interest in genetic counseling among the persons living in these areas.

Local, district, and central taxation provide funding for health care, and the patient only pays for the consultation; investigations associated with the consultation are free of charge. Care for persons with mental retardation is also regulated by law and funded by local, district, and central taxation. For non-Swedish speaking immigrants a free-of-charge interpreter service is available upon request. Private health insurance is practically non-existent.

Clinical genetics centers are found at the university hospitals in the following locations, with the number of specialists in brackets: Gothenburg, East Hospital (3); Linköping (1); Lund (5); Stockholm, Karolinska Hospital (4); Umeå (1) and Uppsala (2). In Linköping, the genetics unit consists of a cytogenetic laboratory and a small counseling unit, whereas the other clinical genetics centers have a broad competence in molecular biology and cytogenetics as well as in genetic counseling. The departments are affiliated with the Universities and provide excellent research possibilities; Lund, Stockholm, and Umeå hold positions for professors in clinical genetics.

Clinical genetics is a recognized medical specialty. The National Board of Health and Welfare regulates specialist training. It takes about five years to complete the specialist requirements after having fulfilled the requirements for the medical license. Today about 20 doctors working in Sweden are specialists in clinical genetics; 16 have positions in clinical genetics departments. About ten physicians are in specialist training. A voluntary specialist examination has been available since 1993 from the Swedish Society of Medical Genetics for those who have fulfilled their specialization training. Genetic nurses and non-physician counselors are found almost exclusively in the cancer genetics units, often combined with research nurse positions. Those who perform genetic counseling have extensive in-house training in a clinical genetics unit and work under supervision of an experienced clinical geneticist.

There are no national codes regarding ethical standards specific to the practice of clinical genetics. However, within the profession there is an ongoing discussion of ethical problems. There is a consensus that counseling should be non-directive and that the profession should have guidelines for pre-symptomatic testing for late onset disorders. A working group of geneticists, oncologists, and other specialists has formulated professional guidelines for cancer genetic counseling.

Partly due to the long distances to the genetics centers, pediatricians, gynecologists and midwives provide much of the genetic counseling. The clinical geneticists mainly work to provide the counselor with up-to-date information. With increasing costs for medical services there is a tendency among clinicians not to refer patients to specialists for genetic counseling. This means that sometimes the information given is not up-to-date, but at least for the classical monogenic Mendelian disorders the system functions adequately.

In the Northern health care region, which covers almost half of Sweden, but has only about 10% of the population, the clinical genetics department provides genetic counseling at two of the district hospitals three to four times a year.

Acceptance

The public generally accepts and approves of medical genetics. People are aware that prenatal testing for several severe hereditary disorders is available. Clinical genetics is regarded as a service to individuals and their families. There is an ongoing public debate on the impact of the advances of molecular genetics on society.

Prenatal cytogenetic screening is offered to all women above 35 years of age. About 90% of the investigations performed are amniocenteses, usually at 13–15 weeks of gestation. In 1995–1996, 11% (6,310 of 103,326) of all liveborns had been investigated prenatally by amniocentesis or chorionic villi biopsy. Prenatal diagnosis has reduced the number of chromosomally abnormal children born to women older than 35 years of age. However, because only 50–70% of all women who are offered it want prenatal cytogenetic analysis. Only 20% of all children with trisomy 21 are born to women over 35; those under 35 are not offered amniocentesis. Therefore, the total impact on births of children with this condition is low. Also, women who decide to have many children are not interested in prenatal screening. Prenatal chromosome analysis has had no impact at all on the risk of having a child with trisomy 21 among women with more than five children.[1]

In principle, all Swedish women whose prenatal investigation revealed a severe malformation or chromosome abnormality chose to terminate the pregnancy. Some exceptions have occurred when screening for trisomy 21 has detected a sex chromosome abnormality; in these cases the woman, after proper genetic counseling, has sometimes decided to continue the pregnancy.

Midwives and gynecologists at the local or district hospitals usually provide information about prenatal diagnosis. This information is usually provided only to women of advanced age (35 years or over) or with a family history of a specific disorder to discuss, but in the latter cases the pregnant woman most often is aware of the availability of prenatal diagnosis. Also, all women are informed that ultrasound can be a potent tool to detect malformation, and not only useful for pregnancy dating. A law stating that all women, regardless of age, should be informed of the advantages and disadvantages of prenatal diagnosis was passed in 1996. The Board of Health and Welfare has issued guidelines about how the information should be provided. They have also issued guidelines on ethical contributions with regard to use of genetics and gene technology in health care. Counseling for genetic disorders is often provided by the doctor in the district hospital and not by a specialist in clinical genetics. There are no national standards for how this type of information should be provided. For prenatal diagnosis, all the clinical genetics units have written information, which is used by midwives and doctors in their respective regions. Some lay organizations provide information sheets on the specific disorders they represent.

Quality Assessment and Patient Data Registration

The Swedish Society for Medical Genetics has brought forward a quality assessment document for clinical genetics units, including guidelines for cytogenetic and molecular procedures and analysis as well as for genetic counseling. All the university clinical genetics departments have, as a minimum standard for quality, adopted this document. The Society also provides external quality assessment for cytogenetic and molecular genetic investigations.

Göran Hermerén and Ulf Kristoffersson

Reference

1. Bui, T.H. and Kristoffersson, U. Prenatal diagnosis in Sweden. Organization and current issues. *European Journal of Human Genetics* 1997; 5(Suppl. 1): 70–76.

SOUTHERN EUROPE

Greece

Greece is a small country (131,957 km^2) with a population of 10,450,000 inhabitants. Four-fifths of the land is on the mainland peninsula and the rest on 1,400 islands. Greece is a mountainous country with a population density of 79 people per km^2. Sixty-four per cent of the population live in urban areas with a population greater than 7 million. Greece is a member of the European Union, a parliamentary republic and a country with social benefits to its people, such as a free elementary and higher education and socialized medicine. The main religion is Greek Orthodox.

Physicians in Greek culture hold a very high social status. Their opinion and advice is sought not only on medical matters, but on family problems as well. This is the reason why only a few cases of malpractice have been brought into court.

Since 1985, Greece has had a National Health Service (NHS) under which medical inpatient care is covered entirely by health insurance, while outpatient services are provided at no cost or at a minimal fee. Greece has never experienced a purely market oriented type of health delivery system. Although all individuals are covered by health and pension insurance according to the profession or type of work, many families also have private insurance coverage, which allows them a wider use of medical services in Greece or overseas. "Professional" insurances are semigovernmental institutions providing coverage mostly in their own or public hospitals and clinics. Private genetic care is offered in maternity hospitals, most of which are privately owned. There are two private labs in Athens, one offering mostly cyogenetic studies, the other molecular techniques. Almost 99% of the population has insurance coverage. The costs of genetic services, prenatal diagnosis and abortion if needed, are covered by health insurance. Physicians not working full-time in the NHS are free to practice privately and charge fees. Private hospitals exist throughout the country, with quality varying from poor to excellent. There are 40 doctors and 49 hospital beds per 10,000 people.

Genetic clinics or units appeared in the early 1960s in the "Aghia Sofia" Children's Hospital and the "Alexandra" Maternity Hospital. In 1965, the Institute of Child Health started an Amino Acid Laboratory and neonatal screening was applied nationally. Prenatal diagnosis of chromosomal disorders by amniocentesis was initiated by the First Department of Pediatrics of the University of Athens and the "Alexandra" Maternity Hospital in 1976, followed by prenatal diagnosis of hemoglobinopathies by the Mediterranean Anemia Laboratory of the First Department of Internal Medicine and the First Department of Obstetrics and Gynecology of the University of Athens.

Newborn screening is excellent and is covered by one centralied center. Genetically trained physicians, trained diagnostic lab staff, the integration of clinical and lab genetics services, prenatal biochemical screening, prenatal screening for maternal age and population screening for carriers of recessive disorders are available only in some places. There are no general genetic, hereditary cancer, cytogenetic, abortion, or congenital anomalies registers.

For almost 20 years genetics labs, initially offering cytogenetic tests, have maintained excellent standards and contributed to the prevention, through early diagnosis, of the birth of malformed babies or those with hemoglobinopathies or hemophilia. New centers appeared in the 1980s and 1990s in the Universities of Thessaloniki, Ioannina, Patra and Crete. These extended prenatal diagnosis services to diseases such as cystic fibrosis, muscular diseases (Duchenne and Becker muscular dystrophy, spinal muscular atrophy, myotonic dystrophy) polycystic kidney disease, fragile X syndrome, Prader–Willi syndrome, Wilson disease, and certain neurological (neurofibromatosis, Huntington disease) and mitochondrial diseases.

There are six molecular genetics labs in Athens (one private) and two in the rest of the country. There are 12 cytogenetic labs, including two private, and five biochemical genetic labs. There is, however, only one for neonatal screening of PKU, hypothyroidism, G6PD deficiency and galactosemia (Institute of Child Health) covering more than 95% of newborns.

There is no formal list of specialists in Greece, but there are 60 members of the Hellenic Association of Medical Genetics (HAMG), which includes physicians mostly trained in genetics overseas and also those involved in hematological diseases (hemoglobinopathies, hemophilia). The number of units offering genetic counseling is 11. It is impossible to obtain patient figures.

Unfortunately, medical genetics in Greece has not achieved the same status of recognition as it has in other European countries (1). The Greek state does not recognize medical genetics as a separate medical speciality. According to a recent study of European genetics services (2), there is evidence that Greece lacks the number of appropriate genetic services to serve its population. Almost all the large genetic services are located in Athens and the number of physicians in full time clinical genetics is extremely limited. The ratio of physicians to technical personnel is disproportionally small, un-economic and basically dysfunctional (2).

Since 1980, prenatal testing has been widely accepted and the number of tests has gradually kept increasing. For 100,000 births per year, 6,500 amniocenteses were performed in 1995. Another 700 pregnancies were tested for hemoglobinopathies. The diagnostic procedures currently available include fetal karyotyping, serum screening during the first and second trimester of pregnancy, ultrasonography for congenital malformations and molecular techniques for the diagnosis of hemoglobinopathies, cystic fibrosis, hemophilia, fragile X syndrome, Duchenne and Becker muscular dystrophies, Charcot–Marie–Tooth disease, Wilson disease, adult polycystic kidney disease, neurofibromatosis, Huntington disease, spinal muscular atrophy, myotonic dystrophy and Prader–Willi and Angelman syndromes (3). Only one hospital ("Alexandra" Maternity Hospital in Athens) reported figures for 1995: 3751 amniocenteses and 999 CVS, which resulted in 92 and 160 abortions.

Access to/Demand for Services

Four million people live in Athens and the majority of the services are located there. Due to transportation problems, an NHS emphasis on hospitals, a lack of information and inadequate genetic care may be found in the remote mountainous regions and island populations.

The following factors are very important in determining access to genetic services in Greece: administrators are poorly informed about genetics, there are too few genetic specialists, there is difficulty in funding new posts for genetically trained physicians and specialist lab geneticists and no continuity of care for patients/families with long-term genetic problems.

Competition between the private and public sectors of health care has tended to create mistrust among health care consumers. Usually medical expenses incurred abroad are compensated by national insurance if local resources are considered inefficient or non-existent for the particular disease, provided that these expenses are approved by a local medical board.

Targets of Genetic Services

Initial targets were the prevalent hemoglobinopathies. The establishment of thalassemia units throughout the country, where interested persons could have screening free of charge and with the offer of genetic counseling led to satisfactory results. Heterozygote carriers were diagnosed and couples were offered options, including prenatal diagnosis. Newborn screening was initiated 30 years ago to include the entire newborn population and this assisted in eliminating complications of PKU and hypothyroidism and also helped those with G6PD deficiency to avoid contact with harmful substances. Infant mortality rates are diminishing, as the creation of Newborn Special Care Units is expanding throughout the country. Health care for children is good and the principal cause of death is accidents. Adult life expectancy at birth is 75 years for men and 80 years for women, among the highest in Europe.

Genetic Research

Limited funding for clinical and laboratory work is an important block to research. National guidelines exist only for hemoglobinopathies. Nonetheless support through grants (public or private) allows the purchase of modern equipment. Most laboratories are well equipped for the services they are offering. Genetic personnel are well trained because of scholarships from the State Scholarship Foundation and the Onasis Foundation, with the NHS even allowing access to excellent training at the best foreign centers. Contacts and collaboration with these centers following return to Greece allows continuity and support for services not fully offered locally. Seminars and congresses are also important opportunities for continuing education.

Genetic teaching in medical schools is insufficient: There are elective courses at the University of Athens Medical, Dental, and Nursing Schools. The Faculty of Nursing offers genetic counseling as an elective postgraduate course. Nonetheless, the time allotted is insufficient. Although Human Genetics is taught to Health Sciences students as an elective course only, their interest is remarkable. The knowledge acquired may be considered as more than adequate. It is remarkable that the applications for taking the Genetic Counseling course offered through the Master's level Graduate Programme exceed by far the available positions. Graduate nursing students have an opportunity for practical application of counseling.

Quality Assessment and Effectiveness of Genetic Services

Quality assessment of genetic counseling is non-existent. Quality assessment of lab methods is practiced on a voluntary basis by the Hellenic Association of Medical Genetics and the "Demokritos" Research Center. There are no national schemes or publications that assess counseling. Local review boards exist in hospitals, but these limit their jurisdiction of approval of the safety and ethical issues of new therapies or research projects. Use of computerized systems for hospital records is in its infancy. There are no national registries of hereditary diseases. So far, the state has not issued any regulations or guidelines for prenatal diagnosis or genetic testing.

Effectiveness of medical genetic services has not been formally evaluated, but there is evidence of decreasing birth rates of patients with thalassemia, longer survival of patients with genetic disorders such as hemophilia, thalassemia, cystic fibrosis, diabetes mellitus, and the absence of mental retardation in patients with phenylketonuria. Of course, some may claim that these improvements are not necessarily attributable to geneticists. Nevertheless, progress is almost entirely the result of referrals of patients to specialized clinics, better and wider screening applications, and improved diagnostic methods and therapies.

Measures of desirable outcomes are not in use in Greece at present. It is hoped that their development will help in choosing methods, i.e., chorionic villi sampling vs. amniocentesis and maternal serum sampling for certain particular chromosomal disorders. These measures would assist the physician to apply the most specific and effective methods, and patients should benefit in the long run.

Although there is a lack of evaluation of quality assurance and patient satisfaction about genetic services, the overall provision of services may be considered as satisfactory. Collaboration with foreign centers and labs contributes to the expansion of these services to less common or rare disorders.

Christos Bartsocas and Ariadne Malamitsi-Puchner

References

1. Vasilopoulos, D. *Genetic Diseases and Community*, Athens: Medipress, 1985.
2. Concerted Action on Genetic Services in Europe (CAGSE) Harris, R. (ed.). Genetic services in Greece. *European Journal of Human Genetics (Special Issue)*, 1998.
3. Metaxotou, C., Mavrou, A., and Antsaklis, A. Prenatal diagnosis services in Greece. *European Journal of Human Genetics* 1997; 5(Suppl. 1): 39–41.

Italy

Health Services

In Italy health is a primary value, a human right and a community interest guaranteed by the Italian Constitution. A National Health Service (NHS) operates throughout the country, guaranteeing access to medical services to all Italian citizens as well as to immigrants. Private health insurance companies and medical facilities are also available and widespread.

Genetic Services

Genetic services in Italy often stem from the initiative of medical staff in public or private institutions rather than from regional or national planning within the NHS. This results in a plethora of unevenly distributed genetic services throughout Italy. A recent survey (Dallapiccola *et al.*, 2000), showed that Italy, with about 57 million people, had, in 1997, 96 laboratories providing molecular genetic tests, i.e., about twice the number available in France and England with a similar population and three times that in Germany, with its 87 million people. In Northern Italy, with a greater economic development, there were 108 labs offering cytogenetics or molecular genetics tests in 1997 (Dallapiccola *et al.* 2000), while in the South, with a similar population but a much lower economic development, the number of labs was less than half. In some areas, such as Rome, several centers offer tests for very rare disorders (e.g., Huntington disease), for which just one center would be enough, while in other areas this kind of service is completely lacking.

The absence of national planning for genetic services also implies that genetic counseling (g.c.), when offered, is usually not provided by specifically trained professionals. Qualification programs in medical genetics for a limited number of medical doctors and biologists are offered by several universities, and short (about 1 week) training programs in genetic counseling open to non-medical professionals (psychologists, nurses, etc.) have recently begun. However, trained professionals, at any level of preparation, are insufficient when confronted with the ever-increasing number of genetic tests performed (Dallapiccola *et al.*, 2000). All too often, genetic tests are provided without the appropriate counseling, as if they were usual blood or urine tests, and within the context of a paternalistic doctor–client relationship, far from the ideal nondirective counseling promoting the counselee's autonomous decision-making process.

National Guidelines for genetic testing have been available since 1999 (see also Chapter 11), but so far they have not been widely distributed among health professionals and are not enforced by the Ministry of Health.

Another consequence of the lack of planning is the absence of quality controls for testing or counseling provided by private or public institutions, although a pioneer project in this area is supported by the Ministry of Health. The lack of quality controls and of regulations for labs providing genetic tests. Sometimes results in a poor quality of genetic services.

Screening Programs

Neonatal screening for PKU, hypothyroidism, galactosemia, and CF are performed according to regional laws. Screening for carriers of thalassemia covers in particular the coastal areas of Southern Italy (including Islands) and other formerly malarial regions where the carrier frequency is very high, but is also widespread in other regions. The screening programs together with the prenatal diagnosis and selective abortion of thalassemic fetuses have drastically reduced the number of thalassemic patients, but have not completely prevented their birth (Bianco Sivestroni 1998).

Marina Frontali and Gioia Jacopini

References

Bianco Silvestroni, I. Le Talassemie. Un problema medico-sociale oggi. Ist. It. Medicina Sociale, Roma (Italia) 1998.

Dallapiccola, B., Torrente, I., Mingarelli, R., and Novelli, G. Censimento della diagnosi genetica in Italia 1997. *Tendenze* 2000; (Suppl. 3).

Spain

Geographic and Demographic Background

Spain had 39 million inhabitants in 1995 (INE, 1997). It is divided into 17 autonomous regions with their own parliaments. The density of population is variable, ranging from rural areas (Castilla La Mancha) to urban regions (Madrid). The Spanish birth rate is the lowest in Europe (9.2 per thousand in 1995) and has been progressively declining since 1975. The fecundity rate has dropped mainly among mothers aged 15 to 29 years, whereas a slightly contrary effect has been observed in mothers from 30 to 44 years. In 1995, there were 363,469 births, the mean maternal age was 27.7 years, and 13.8% of mothers were 35 years or older. In 1995, population growth was 17,242 people (0.44%).

Health Services and Medical Genetics

The National Health Service, regulated by the General Health Law (1986) is supported by different public institutions, which are INSALUD (Instituto Nacional de Salud) and 7 autonomous Regional Health Services in Andalusia, the Basque Country,

Canary Islands, Catalonia, Galicia, Navarra and Valencia. The whole system covers almost all the population. Genetic services were incorporated into medicine around 1964, mostly as a sub-area of other medical specialties such as obstetrics, pediatrics, hematology, biochemistry, etc. Since then, they have undergone irregular development around the country.

At present, there are 38 public and 19 private genetic units/centers performing diagnosis, excluding those that are involved exclusively in research (AEGH data, Ramos-Arroyo *et al.*, 1997; Gabarrón and Ramos, 1997). They are located mainly in hospitals; few belong to universities. Almost 65% are functionally independent, while the remaining 35% are part of other departments. Their personnel include 150 graduates specifically trained in genetics. Half are physicians and the other half are non-medical scientists. There are also about 150 technicians and 100 students. In 1994, about 19,000 cytogenetic analyses (including prenatal diagnosis) and 15,000 consultations were performed. However, resources in medical genetics are not uniformly distributed. There are some extensive regions like Extremadura and Castilla-León where there is no genetic unit at all, because Spanish Genetic Services are mainly placed in large cities (Madrid, Barcelona in Catalonia, Valencia in Valencian Country, Sevilla in Andalucia, Bilbao in Basque Country, etc.). In spite of this, access to centers in other regions is easy for people without a unit in their own region.

Genetic Counseling

100% of the genetic units provide genetic information services related to prenatal diagnosis or to specific diagnoses that they perform, but only 27% offer comprehensive genetic services including clinical diagnosis, counseling and follow-up.

Prenatal Diagnosis

Prenatal diagnosis is well developed, but the centers are irregularly distributed, mostly concentrated in large cities (Gabarrón and Ramos, 1997); 83% of the genetic units perform cytogenetic analysis for prenatal diagnosis, including amniocentesis, transcervical and transabdominal chorionic villus sampling and chordocentesis, although the first procedure is the most extensively used. In 1994, a total of 13,780 amniotic fluid, 1,089 chorionic villi and 90 fetal blood cytogenetic analyses were performed. Between 30 to 40% of fetuses at risk because of maternal age (more than 34 years) were studied. At present, about 40,000 amniotic fluid analyses per year are performed. In several centers, cytogenetic analysis techniques include fluorescence *in situ* labeling (FISH) on interphase nuclei and metaphase plates. Molecular tests for a number of genetic diseases are also available, and if needed, diagnosis can be performed in another country by sending the appropriate sample to the referral center. There is a very fluent communication between different Spanish genetic and prenatal diagnosis services, allowing exchange of information, availability of different techniques and discussion about diagnostic dilemmas.

Routine ultrasound examination (type II) in the second trimester is well established. However, screening of some morphological markers (like nuchal thickness or choroidal plexus cysts) in the first trimester is not yet very developed. Although there is no universal program for prenatal screening for Down syndrome and neural tube defects, maternal serum screening of alpha-feto-protein and beta-HCG is very extensive and there are local population-based programs in some small areas in Madrid,

Navarra, Castilla-La Mancha and Asturias. However, not all screening programs have a cytogenetics laboratory to which to refer women with abnormal biochemical results among this low-risk population of pregnant women. Funded by the Ministry of Health, a research project on analysis of fetal cells in maternal circulation is in progress. Other techniques under investigation are amniofiltration, analysis of cervical samples and early amniocentesis.

Neonatal Screening

Neonatal screening for PKU and congenital hypothyroidism is performed by 21 different laboratories around the country and covers practically 100% of the population. There are also 10 laboratories located mainly in large cities, which also perform other types of biochemical screening, such as 21-hydroxylase deficiency.

Registry of Malformations

ECEMC (Estudio Colaborativo Español de Malformaciones Congénitas) is a national service for epidemiological control of congenital malformations, implemented by the Minister of Health and Madrid Regional Health Services. This is a program of clinical and epidemiological research on congenital defects, based on a newborn register (mainly based in hospitals) and case-control design, which has been in process since 1976. The study has examined 1,479,354 newborns and has been able to measure the impact of prenatal diagnosis on neural tube defects and Down syndrome, whose prevalence among the Spanish population has declined significantly since prenatal diagnosis and subsequent termination became available (Bermejo and Martinez-Frías, 1998). The prevalence of anencephaly (both syndromic and isolated) has been progressively reduced by a mean of 3.5 cases per year (from around 4.5 per 100,000 newborns in 1980 to around 0.5 per 100,000 in 1997; $p < 0.000001$). Similar progressively decreased frequency has been observed for spina bifida (2 cases less per year; $p < 0.000001$). Down syndrome showed a decrease during this period, with a prevalence at birth reduced from 15/10,000 newborns in 1980 to 11/10,000 in 1997 ($p = 0.000001$). This was due mainly to a decrease in Down syndrome among mothers aged more than 34 years, while in women under 35 years the prevalence of Down syndrome remains unchanged.

Access to Medical Genetic Services

Prenatal diagnosis, genetic tests and genetic counseling services are part of national public health care. Access to genetic services that are medically indicated, including prenatal diagnosis, is free and the costs are covered by public funding. All procedures are done in public as well as in private settings.

In practice, the current overload of medical genetic services, due to the scarcity of posts, is an important limiting factor. On the other hand, sometimes general practitioners and obstetricians do not have enough information about the existence and availability of genetic services. There is insufficient education in medical genetics in medical schools. Therefore many doctors do not refer patients to medical genetics services.

Carmen Ayuso, Carmen Ramos, and Isabel Lorda

References

1. Vigilancia epidemiológica de anomalías congénitas. Bermejo-Sanchez, E. and Martinez Frías, M.L. Bol ECEMC (Estudio Colaborativo Español de Malformaciones Congénitas) 1998; Ser IV(3): 37–116.
2. Ramos-Arroyo, M.A., Benitez, J., and Estivill, J., Genetic services in Spain. *European Journal of Human Genetics* 1997; 5(Suppl. 2): 163–168.
3. Gabarrón, J. and Ramos, C. Prenatal diagnosis in Spain. *European Journal of Human Genetics* 1997; 5(Suppl. 1): 64–69.

EASTERN EUROPE

Czech Republic

Czechoslovakia was established as an independent state in 1918 and developed successfully as a democratic country for two decades. After World War II, under the influence of the Soviet Union, Czechoslovakia adopted a communist regime that lasted for more than forty years. This period finally ended in 1989, with the Velvet Revolution, when Czechoslovakia shifted again toward a democratic system and a private economy. In 1993, the Czechs and the Slovaks decided to divide the country along linguistic and cultural lines, a division that took place peacefully. Despite the similarities of language and culture, the Czech and Slovak Republics developed differently. Major changes had already been made in the economic transition from communism to capitalism. The health care system in the Czech Republic is now undergoing major reforms.

Health Care System

During the communist era, the state financed the health care system and owned all medical institutions. Health care was "free-of-charge" for everyone and all services were easily accessible. A relatively dense net of hospitals and primary care offices was established. Today, the Czech Republic has one of the highest ratios of physicians (one per 319 persons) and beds (one per 98 persons) in the world. Under the new health care system, everyone must be insured, and no patient may be denied insurance because of any pre-existing conditions. Currently, 66% of medical expenses are covered by autonomous insurance funds, and 34% by local governments. Only a very small part is paid by the patient.

There is a General Health Insurance Company created in 1992 by the government. In the beginning, it received over 66% of its budget from the Ministry of Finance. Since 1993, the General Health Insurance Company has become an autonomous insurance fund that collects premiums directly from the insured, without state intervention. Also since January, 1993, other health insurance "companies" have been founded, which resulted in the initiation of 28 new health insurance funds.

The majority of primary care offices are now in private hands; however, many hospitals are still waiting for privatization and their status is not yet clear. The system of financing medical expenses through autonomous insurance funds is not working well, and many hospitals and private physicians are experiencing financial troubles. Creating a well organized and functioning health care system in the Czech Republic will most likely be a long process, and cooperation with other countries will be necessary, especially in legal and ethical issues.

Demographic and Social Characteristics

The population of ten and a half million in the Czech Republic is ethnically homogenous, with little social or economic stratification. There are few problems related to ethnic groups and geographic regions. Health services are fairly accessible for patients in all parts of the country. The birth rate in the Czech Republic is one of the lowest in the world. In 1993, the number of newborns per 1000 inhabitants was the lowest in 200 years. The mean age of a woman at the birth of her first child (and also at the time of her marriage) is 21.5 years. This is the youngest in Europe. The proportion of children born in single-parent families was 12.7% in 1993. This proportion is increasing, though it is still lower than in Western European countries.[1]

The Czech Republic has a very high abortion rate. However, from 1992 to 1994, this rate decreased by 41%[2] (Table 3). The reason for this may be financial (abortion is not covered by insurance), or it may be related to the more responsible behavior of people under capitalism and privatized medical services. The number of women using medically effective contraception has significantly increased.

Most families have only one or two children, probably due to traditional attitudes and financial reasons. According to the law, women have the right to three years of maternity leave with financial support from their employers, and most women take advantage of the opportunity to stay at home with the children. The vast majority of women return to work after the three years leave of absence. In general, families depend on the income of both parents in order to maintain a sufficient standard of living. It is expected that in the future more women will pursue their careers and will not be willing to take such a long leave of absence. It is unlikely that private businesses will guarantee employment to mothers after such long maternity leaves.

Day care for children older than three years and after-school programs are provided by the public school system. The assortment of other public services is still much less in the Czech Republic than in Western European countries, and these must be improved. Most women are employed, and some are also active in public and political life. The feminist movement is very weak, and women's issues are not discussed publicly.

Table 3. Abortions in the Czech Republic, 1987–1994[2]

Year	Spontaneous abortions	Induced abortions total	Induced abortions for medical reasons*
1987	15,081	107,717	11,095
1988	15,282	108,350	10,087
1989	14,690	107,380	11,811
1990	14,656	107,130	9,428
1991	13,892	103,124	8,836
1992	13,324	93,435	10,270
1993	13,076	69,398	15,738
1994	10,958	53,674	13,046*

*2–3% were abortions for genetic conditions.

Support for people with disabilities has always been provided by the government. Financial aid is given to families with disabled children. The level of support is not high, but it may be sufficient for a modest lifestyle. The "Association of the Disabled", which worked for rights, benefits, and services for people with disabilities, no longer exists and has been replaced by many small special interest groups. There is a great need to incorporate people with disabilities more fully into society, to protect their rights, to improve services, to create an environment without barriers and to assist them in finding employment.

During the socialist regime, religion was restricted. One-third of the population consider themselves "believers," but only 5% of the people actively practice their religion. Catholicism is the predominant religion. Acts of charity were considered as insults to human dignity in the past and are being revived very slowly.

Medical Genetics

Throughout the world, Mendel's laws of heredity are well known. In the nineteenth century, Johan Gregor Mendel lived and worked in Brno, the second largest city of the Czech Republic. In 1967, the Society of Medical Genetics was founded and in 1969 medical genetics became an independent medical discipline. Becoming a medical geneticist requires three years of training and postgraduate education after specialization in a basic field of medicine (gynecology, pediatrics, internal medicine, etc.). Postgraduate training in medical genetics is provided for medical and non-medical laboratory personnel. In 1969, the first counseling centers were established, and since then, the network of genetic services has grown. Currently, there is approximately one genetic center per one million people, and about 150 specialists with university degrees working in medical genetics. Prenatal diagnostics have been available since 1970, and DNA analysis was introduced in 1986.[2]

Medical genetics is used as preventive medicine and genetic services are accessible in all parts of the country. Biochemical screening (maternal serum alpha-fetoprotein, human chorionic gonadotropin and unconjugated estriol in the sixteenth week of pregnancy), and ultrasound usually in the twelfth, twentieth, and thirtieth week of pregnancy are offered to most pregnant women. Invasive prenatal diagnosis is performed in indicated cases, that is, for women whose biochemical or ultrasound screening, personal history of miscarriages, or family history of chromosomal aberration or other genetic diseases indicate that they are at high risk. For women older than 35 years, amniocentesis or chorionic villus sampling are highly recommended. In all procedures, except biochemical screening and ultrasound, the pregnant woman must sign an informed consent form. Phenylketonuria screening of all newborns began in 1975, and hypothyroidism screening of all newborns in 1984.[3]

Despite the fact that the economy and health care system are in transition, genetic services are working well to provide people with standard care that is financially affordable.

Ethics in Medicine

In the 1920s and 1930s, studies of moral problems were part of medical training. In 1946, the Deontological Institute of the Medical Faculty of Charles University in Prague was founded. In 1989, several Departments of Ethics were established within medical schools, as well as the Ethical Committee of the Czech Ministry of Health. The ethical issues of genetics were included on their list of priorities.

Table 4. Abortions Indicated for Genetic Reasons from 1985 to 1995[4]

Year	Total number of abortions	Abortions per 10,000 live newborns
1985	67	4.9
1990	315	24.1
1995	360	37.5

In 1992, the Association of Geneticists promulgated a very general Code of Ethics in medical genetics, which was based on the Hippocratic oath. From the beginning, it stressed a professional standard of services, collegial cooperation, non-directive counseling, respect for human life, and dignity of the patient. Geneticists are more aware than other physicians of the psychological, social, and moral factors that might influence reproductive and other decisions. They are more willing to accept clients' decisions even if the geneticists themselves deem the decisions "irrational" and "irresponsible".

Birth Defects and their Registration

The Czech Republic is one of the founding members of the International Clearinghouse for Birth Defects Monitoring System. Birth defects in newborns have been registered since 1965.[3] According to the new system implemented in 1994, there are 63 birth defects registered in children up to 15 years of age. Unfortunately, this system became more difficult to use and monitor with the collapse of a centralized health care system.

The ratio of pregnancies terminated for birth defects per 1000 newborns was 0.49 in 1985; this ratio has increased each succeeding year, reaching 3.75 in 1995[4] (Table 4). Increasing numbers of affected pregnancies are not terminated. In 1994, 95 out of 394 affected pregnancies were carried to term. Some had mild defects or late findings of conditions difficult to detect by early ultrasound. There were 3.7 birth defects per 1000 deliveries (2.7 were terminated and the remainder were carried to term). Parental desire not to terminate a pregnancy in the case of a severe birth defect is very rare.[5,6] In 1994, the detection rate of Down syndrome was 43%; the pregnancies were not terminated in four out of 62 cases. It was the desire of the parents knowingly to carry to term a pregnancy diagnosed with Down syndrome in only one of the four cases. The other three cases were detected too late for termination.

Jana Zidovská and Lenka Foretova

References

1. Czech Health Statistics 1994, UZIS, Prague [Czech].
2. Kapras, *et al.*, 1992, Advances in Medical Genetics [Czech], Avicenum, Prague.
3. Gregor, Vladimir 1996, Secondary prevention of abnormal phenotypes in the Czech Republic in 1995 [Czech], Inborn Errors, Health Statistics, UZIS, Prague (in press).
4. Sipek, A., Gregor, V., and Chudobova, M. The role of prenatal diagnostic in decreasing occurrence of birth defects in children in the Czech Republic [Czech]. *Ceska Gynekologie* 1996; 61: 59–70.

5. Gregor, V. 1994, Report on the performed prenatal diagnostics and prevention of abnormal phenotypes in the Czech Republic (Annual Report for Medical Geneticists, Prague) [Czech].
6. Sipek, A. Birth defects in the Czech Republic in 1993 [Czech]. *Ceska Gynekologie* 1995; 60: 241–249.

Hungary

At present there are 16 genetic counseling clinics. Nearly all have a suitable cytogenetics lab and auxiliary personnel (technical assistants). The number of new counselees continues to increase; in 1994, 118 consultations per 1000 liveborn infants were recorded, which is very high. These clinics also provide advice concerning teratogens to about 15% of counselees. Since 1987, there has been an attempt to lower this proportion by providing a centralized teratology telephone counseling service. Of 16 medical cytogenetics labs, 12 either work directly with or are closely connected to genetic counseling clinics. Three others are involved with tumour cytogenetics in Budapest and Debrecen, and another works in a forensic institute in Budapest.

Nine prenatal diagnostic centers function within obstetric in-patient clinics. Both chorionic villus sampling and transabdominal amniocentesis are used after ultrasound scanning on pregnant women admitted to hospitals. The national policy on indications for prenatal diagnosis has been determined by official guidelines of the Board of Medical Geneticists and the National Institute of Obstetrics and Gynecology. The general principle is that prenatal diagnosis is reasonable in cases where there is $\geq 2\%$ risk of severe fetal disorders that cannot be effectively treated. In general, prenatal diagnosis is offered by geneticists; however, the heads of prenatal diagnostic centers can decide to do it and to select amniocentesis or chorion biopsy. All prenatal diagnostic centers are able to perform chromosomal analysis, although availability of biochemical and molecular genetic methods is limited to only five centers.

There are 14 molecular genetics labs in Hungary, including two well-established biochemical labs that are mainly involved in medical genetics (pediatric clinics in Pécs and Szeged) and some hematological-immunological labs in the National Institute of Hematology, Bloodtransfusion, and Immunology. DNA probes are available for clinical and fetal diagnosis of cystic fibrosis, Duchenne muscular dystrophy, spinal muscular atrophy type I (Werdnig–Hoffmann disease), hemophilia A, Huntington chorea, and some other disorders.

There are two prenatal screening programs. Prenatal chromosomal examination of the fetus is recommended for pregnant women ≥ 38 years of age; however, it is done for women aged 35 and over on request. Since 1985, mandatory maternal serum AFP (MSAFP) screening has been undertaken in the sixteenth week of gestation with ultrasound examination.

There is good cooperation between the secondary care centers and genetics centers, as well as between prenatal care clinics and prenatal diagnostic centers. Almost all pregnant women attend prenatal care clinics and participate in the mandatory MSAFP screening.

Of the four medical and one postgraduate medical universities, only one (Szeged) has an independent medical genetics institution. The Institute of Medical Genetics, Szent-Györgyi Medical University, Szeged includes a genetic counseling clinic, a prenatal diagnostic center working in collaboration with the Obstetric In-patient Clinic in the university, and also provides medical genetics education for students. Three pediatric in-patient clinics (Budapest, Debrecen and Pécs) and one obstetric in-patient clinic (Budapest) have clinical genetics units. The Department of Human Genetics and Teratology, National Institute of Public Health in Budapest, has three units. The Hungarian Centre for Congenital Anomaly Control is concerned with public health activities. The task of the Epidemiological Unit is research. The Family Planning Clinic provides periconception care for couples, coordinates the work of other centers, and includes a Genetic Counseling Clinic.

There are some links between the oncological in- and out-patient clinics and cancer genetics services, mainly for childhood cancer. The National Registry of Childhood Cancer is located in the Second Pediatric Clinic of the Semmelweis Medical University, Budapest. The Pediatric Clinic of Debrecen Medical University is interested in the cytogenetic aspects of childhood cancer. The National Institute of Oncology and National Institute of Hematology, Bloodtransfusion and Immunology include molecular, cytogenetic and other units oriented towards the study and practice of genetics.

There is no appropriate link between primary health care and genetic counseling clinics in Hungary. The majority of patients visit genetic counseling clinics spontaneously, without being referred by general practitioners or other medical doctors. This is one of the reasons that prompted the establishment of a new health care service system, i.e., the periconceptional service.

Andrew E. Cziezel and Laszlo Timár

References

Czeizel, A.E. Medical genetics in Hungary. *Journal of Medical Genetics* 1988; 25: 2–8.

Czeizel, A.E. First 25 years of the Hungarian Congenital Abnormality Registry. *Teratology* 1997; 55: 299–305.

Czeizel, A.E., Dobó, M., Dudás, I., Gasztonyi Z., and Lantos, I. The Hungarian periconceptional service as a model for community genetics. *Community Genetics* 1998; 1: 252–259.

Papp, Z. *Obstetric Genetics*. Budapest: Akadémia Kiadó, 1990.

Poland

For a long time after the Second World War, the science of genetics in Poland, as in the Soviet Union, was virtually banned in medical schools. This situation lasted until 1956. Then there was some improvement, but for many years to come, the teaching of medical genetics was almost nonexistent. Genetics services in Poland were initiated in the early 1960s in the form of a few cytogenetic laboratories and genetic counseling units. There was a considerable delay in the development of medical genetics in Poland, and even today human genetics is not a regular public health activity, but rather a service to individuals and families. Specialization in medical genetics does not currently exist in Poland, but there are plans to initiate it soon.

There are no codified ethical standards for Polish geneticists, but some ethical problems pertinent to medical genetics appear in a few articles of the Code of Ethics of Physicians accepted by the 3rd National Congress of Physicians in 1993. Two of these are included in a section entitled "Procreation". Article 37 states that "A physician is obliged to inform patients who belong to a group at increased risk about diagnostic and therapeutic possibilities of contemporary medical genetics, including prenatal diagnosis. When providing the above mentioned advice a doctor has to inform interested patients about the risk involved in prenatal diagnosis".

Article 38 states that "A physician performing medical procedures in pregnant women is also responsible for the health and life of the child". Opponents of the termination of pregnancy after prenatal tests indicating a disorder may interpret these articles as they see fit.

The privacy of patients is an important consideration in the code. An example is article 28, which says that "A physician and collaborating personnel are obliged to protect confidentiality of stored samples of DNA obtained from patients and members of their families".

Polish geneticists in general observe the rules accepted by most medical genetics centers in Europe and the United States, such as nondirectiveness in genetic counseling and protection of the autonomy of patients. International recommendations such as those concerning predictive testing for Huntington disease (Guidelines for the molecular genetics predictive test in HD, 1994) are observed by Polish medical geneticists.

The percent of women who choose abortion when they receive positive prenatal test results that indicate the presence of a genetic condition has recently been analysed by one of the larger centers that does prenatal diagnosis. 84% of patients chose termination for a chromosomal aberration in the fetus, whereas 16% decided to continue the pregnancy. The choice in most instances depended upon the perceived severity of the chromosomal aberration. The total number of tested women was 5,623, and the number of chromosomal aberrations detected was 80 (Pawlowska *et al.*, 1997).

Recent data on the proportion of abortions done for genetic reasons in the context of the total number of abortions have not been analysed. Such data, however, were available for a survey carried out over 10 years ago in Warsaw. Selective terminations following prenatal test results indicating a disorder constituted at that time 3.35% of overall induced abortions, i.e., 1 in 30 (Zaremba, Wisniewski and Mazurczak, 1986). At that time, however, artificial abortion was a major method of family planning and this is presently not so much the case. The demographic impact of genetics services on the frequency of genetic diseases has been negligible, mainly because screening tests such as biochemical maternal serum screening or ultrasonographic screening are not available. Screening programs for PKU and congenital hypothyroidism are exceptions. These programs have certainly reduced the number of cases of intellectual disability caused by these two conditions.

A prevention program for neural tube defects is being planned by the Ministry of Health and Welfare, though there are no plans for a screening program. The prevention program will be based on supplementation of the diet with folic acid.

Jacek Zaremba and Elzbieta Fidzianska

References

CBOS, Center of Studying Public Opinion: attitudes towards legal regulations of abortion and possible change of law, Report of the Survey, Warsaw, April 1996, p. 6.

Guidelines for the molecular genetics predictive test in Huntington's disease. *Neurology* 1994; 44: 1533–1536.

Pawlowska, B., Ilnicka, A., Czylewska, J., Szirkowiec, W., Zdzienicka, E., Mierzewska, H., and Zaremba, J. Abnormal result of prenatal test – type of chromosomal aberration and patients' decisions (in Polish). Ginekologia Polska, 1997.

Zaremba, J., Wiœniewski, L., and Mazurczak, T. Results of prenatal diagnosis programme – medical part. In: Zaremba, J. (ed.), *Programme of Prenatal Diagnosis for the City and Province of Warsaw*. Warsaw: Institute of Psychiatry and Neurology, 1986, pp. 13–67.

Russia

The Russian Federation occupies an area of 17 million km^2, about half of which is in arctic and subarctic regions, with about 146 million inhabitants belonging to more than 120 national and ethnic groups, some of which differ essentially in language, life style, culture, and religion. About three quarters of the total population are urban residents. The overall birthrate is about 1.2 million per year. There is no separate entry for genetic diseases in general health statistics, but by summarizing the results of regional surveys a rough estimate may be made that the load of genetic and inborn diseases is somewhere within 4–8% of live births (Ginter, 1994).

Current medical genetic services are provided by some 600 physicians trained in genetics, including about 300 genetic counselors, who have had special post graduate training, about 100 biochemists and molecular geneticists and up to 200 cytogeneticists. The service is co-ordinated by the Mother and Child Care Department of the Russian Federal Ministry of Health through a Council of 18 experts headed by Dr. E.K. Ginter.

Primary genetic care, i.e., preliminary counseling based on visible morbid traits in patients and ultrasonographic surveys of pregnant women, is provided by genetic units in 70 local hospitals. The bulk of routine medical genetic procedures (detailed counseling, post- and prenatal cytogenetic, biochemical, and ultrasound testing) is provided by ten of the genetic units, which have the status of Regional Medical Genetic Centers. The topmost level of the Federal Service includes seven Federal Medical Genetic Centers, the activities of which cover molecular analysis of genetic disorders, allele and mutation identification in monogenic diseases, cytogenetic analysis of complicated chromosome rearrangements in families at risk, prenatal sampling and testing, teaching personnel for all levels of the service, and quality control.

Average data on the yearly amount and level of genetic care recently provided in Russia have been reported by Zelinskaya and Novikov (1997), responsible officers of the Russian Ministry of Health. Within the last three years about 80,000 families asked for genetic help. Routine karyotype analyses were performed for 23,000 and chromosome disorders were found in about 9.5%. Seventy-five percent of newborns were screened for PKU and 50% for hypothyroidism, as well. Selective screening for metabolic disorders included about 22,000 newborns. Up to 300,000 out of a million pregnant women in 60 of the 89 regions of Russia had ultrasound. On the basis of the ultrasound results, 1,590 of the examined women were selected for karyotyping or DNA-testing of fetuses for Mendelian disorders. In 3.6% of the examined fetuses serious anomalies were found, including 0.72 % with chromosome disorders. In addition to the ultrasonography, molecular genetic, biochemical and cytogenetic diagnoses of genetic disorders based on chorion villi sampling, amnio- and cordocentesis are also available in some Medical Genetic Centers, mainly in Moscow and St. Petersburg, and also in Tomsk and Novosibirsk.

In 1996, prenatal diagnostic procedures were carried out on 1743 fetal samples, including 1483 cytogenetic, 279 biochemical, and 364 molecular investigations. A list of diseases diagnosed by DNA investigation in the four above mentioned Centers, either pre- or postnatally, includes 25 Mendelian disorders (Baranov *et al.*, 1996). Cytogenetic studies are widely applied in all the genetic units, and in a few centers FISH analysis of chromosomes 13, 18, 21, X and Y has been available since 1996 (Kuleshov *et al.*, 1997). Dozens of metabolic diseases are diagnosed in all regional genetic centers, but elaborated biochemical programs are carried out only in some Federal Centers, especially in the Research Center for Medical Genetics in Moscow, where about 40 lysosomal storage diseases are accurately diagnosed, both prenatally and postnatally (Krasnopolskaya *et al.*, 1997).

To help the clinical diagnosis of multiple malformation syndromes, a set of Russian language computer software has been developed, which is available now in regional genetic centers (Ivanov *et al.*, 1994, 1996, 1997, 1998). Under- and postgraduate training in medical genetics is available now at a number of universities and academies in Moscow, St. Petersburg and Tomsk (Bochkov, 1997; Ivanov and Khrennikov, 1996; Kozlova *et al.*, 1996; Puzyrev and Stepanov, 1997).

Funding of the Medical Genetic Service in Russia is now provided via the Federal Foundation for Invalids and Handicapped Persons. Additional support comes from regional and local budgets and voluntary donations. Multisource funding permits only a rough estimate of the overall expenses, as about US $6.5 million (including services for PKU). This sum is really insufficient. Therefore, essential help comes from some federal research foundations, such as Human Genome, Peoples' Health, Fundamental Research and Priorities in Medicine and Genetics, especially for complex technological studies.

Vladimir Ivanov and Vera Leonidovna Izhevskaya

References

Baranov, V.S. Early diagnosis of hereditary diseases in Russia: current status and prospects. *International Medical Review* 1994; 2(4): 236–243 (Russian).

Bochkov, N.P. *Clinical Genetics*. Moscow: Meditsina, 1997, 288 pp. (Russian).

Evgrafov, O.V. Prenatal DNA diagnosis in central Russia. PECO EUCHROMIC Congress, Prague, 1996, September 4–7, Abstract p. 105.

Ginter, E.K. Load and diversity of hereditary disorders in Russian population: influence of genetic structure. *Medizinische Genetik* 1994; 6: p. 101.

Ivanov, V.I. Bioethical aspects of medical applications of human genome and gene therapy projects in Russia. In: Fujiki, N. and Macer, D.R.J. (eds), *Bioethics in Asia. The Proceedings of the UNESCO Asian Bioethics Conference and the WHO-Assisted Satellite Symposium on Medical Genetics Services, 3–8 November, 1997 in Kobe/Fukui*. Tsukuba, Japan: Eubios Ethics Institute, 1998, pp. 268–270.

Ivanov, V.I. Gene therapy and medical genetic services: biological, safety and ethical aspects. In: Mueller, S., Simon, J.W., and Vesting J.W. (eds), *Interdisciplinary Approaches to Gene Therapy. Legal, Ethical and Scientific Aspects*. Berlin: Springer, 1997, pp. 85–87.

Ivanov, V.I. Medical genetics services in Russia. In: Fujiki, N. and Macer, D.R.J. (eds), *Bioethics in Asia. The Proceedings of the UNESCO Asian Bioethic Conference and the WHO-Assisted Satellite Symposium on Medical Genetics Services, 3–8 November, 1997 in Kobe/Fukui*. Tsukuba, Japan: Eubios Ethics Institute, 1998, pp. 381–383.

Ivanov, V.I. and Izhevskaya, V.L. Human genetics: current and prospective ethical problems; eugenics. In: Pokrovsky, V.I. (ed.), *Biomedical Ethics*. Moscow: Meditsina, 1997, pp. 98–115 (Russian).

Ivanov, V.I. and Khrennikov, V.Yu. *Basic General Genetics*. Moscow: Russian State Medical University, 1996, 128 pp. (Russian).

Ivanov, V.I., Konstantinova, L.M., Levina, L.Ya., Tychinin, S.P., and Antonenko, V.G. SYNGEN: an electronic manual for clinical diagnosis of malformation syndromes. *European Journal of Human Genetics*. 1996; 4(Suppl. 1.): p. 149.

Ivanov, V.I., Levina, L.Ya., Konstantinova, L.M., and Agapova, R.K. Computerized analysis of chromosomal dysmorphies in man. *Abstract of 26th Annual meeting European Society of Human Genetics Paris* 1994, p. 69.

Ivanov, V.I., Levina, L.Ya., and Smolyanova, G.G. Human chromosome 2: database for cytogenetic mapping and phenotypic characteristics of structural aberrations. *Russian Journal of Genetics* 1997; 33(12): pp. 1436–1441.

Kozlova, S.I., Seemanova, E., Demikova, N.S., and Blinnikova, O.E. *Hereditary Syndromes and Medical Genetic Counseling*. Moscow: Praktika, 1996, 470 pp. (Russian).

Krasnopolskaya, X.D., Mirenburg, T.V., Akhunov, V.S., and Voskoboeva, E.Yu. Postnatal and prenetal diagnosis of lysosomal storage diseases in the former Soviet Union. *Wien. Klin. Wochenschr*. 1997, Bd. 109/3. pp. 74–80.

Kuleshov, N.P., Modern cytogenetic diagnosis and medical genetic counseling. *Russian Conference: Medical Genetic Counseling in Prevention of Hereditary Diseases: Theory and Practice. Abstracts and Reports*. Moscow: 1997, pp. 67–68. (Russian).

Puzyrev, V.P. and Stepanov, V.A. *Morbid Anatomy of the Human Genome*. Nauka. Novosibirsk. 1997, 223 pp (Russian).

Zelinskaya, D.I., Kuleshov, N.P. and Novikov, P.V. The status of and trends in development of medical genetics services in the Russian Federation. *Russian Conference: Medical Genetic Counseling in Prevention of Hereditary Diseases: Theory and Practice. Abstracts and Reports*. Moscow: 1997, pp. 13–18 (Russian).

NEAR EAST

Israel

Genetics services, which began almost 30 years ago, are currently offered throughout the country. Genetics institutes and clinical genetics centers belong mostly to high and medium complexity hospitals that have academic affiliations with each of the medical schools from all four major universities. This allows for close interaction with research, teaching and training at the pre- and postgraduate levels. General clinical genetics services basically deliver genetic counseling and routine cytogenetics for pre- and postnatal diagnosis, without any regional, ethnic, or social restrictions. Molecular cytogenetics, which has been incorporated more recently, is offered in an increasing number of centers. There is no social consensus yet as to whether coverage of these more sophisticated techniques should be paid for by the individual or by the national health coverage.

Geneticists in Israel have a unique opportunity to study clusters of patients with specific disorders and mutations.[1] This is due to the fact that socio-cultural and historical factors played a role in the development of genetic isolates, allowing for the higher frequency of some genotypes. Prenatal and postnatal biochemical and molecular biology tests for specific diseases are available to the entire population.

Prenatal diagnosis for chromosomal anomalies is delivered by hospital-based cytogenetic laboratories. Genetic counseling and a thorough explanation are given before performing the tests, and an informed consent form is signed by the consultand. The Social Security Institute and the Ministry of Health sponsor cytogenetic prenatal diagnosis related to advanced maternal age for every woman who is 35 years old at the beginning of the pregnancy. The Ministry of Health, through its Program for the Prevention of Congenital Malformations, supervises and centralizes the data on Down syndrome. It is also responsible for the overall screening programs for Tay–Sachs disease, phenylketonuria and congenital hypothyroidism. In the beginning, screening for Tay–Sachs carriers targeted the Ashkenazi Jews, where the heterozygote frequency is about 1/30. It has now been extended to the entire population, due to the finding of mutations in other groups.[2] Screening for phenylketonuria, and hypothyroidism is performed on every neonate before discharge from the hospital, ensuring complete coverage.

In the second trimester, the “triple test” (maternal serum alfa-fetoprotein, chorion gonadotrophic hormone, and unconjugated estriol) is offered to all pregnant women irrespective of age, and it is completely or partially covered by all the health insurance funds. When results indicate a possible genetic condition, genetic counseling is provided. It may include recommendations for further studies such as amniocentesis or ultrasonography. These tests are partially or totally covered by health funds or the government, depending on the type and level of risk. Compliance with the triple test is satisfactory in almost all social and ethnic groups.

The willingness and readiness to accept invasive procedures, either because of advanced maternal age or following abnormal ultrasonographic/biochemical findings, depends on several factors, including the following:

(a) Religious beliefs. The Orthodox and Conservative Jewish families, Muslims, and many Christians do not accept interruption of pregnancy. Therefore, most of them will not undergo invasive procedures during pregnancy.
(b) Lack of awareness of birth defects. Patients’ lack of recognition of the phenotype of Down syndrome and the counselor’s difficulty in conveying the concept of developmental delay or mental retardation[3] may express a cultural pattern[4] rather than a denial mechanism.[5]

In recent years, health professionals generally have recommended carrier screening for disorders such as fragile X, cystic fibrosis, Gaucher disease or Canavan disease. The information is usually delivered to the pregnant woman by the gynecologist or the nurses at the maternal clinics belonging to the net of maternal-child care centers. Compliance with these tests, which are performed on a private basis, varies widely among women from different cultural-economic backgrounds. In general, the number of people desiring newly developed tests increases with time, due to increased awareness resulting from the coverage of medical issues by different informative channels.

In order to avoid the consequences of indiscriminate testing, the Society of Human Genetics in Israel prompted discussions among its members on the ethical aspects of carrier screening for several genetic disorders. The conclusions include recommendations not to test for disorders such as Gaucher disease in those families where there are no affected individuals, nor to test for cystic fibrosis in those ethnic groups where specific mutations have yet to be identified.[6,7]

As in most other countries, ethical and legal dilemmas have arisen in Israel regarding presymptomatic testing for disorders such as Huntington disease,[8] adult polycystic kidney disease, mutations associated with specific malignancies, and assisted reproduction. These problems are generally discussed within working groups that consider basic ethical principles and the legal framework.

Postgraduate training in medical genetics in Israel was approved by the Israeli Scientific Council at the end of the past decade and is currently provided in a limited number of institutions. This accounts for the small number of professionals who have already finished the 2-year program. Despite the increasing need for formal training for genetic counselors, due to the growing diversity of medical genetics services and the number of interested individuals, such a program is still not available.

Cultural Climate Influencing Genetic Services

Medical genetics, as an integral part of health services, is influenced by the attitudes and beliefs of many different sectors. This holds true for the conservatively-oriented religious communities (whether Jewish, Muslim, or Christian), who will accept neither tests nor interventions that do not have the approbation of their religious counselors. Even though these groups represent a minority, their disapproval may interfere with the provision of preventive measures.

The detection and treatment of affected individuals and their families raise ethical concerns in relation to confidentiality[9] and respect for privacy, especially in small communities where people are more likely to be known to each other. Health is valued as a right and an obligation for each person and the society. As a fundamental factor in the well-being of the individual at all stages of life, health care includes developing pre- and postnatal preventive measures, health education for the mother and the child, the organization of a protected environment and adequate support for those with disabilities and their families. The latter is achieved by the inclusion of social workers as an integral part of the professional team, and by support groups for different genetic disorders, many of them in conjunction with similar organizations outside Israel. Programs for individuals with mental retardation and physical disabilities include protected work-units and hostels where children and adults work and live, with the help and supervision of trained personnel.

Ethical issues related to everyday life, interpersonal relations, sexuality and parenthood may have to be discussed by a multidisciplinary team, including a geneticist. When a request for child-bearing comes from a disabled couple, counseling should also take into account the right of the unborn child to be healthy.

The following experience may be a good example: A 32-year-old woman with clinical features of Down syndrome was married to a 40-year-old man who had a psychiatric disorder. He was under supervision and treatment, having been diagnosed as schizophrenic at an early age. They were living alone without any direct family contact, but with community support. Contraception with gynecological supervision was effectively implemented in the woman. The couple expressed a strong desire to have a child. The man argued that he would divorce his wife if she could not get pregnant, causing her a lot of distress.

Cytogenetic studies in both consultands confirmed a mosaic trisomy 21 in the woman (40% of fibroblasts from a skin biopsy culture showed a trisomy 21 while a normal karyotype was found in peripheral blood lymphocytes). The husband's karyotype was normal. The couple received a simplified explanation about the possibility of detecting Down syndrome prenatally, but were told that it would be difficult to determine the status of the fetus with respect to the paternal illness. Discussions on the ethical aspects included the social worker and the family doctor, who were in permanent contact with the couple and who eventually had to orient and support the couple afterwards.

The last 10 years have been a very productive period in the field of molecular genetics, concomitantly raising ethical issues connected to the handling and storage of DNA, the disclosure of test results, and the possibility of presymptomatic testing of fetuses and children for late-onset diseases.[10] There is no consensus yet about the extent of permissible uses of stored material and the type of written consent to be signed by the individual. Certain groups, such as those working in genetic oncology, have been discussing a written consent form for common use that may overcome ethical and legal problems.

Zully Gelman-Kohan

References

1. Zlotogora, J. and Chemke, J. Medical genetics in Israel. *European Journal of Human Genetics* 1995; 3: 147–154.
2. Drucker, L., Proia, R.L., and Navon, R. Identification and rapid detection of three Tay–Sachs mutations in the Moroccan Jewish population. *American Journal of Human Genetics* 1992; 51: 371–377.
3. Gelman-Kohan, Z., Merlob, P., Akstein, E., Nisani, R., Demb, N., and Moguilner, M. Low perception of MR and malformations in Ethiopian immigrants in Israel (personal communication)1996.
4. Christianson, A.L. and Kromberg, J.G.R. Maternal non-recognition of Down Syndrome in Black South African infants. *Clinical Genetics* 1996; 49: 141–144.
5. Lubinsky, M.S. Bearing bad news: dealing with the mimics of denial. *Journal of Genetic Counseling* 1994; 3(1): 5–12.
6. Conclusions of the Evaluation Commision on Population Screening in Israel-Chairwoman: Dr. D Abeilovich. Society of Human Genetics in Israel – July, 1996.
7. Abeliovich, D., Pashut-Lavon, I., Lerer, I., Cohen, T., Springer, C., Avital, A., and Cutting, R. Screening for five mutations detects 97% of CF chromosomes and predicts a carrier frequency of 1 : 29 in the Jewish Ashkenazi population. *American Journal of Human Genetics* 1992; 51: 951–956.
8. Gelman-Kohan, Z. and Chemke, J. Nisani. Huntington disease: ethical aspects of presymptomatic testing (in Hebrew – Abstract in English). *Harefuah* 1993; 124(4): 196–197.
9. Shaap, T. Confidentiality in counseling for X-linked conditions. *Clinical Genetics* 1995; 47: 155–157.
10. Report of a Working Party of the Clinical Genetics Society (UK). Chairman: A Clarke. The genetic testing of children. *Journal of Medical Genetics* 1994; 31: 785–797.

Turkey

Turkey has a young population, as a result of a high birth rate. A third of the population is under the age of 15 years and the proportion of older people is quite low. Health services receive only 3% of the national budget of Turkey. The falling infant mortality rate is an important indication of heath care. The infant mortality rate was around 200 per thousand newborns in the late 1950s, fell to 130 per thousand in the mid 1970s, and at present is about 50 per thousand. This figure is not constant all over Turkey. There are differences from east to west, which parallel the richness of the land and the percent of mothers receiving prenatal care. In the west, 66% receive prenatal care, in contrast to 14% in the east. Immunization rates for children are 76% in the west and 41% in the east, where malnutrition is also a problem.

Genetic Services

Genetic services have been offered by some medical schools since the late 1960s. The number of medical schools has increased, and more are providing services. In 1984, genetic services were provided at 7 out of 19 medical schools, but by 1997 the number had reached 26, according to the National Association of Human Genetics. Most survey respondents (75%) were practicing at University centers. These centers mainly carry out cytogenetic analyses for postnatal and/or prenatal diagnosis, but some also do molecular analyses. Molecular tests are also performed at molecular biology departments of some Science Faculties (such as Bogazici and Bilkent).

Private practices in Ankara and Istanbul have been started during the last five years. In June 1998, the Ministry of Health prepared regulations for diagnostic centers in the field of genetics; these regulations led to some discussion and opposition. Private services cost more and provide the possibility of greater gains. In private practice it is difficult to control the quality of services and qualifications of persons providing services. Persons without experience may aim to increase their incomes and may not care about ethical rules and quality controls. The possibility of financial gain may prompt scientists working at the universities to open such private labs. These will bring more inequalities of access to services and will do harm to the labs at the faculty hospitals and in the end to the public. Authors of this chapter are in favor of regulations when the consequences of practice without guidance and control are considered.

Existence of over 26 services for 62 million people seems appropriate for the size of the population. In reality this is not the case; lagging staff and poor equipment make the units far from expectations. In many there are no medical doctors trained in the field (75% of survey respondents were M.D.s, but only 25% had been trained formally in genetics).

It is not easy for people who need genetic services to obtain them in Turkey. One reason is the unequal geographical distribution of services, which are in the largest cities. Another reason is the high cost. Primary health care services can be obtained from facilities of the Ministry of Health without payment if needed, but if there is a referral, patients have to pay their own expenses. If they are not covered by the social security system (an estimated 50% are not) patients have to pay out of pocket. "Health insurance for all" is on the waiting list. This means that services can only be obtained by patients who can afford them.

Tests for hemoglobinopathies and PKU are carried out in Turkey as research projects or by cooperation between The National Genetics Society (NGS) and the Health Ministry. The Ministry of Health has tried to add some other tests to newborn screening such as hypothyroidism. Newborn screening for PKU is increasing each year through one of the medical schools in Ankara. Huntington disease predictive tests are carried out for a small group of patients and their relatives, with the cooperation of neurology, psychiatry and medical genetics departments under a project (SBAG-159) supported by the Turkish Scientific Association.

In Turkey some, but not all, research is carried out only after informed consent.

Işik Bökesoy and Berna Arda

LATIN AMERICA

Argentina

Global indicators of health in Argentina show some improvement, particularly in the last two decades (longer life expectancy, lower infant mortality). The level achieved is still far from that in industrialized nations, the pace of progress is slower than in other countries in the region, and there are some deficiencies in the quality of public health care.

The Argentine health sector is characterized by the coexistence of three types of services that supplement and sometimes overlap each other: the public sector, the private sector, and what are known as "social health care organizations." The latter include organizations with compulsory membership, traditionally managed by labor unions and workers' associations. The Health Ministry is currently considering their deregulation and restructuring. The private sector began with individual practitioners, complemented by the services offered by private clinics, a system that led to the consolidation of health care management systems that provide multiple alternatives and systems of coverage. The public sector provides medical care in hospitals and health care centers under national, provincial or municipal jurisdiction, mainly to low income families. The Constitution states that the responsibility for providing health care lies with the provinces rather than the central government, although the central government has slowly begun providing services. Financing is provided by the provinces and the national government.

World Bank estimates indicate that there are three physicians per one thousand inhabitants. Approximately 5000 students graduate from medical schools every year. Nevertheless, the current geopolitical situation leads physicians to concentrate in urban and suburban areas, causing a relative shortage in rural areas. There is an uneven distribution of technological resources in the different sectors: during the last decade, the private sector has invested more in highly sophisticated medical equipment than the public sector, which endured budgetary cuts and consequently a relative deterioration.

In 1989, approximately 70% of the total population had entered a private or "social" health care system, whereas the remaining 30% were still in the public system. These percentages are currently changing and more people are entering the public health care system. The current government's growth and development programs include restructuring of the expenditures in the health system, the organization of existing resources, and the creation of a global health care system. None of these strategies assigns importance to or specifies a policy for medical genetics. The public, "social," and private sectors offer only some genetic services, (e.g., familial genetic counseling), and are only partially involved in identifying high-risk populations. Services for treatment and rehabilitation of people with genetic disorders is another critical issue.

There are very few public health care centers, all in urban areas, with a genetics department, and their laboratories are dedicated almost entirely dedicated to cytogenetic studies. Molecular studies are generally performed by other health care sectors or by the universities. Public hospitals do not perform prenatal diagnosis (amniocentesis, chorionic and placental biopsy and cordocentesis), but these services are provided at private health care centers. In general, private centers do cytogenetic diagnosis and alphafetoprotein determination.

At the three levels of health care, genetic care is in the hands of health professionals such as physicians, medical assistants, chemists, biochemists, biologists, etc. They meet in discussion and training for a such as genetics societies and other medical associations (Society of Paediatricians, Society of Gynaecology, etc.). Women manage most of the genetic centers.

The population with genetic care coverage is far from numerically satisfactory, and those with greater access have higher incomes. As in other countries in the region, genetic care has developed as a specialty mainly in urban areas. As a result of the lack of information among many professionals, health authorities, and to some extent the community, it is difficult to create an awareness of the preventive role that an adequate official policy for genetics should have at the primary care level.

Elba Giorgiutti

Brazil

Brazil is a large nation. One hundred and seventy million persons are distributed in 8.5 million km^2, which present a vast array of climates and ecological characteristics. The socioeconomic conditions of the people from these regions are also highly diversified, with extensive poverty areas, especially in the Northeast and North. In the Southeast and South, on the other hand, better economic conditions prevail, while the Center-West presents an intermediate picture.

The Europeans arrived in this part of the world in 1500, and found an Amerindian population estimated as having at least one million persons. Subsequent population growth was heavily influenced by voluntary migrations from Europe, forced migrations from Africa, and the living conditions of these ethnic groups. Intermixture between them led to a fantastic array of physical types (details in Salzano and Freire-Maia, 1970; Salzano, 1987).

In terms of development, Curi (1991), after considering 16 variables (which included mortality and fertility rates, life expectancy at birth, percentage of rural/urban populations, percentage of births assisted by health personnel, degree of literacy, gross national product, acquisition of domestic appliances and rate of inflation) applied two multivariate techniques and compared Brazil, first with 25 other North and South American countries, and then with 36 nations all over the world. He distinguished four levels of development (High, Medium Superior, Medium Inferior, and Low). Brazil was classified as representing Medium Superior development in relation to the other American countries, but only Medium Inferior in comparison with the world in general.

There has been a decline in the rates of infant mortality in the past ten years; the national average is now around 45 per thousand. However, there is extreme regional heterogeneity, with rates as high as 200 per thousand in the Northeast and North, and 15 to 20 per thousand in São Paulo and the South (Brunoni, 1997).

Brunoni (1997) has assessed the conditions and level of development of medical genetics in Brazil. Earlier accounts can be found in Salzano and Pena (1989) and Salzano (1992). Medical genetics developed as an outgrowth of studies in general genetics and human genetics, which date back to the early 1930s, but the Brazilian Society of Clinical Genetics was formed only in 1986. Its main aim was to establish and regulate the title of Specialist in Clinical Genetics according to the norms of the Brazilian Medical Association and of the Federal Council of Medicine. This was readily achieved. In addition, the Society has had a very active program, which included 12 national meetings. Total membership is now 300, but the number of titles of Specialist awarded is still modest: 34.

The information obtained by Brunoni (1997) through a mailed questionnaire indicated the existence of 33 centers of research in medical genetics, distributed among nine States and the Federal District. A total of 118 clinical geneticists work in these centers, as well as 85 biologists and 109 other personnel, which includes social assistants, nurses, and psychologists. In 1995, around 35,000 families received care in these centers. The facilities available included tests for a wide variety of genetic conditions, at the DNA and protein levels; karyotype determinations; chorionic villus sampling and amniocentesis; family and population screening for inborn errors of metabolism; services for teratogen information; monitoring and registration of congenital malformations; and management and/or gene therapy for certain genetic conditions.

Besides these centers, there are associations related to specific diseases; three of the most active are for hemophilia, cystic fibrosis, and muscular dystrophy. Some of these groups work together with genetic services, but others function independently.

Although specific training in genetics is not compulsory, most Brazilian schools of medicine now have courses in medical genetics in their curricula. Presently five Residencies in Medical Genetics are offered (two in Porto Alegre, one in Campinas, one in Ribeirão Preto, and one in Rio de Janeiro). Several graduate courses in genetics also offer options in medical or clinical genetics for the awarding of Ph.D. and M.Sc degrees. Assuming the need of one physician per one thousand inhabitants, and a prevalence of genetic diseases on the order of 5%, Brunoni (1997) estimated that the total number of medical geneticists necessary for a reasonable level of assistance to the Brazilian population would be 8,000. Clearly we are well below this number.

Francisco M. Salzano and Lavínia Schüler-Faccini

References

Brunoni, D. Estado atual do desenvolvimento dos Serviços de Genética Médica no Brasil. *Brazilian Journal of Genetics* 1997; 20(Suppl. 1): 11–23.

Curi, P.R. Agrupamento de países segundo indicadores básicos e econômicos. *Revista de Estudos de População* 1991; 8: 112–124.
Salzano, F.M. Brazil. In: Schwidetzky, I. (ed.), Rassengeschichte der Menschheit, Vol. 12. München: R. Oldenbourg, 1987, pp. 137–175.
Salzano, F.M. History and development of human genetics in Brazil. In: Dronamraju, K.R. (ed.), *The History and Development of Human Genetics*. Singapore: World Scientific, 1992, pp. 228–255.
Salzano, F.M. and Freire-Maia, N. *Problems in Human Biology. A Study of Brazilian Populations*. Detroit: Wayne State University Press, 1970, 200 p.
Salzano, F.M. and Pena, S.D.J. Ethics and medical genetics in Brazil. In: Wertz, D.C. and Fletcher, J.C. (eds), *Ethics and Human Genetics*. Berlin: Springer-Verlag, 1989, pp. 100–118.

Chile

During the long colonial epoch (1536–1810), Chile was the most isolated and poorest of the Spanish-American colonies. The country was slow to develop modern culture, because of the lack of printing presses and technical higher education. After independence from Spain (1810), the first School of Medicine was founded in 1833 under the direction of British, French, German, and Spanish physicians. During the nineteenth century, Chilean medicine and biological sciences developed according to European models. Starting in 1875, professors of the new School of Medicine travelled to Europe for post-graduate training and returned to Chile to develop new specialties.[1]

The teaching of genetics was introduced in Chile in the 1920s by the Italian-born physician Giovanni Noe, Professor of Medical Zoology at the University of Rome. After the Second World War, Chilean biology and medicine fell under the influence of Anglo-Saxon culture. The Chilean pioneers of Mendelian genetics were basic biologists and physicians of the Medical School of the University of Chile who travelled to the United States, England and France in the 1950s.[1]

The Chilean Genetics Society was founded in October 1964, joining together all the Chilean geneticists from the areas of medicine, zoology, and agriculture. The first courses in modern human genetics started in 1963. Chilean participation in International Congresses of Human Genetics began in Chicago in 1966. In 1969, Chilean geneticists participated in the foundation of the ALAG (Asociacion Latinoamericana de Genetica). In 1970, Mendelian genetics had been firmly established in the academic and scientific community in Chile.[1]

Ethics and Society

In the nineteenth century, liberal European immigrants to Chile transformed their new fatherland into a parochial state of western civilization, using the moral and imperialistic standards of classical European bourgeois culture. Therefore, ethical standards of Chilean society have been modelled by the interaction of the Republican ideals of the French Revolution and the strong influence of the Roman Catholic Church. By the end of the nineteenth century, liberal reforms secularized the civil and penal codes, but the influence of the Church remains strong. There is pluralism in politics, religion, and sociology. Penal codes and civil law have condemned and repressed abortion, euthanasia, infanticide, and homosexuality.

The Chilean Medical College, founded in 1948, strictly followed the ethical standards of the World Medical Association inscribed in the codes of Helsinki and Sydney. These standards are accepted and applied by the ethical committees of Chilean hospitals and universities. All Chilean medical doctors swear to the Hippocratic Oath at their graduation.[2]

Vital Statistics

During the last three decades, economic progress and the drop in morbidity and mortality from infectious and nutritional diseases has changed the epidemiological profile of a country in transition to an industrialized nation.[3,4] In 1998, the infant mortality rate was 10.2 per 1,000 births and primary health care covered 98% of perinatal mother and child health care needs. Life expectancy at birth is now 75 years of age.[5]

Vital statistics show that Chile has a relatively low level of mortality from abortion. During the last 25 years (1974–98), women's mortality rate from abortion has decreased from 0.5 to 0.05 per 100,000 population (from 111 down to 38 cases per year).[5] Hospitalizations for all types of abortion, including spontaneous abortion have decreased from 44,000 cases to 22,200, with a population increase from 10 to 15 million.[5] The number of illegal criminal cases of clandestine abortion are unknown. Teenage mothers of low social strata have high pregnancy rates.

Genetic Services

Chile was the leader among Latin American nations in pioneering social and public health reforms, with the enactment of the Sanitary Code Act (1918), Social Security (1924), and the Preventive Medicine Act (1938).[6] As a Minister of Health, Dr. Salvador Allende contributed to the design of the future National Health Service. During the 1960s, Chile developed the characteristics of a welfare state with socialized medicine under a centralized National Health Service (33,000 beds for 10 million people).[6]

The Ministry of Health does not have an official policy for providing genetics services at public and private hospitals and consultories. Foundation and provision of genetics services are due to the pioneering work of geneticists from the University of Chile.

Starting in the 1960s, human and medical genetics has been taught at the University of Chile School of Medicine, and at the present time is taught in all six Chilean schools of medicine. There are 50 trained medical clinical geneticists and about 80 genetic technicians distributed in 12 genetic centers in public and private hospitals. The nine leading Chilean hospitals developed the Latinoamerican Program for Monitoring Congenital Malformations. Newborn screening for PKU and hypothyroidism is currently performed in the public and private hospitals in Santiago and Valparaiso and covers 95% of Chilean births.

The distribution of genetic services is unequal, and influenced by social class, ethnicity and geographical region. Due to the present decline of the welfare state, there is a medicine for the rich and another medicine for the poor. The upper economic classes have wide access to genetic services. Only a fraction of middle and low income people receive genetic counseling. Northern and southern regions have few genetic services; services are concentrated in urban metropolitan areas of central Chile. The ethnic minorities (1 million people, in a total population of over 14 million) distributed in rural areas and on frontiers have no access to these services, only to primary health care. Although this unequality exists, the ethnic minorities are covered by primary health care, which includes nutrition, vaccination, sanitation, and dental care. In 1998, vital and health statistics showed low rates of morbidity and mortality, very close to those of industrialized countries.[4] This indicates how successful the official primary health care policy has been. The inequality of access to genetics services is due to the absence of an official genetics policy of the Ministry of Health.

During the last thirty years, ethnic minorities in Chile have been surveyed by population geneticists to study genetic polymorphisms.[1] The description and discovery of genetic isolates helped to introduce primary health care. The new Law CONADI (Comision Nacional de Indigenas), passed in 1993, is now improving the economic and social situation of these minorities.[7]

Due to the absence of state policy to provide genetic services, these are oriented to the private use of individuals and families, discarding public health or eugenic purposes. There are no special codified ethical standards for interaction between genetics services providers and their patients. Guidelines are based upon standards for patient-physician relations set forth by the Chilean Medical College.

The ethnic minorities speak fluent Spanish and receive primary health care from the National Health Service. However, they have no access to genetic counseling, due to the absence of counselors at the primary health level.

Consequences of Genetic Counseling

Statistical data about the population receiving genetic counseling in Chile is scarce and limited. Only three genetic centers have regular statistical data available, and only for a few thousand cases.

Prenatal diagnosis has limited use in Chile, because legal abortion is not available and there are no commercial clinics. In 1998 the birth rate was 18.3 per 1,000 population, some 270,600 liveborns. The largest clinical series of prenatal diagnoses extends to 500 cases with 5% positive results. The percent of women who choose abortion is unknown, because induced abortion is illegal. A small fraction are referred outside the country. A significant group has spontaneous abortions. Probably a majority of the women affected, under social and religious pressure, carry to term.

We do not know the proportion of abortions done for genetic reasons. In Chile there are wide differences in how people define health and disease. The decision to raise the defective child belongs to the family, who classify the "serious" and "less serious" levels of disease. The general trend of cultural beliefs of the Chilean people is to follow the "culture of life", maintaining the survival of the fetus and carrying to term.

The influence of genetic counseling in the reduction of births with congenital malformations and single gene defects is negligable. On the contrary, birth defects are increasing in vital statistics of newborns, and the numbers of new cases of foreign genetic diseases (previously unknown in Chile) are also increasing. This fact moves us to review the structure of morbid genomes among the Chilean people.

Morbid Genomes in the Chilean Population

The Chilean population is approximately two-thirds white and one-third Amerindian, with traces of other admixtures (black and oriental). Fewer than 4% are foreign born. Classical ethnic diseases such as those of Jews (Tay–Sachs, Nieman–Pick, Gaucher, etc.) and blacks (G6PD, sickle cell anemia, thalassemia, etc.) are very rare. Only

30% of known inborn errors of metabolism have been diagnosed in the Chilean people.[8] No local hemoglobin or haptoglobin genetic variant has been detected. Autosomal recessive disorders show lower percentages than classical figures for European populations and have a wide range of variation. Cystic fibrosis is estimated at 1 per 3500 births, with a low frequency of the Delta F-508 allele mutation (0.30), compared with 0.70 in Europe. Consequently, Chile seems to be a fortunate country, without the genetic load of Europeans and North Americans.[8]

Notwithstanding, every year new genetic diseases are introduced into Chile by immigrants and are published in Chilean medical journals.[9] Moreover, with the drop in the infant mortality rate (under 1 year of age) from 75 to 10.3 per 1000 live births,[5] the relaxation of the law of natural selection has increased congenital malformations from 112 to 151 per 100,000 in the population.

Ricardo Cruz-Coke

References

1. Cruz-Coke, R. Influence of Mendelism in the development of Human genetics in Chile. In: Dronamraju, K.R. (ed.), *History and Development of Human Genetics*. London: World Scientific, 1991.
2. Colegio Médico de Chile. Normas de Etica Médica, Santiago, 1986.
3. Cruz-Coke, R. *Cardiovascular Diseases in Chile Preventive Medicine* 1985; 14: 541–547.
4. Pan American Health Organization. Health Conditions in the Americas 1994. Vol I. Scientific Publication No 549, Washington, 1994.
5. Annuario de Demografia. 1998. INE (Instituto nacional de Estadisticas), Santiago, Chile, 1999.
6. Medina, E. and Cruz-Coke, R. Chilean medicine under social revolution, *New England Journal of Medicine* 1976; 295: 193–197.
7. Republica de Chile. Codigo Sanitario, Edit Jurídica, Santiago, 1993.
8. Cruz-Coke, R. Structure of the morbid genome of the Chilean population. *Reviews in Medicine* Chile 1985; 113: 702–706.
9. Castillo, T.S. Los servicios de genetica medica en Chile. *Brazilian Journal of Genetics* 1997; 20(Suppl. 1): 25–31.

Mexico

For a population of 90 million, only 155 professionals were certified by the Mexican Board of Human Genetics in 1995. Of these, 111 were physicians and the remaining 44 were biologists or chemists working in human genetics. They are distributed among 40 genetics departments, mainly located in public hospitals attending patients of low socioeconomic level.[1] Only a few groups are located in the main public universities. The departments that are capable of offering a relatively wide range of modern services are located only in the three largest cities: Guadalajara, Mexico City, and Monterrey. Small groups, capable of offering mostly clinical and cytogenetic services, are located in another 12 cities. The rest of the country has no genetic services available, and patients in need must travel to obtain them. Fifteen of the 40 human genetics departments are involved in research in the following areas: clinical genetics, cytogenetics, biochemical genetics, mutagenesis, cancer genetics, and population genetics. Only ten departments are proficient in the use of modern molecular biology techniques. Private health institutions offering genetics services are scarce. At the present time, human genetics is regarded as a service to individuals and families. Screening programs for newborns are very limited; they are available almost exclusively for PKU[2] and hypothyroidism; the latter was made mandatory a few years ago, but it is restricted to certain geographical areas.

Services are practically free in the public institutions, but the patient must pay in the private institutions. There is a general acceptance of the practice of human genetics, except when it has to do with abortion, which is tabu for our legislators because opposed by a small, but very vocal minority. Little has been done in the area of services for people with disabilities; however, private groups and schools for people with Down syndrome or children with mental retardation exist in some cities.

Most medical genetics facilities are located in public hospitals, where attention is practically free and payment is not an issue. The pessimists (or realists) probably know that the departments in the public hospitals are unable to provide many genetic services. Some genetic services are almost nonexistent. Prenatal diagnosis (PD) is mainly available only in the private sector, which is very limited because of cost and number of providers. Screening programs are also very scarce and cover possibly less than one percent of newborns, with the possible exception of hypothyroidism. The facilities for genetic studies are located only in the largest cities, and people living in rural areas have restricted access, which indirectly discriminates on the basis of social class.

Rubén Lisker, Salvador Armendares, Alessandra Carnevale, and Osvaldo Mutchinik

References

1. Lisker, R. and Carnevale, A. Diagnóstico y perspectivas de la medicina. In: Fortes, M. (ed.), *Retos y perspectivas de la Ciencia en México*. Mexico City: Academia de la Investigación Científica, 1995.
2. Nicolini, H., Cruz, C., Camarena, B., Merino, M.S., Bilbao, G., Vela, M., Velazquez, A., Pérez, B., Desviat, L., and Ugarte, M. Molecular analysis of the phenylalanine hydroxylase gene in Mexican phenylketonuric patients. *Archives of Medical Research* 1995; 26: 53–57.

ASIA

Thailand

Thailand is a developing country with a constitutional monarchy. It covers 198,404 square miles, in Southeast Asia, surrounded by Myanmar, China and Vietnam at the North, Laos and Cambodia at the East, Malaysia at the South, and Myanmar at the West, and is divided into five regions, Central, North, South, Northeast, and West. Bangkok, the capital, is in the Central region. Each region has minor differences in culture and language, which pose no problem in communication. Thailand has a population of 60 million people with no ethnic differences except for one minority at the border. Eighty percent of the population is Buddhist; the rest is Christian or Islamic.

Genetic services are limited, due to lack of qualified personnel and qualified laboratory services. Prenatal diagnosis and termination of pregnancy are difficult issues, since religious believers may oppose them. Genetic information through the media helps people recognize how genetics can lead to a better quality of life, and may lead more people to come for the service, especially in the upper socioeconomic group. There is non-invasive routine screening during pregnancy to detect fetal abnormalities. The Ministry of Health has planned a national project to prevent thalassemia and Down syndrome, which are the disorders of most concern in Thailand, by research on cost-effective and easy-to-perform testing to be used nationwide. The Genetic Society of Thailand, together with the medical schools is trying to improve services by working in a team approach and conducting seminars periodically for more organized and uniform genetic services in the future.

There is no registration of genetic data, either in hospitals or at the national level.

Eugenic principles make information to family members more important than confidentiality. No respondent had never counseled a patient with a serious genetic disorder who refused to tell relatives at risk.

Vailarat Dhamcharee, Apiwat Mutirangura, and Orasri Romayanan

CHAPTER 11

LAWS AFFECTING GENETIC SERVICES

ENGLISH-SPEAKING NATIONS

Australia

In Australia there are eight state/territorial jurisdictions, in addition to a federal jurisdiction. Laws affecting genetic services are, in general, the province of state or territorial legislatures. As Australia is a common law jurisdiction, court rulings are an important part of the law. This is particularly important, as the legislatures are often unwilling to legislate on areas that are overtly controversial. For example, the provisions of the New South Wales *Crimes Act* relating to abortion (Sections 82–83) have remained unchanged since they were enacted in 1900; changes in the legal attitude toward abortion have come about through a series of court rulings.

One state, Victoria, has laws that restrict research involving the human embryo; otherwise, in general, the laws of the most populous state (New South Wales) are typical of the approach taken to issues involving genetic services. In regard to abortion, the unlawful procurement of a miscarriage, and in a series of rulings (Wald, 1971; *ES* v. *SuperClinics*, 1995) it has been determined that abortions are lawful when needed for the mental or physical well-being of the mother. This has been interpreted very broadly, so that there is effectively ready access to safe abortions. There are no specific laws or court rulings that restrict the availability or types of prenatal diagnoses that can be performed. The use of abortion following prenatal diagnosis where a fetus has a disorder is, thus far, unchallenged; presumably it could be legally justified by arguing that the birth of such a fetus poses a threat to the well-being of the mother.

The laws in relation to gamete donation are, as yet, unsettled. There are no problems with donating gametes, and legal and administrative practice is to accept that the putative parents are also the biological parents, with the donors remaining anonymous with no parental rights. This has yet to be legally challenged. To date, current practice at fertility clinics has been to restrict the availability of IVF and AID techniques to infertile heterosexual married couples. However, this is being successfully challenged in the courts (*Pearce* v. *South Australian Health Commission, Supreme Court of SA, 1996*) in light of antidiscrimination legislation at state and federal levels, and it is anticipated that these services will be made more widely available to unmarried heterosexual couples, lesbian couples and single women.

Ownership
Medical records belong to the originator of those records, such as a hospital or medical practitioner. The patient has no legal right to access those records (*Breen* v. *Williams*, 1996). However, when the medical record is held by public facilities, such as hospitals, it must usually be made available to the individual concerned upon request under the various *Freedom of Information* laws. Access to all records for third parties is strictly limited under privacy legislation. However, there are proposals to broaden the types of persons who would have access to genetic information about an individual. The life insurance industry has proposed that persons who want life insurance will have to reveal the results of prior genetic testing to the proposed insurer, on the grounds that the insurer needs to have access to the same information as the insured, so that risks are equitably assessed. By the same token, insurers have specifically stated that they will not require potential clients to undergo testing; if the client is unaware of a specific genetic condition, then the insurers are also happy to remain unaware.

Alison Colley and Keith Joseph

Canada

The Royal Commission on New Reproductive Technologies (RCNRT) whose final report, *Proceed with Care*,[1] was released in 1993, was established in 1989 with a wide mandate "to examine current and potential reproductive technologies, considering the social, ethical, health, legal, and economic implications of use of these technologies." The Commission consulted widely among the public and persons from numerous different disciplines and sponsored 130 related research projects. The final report contains 293 recommendations resulting from reviews of the scientific information and surveys of Canadian attitudes, and uses recognized moral principles and evidence-based medicine. These recommendations address many areas of reproduction, including infertility, *in vitro* fertilization, assisted insemination, and surrogacy. Key recommendations relating to genetics and prenatal diagnosis include the establishment of a widely representative national advisory body, which would have regulatory powers but be responsive to changing technology and would be able to approve and revise guidelines. Legislation was also proposed to ban certain practices, such as post-conception sex selection, and to establish a system of licensure for centres providing prenatal diagnosis. In 1996, federal legislators introduced Bill C-47, which dealt with some of the recommended legislation but did not provide for the proposed regulatory body. At that point, a federal election was called and the bill died, but new versions have been reactivated under subsequent governments.

Abortion was removed from the criminal code and a new abortion law passed in 1969. This permitted abortion under certain circumstances – only in hospital by accredited medical personnel and when a committee agreed that continuation of pregnancy would endanger the woman's life or health. Many problems arose as a result of the application of this law, and it was challenged by the establishment of non-hospital abortion clinics in Quebec (1970) and Ontario (1983). The resultant trials and appeals reached the Supreme Court of Canada, which, in 1988, declared the 1969 law unconstitutional as contravening the Canadian Charter of Rights and Freedoms.[2] A subsequent attempt to introduce a new law failed, and Canada is still without any national law relating to abortion. Abortion services are currently provided either through special clinics or at the discretion of individual hospital boards. The trend to regionalisation of hospital facilities means that abortion services are generally available, although there is some provincial variation, and geographic accessibility is still a problem for many rural women.

Elizabeth J. Ives

References

1. Royal Commission on New Reproductive Technologies (RCNRT) Proceed with Care: Final report of the RCNRT, Minister of Government Services, 1993.
2. Roy, D. and Williams, J. *Bioethics in Canada*. Canada: Prentice-Hall, 1994, pp. 199–204.

South Africa

The first Abortion and Sterilization Act was passed in 1975, and it provided for abortion for genetic reasons. Most geneticists who participated in the survey were satisfied with the law as it applied to abortion for such reasons. Other indications for abortion included rape and psychiatric or medical problems in the mother. Abortion for other reasons was not allowed.

The new government has reconsidered the situation and a new law (Choice on Termination of Pregnancy Act, 1996) has recently been promulgated, which allows for abortion on request up to 12 weeks after conception. Access to termination is also allowed, after a medical opinion has been sought, up to 20 weeks, and thereafter with the consent of 2 medical practitioners who agree that the confirmed pregnancy would endanger the women's life, result in a seriously malformed fetus, or pose a risk of injury to the fetus. This will lead to fewer septic backstreet abortions (which previously were a major cause of morbidity in women), and fewer unplanned and unwanted children.

The rate of abortion after genetic diagnosis of fetal defects was high even when there was a more restrictive abortion act. However, the number of genetic abortions in comparison with abortions for psychiatric reasons (which often include psychosocial factors such as unmarried status, low income, no support, grand multiparity, very advanced maternal age, as well as instability, manic-depression or schizophrenia, or drug addiction), is very small. In the total abortion group in a study conducted in 1976, genetic abortions only accounted for about 4%, while abortions for psychiatric and medical reasons accounted for 80% and 16%, respectively (Jenkins and Kromberg, 1976).

Issues surrounding the sterilization of the intellectually disabled are being reconsidered in the light of their basic human rights (as described in the new Human Rights Bill, 1996). New guidelines, under discussion, are concerned with the rights of the parents or guardians of such individuals compared with the disabled persons' own rights to a good quality of life.

New laws regarding gamete donation and research on the human embryo are also under consideration, and meetings of experts and relevant stakeholders have been held to discuss these matters. However, there are no laws regarding prenatal diagnosis and its indications, and most geneticists would prefer this situation to remain as it is.

Previously the law, particularly the restrictive abortion law, might have hindered the delivery of genetic services and the free choices of individuals. However, with the liberalization of the abortion act, after considerable canvassing of public opinion, and the reviewing of many other relevant acts, choices of at-risk individuals are expanding.

The sterilization of persons with intellectual disorders is still somewhat routine, and the rights of affected people may not be fully considered. The new constitution, however, has entrenched human rights, and in the future a legal opinion and patient advocate might be required before the procedure is carried out.

Jennifer G.R. Kromberg and Trefor Jenkins

References

1. Choice on Termination of Pregnancy Act. Act No. 92, 1996. Government Gazette, 22 November 1996, 2–8.
2. Jenkins, T. and Kromberg, J.G.R. Prenatal diagnosis and selective abortion. *South African Medical Journal* 1976; 50: 2091–2095.

United Kingdom

Current legislation treats genetic information as similar to all other types of medical information, but the government has warned the insurance industry that restrictive legislation – banning the use of genetic information in setting insurance premiums, for example – will be prepared if they do not put forward proposals for dealing with developments in genetics in a socially equitable manner.

A legal framework for controlling human embryology and fertility research and associated clinical services has been developed; the Human Fertilization and Embryology Authority (HFEA) is the statutory body created to monitor and regulate this area after the report of the Warnock Committee (Dept. of Health and Social Security, 1984). Research on human conceptions *in vitro* is permitted up to 14 days post-conception but not thereafter. Surrogacy arrangements are permitted in principle but may not be arranged on a commercial basis; the generous "expenses" allowed to compensate the surrogate mother for her costs are likely soon to be heavily restricted. The surrogate mother is also treated in law as the true mother of the child, who cannot be obliged by contract to give up her child but who may do so if she chooses (HFEA, 1998).

Research in the field of gene therapy is similarly tightly controlled, because all research in this area involving human subjects has to be considered by the Gene Therapy Advisory Committee (GTAC), set up by the Department of Health. Research cannot proceed without approval from this body, and separate regulation is not required because all gene therapy remains experimental.

The monitoring of developments in the broader field of genetic testing is the remit of another advisory body established by the Department of Health – the Advisory Committee on Genetic Testing (ACGT). It has produced one report recommending that genetic testing made available directly to the public (e.g., over the counter) should be restricted to tests of adults for carrier status in relation to recessive diseases (ACGT, 1998). Another report sets out a recommended approach to the issue of predictive testing of healthy adults for late-onset disorders (ACGT, 1998). Further reports and guidelines are being developed.

One other public body established to oversee developments in human genetics and their application to society is the Human Genetics Advisory Commission (HGAC), set up jointly by the Departments of Health and of Industry. It has reported on genetic testing and insurance and has issued a consultation paper on the implications of animal cloning technology (HGAC, 1998). The relations between these various bodies remain to be developed in practice, but the HGAC has representation from all the other bodies, two government departments and the biotechnology industry, and is perhaps the senior body. Its role was strengthened at the insistence of the House of Commons Committee on Science and Technology, which had itself reported on Human Genetics: The Science and its Consequences (1995).

The applications of genetic testing to prenatal diagnosis and the termination of pregnancies affected by genetic disorders is covered by older legislation that predates the advances in molecular diagnostics. Recent amendments have restricted the period of gestation in which termination is permitted to 20 weeks, except in the case of a fetus affected by a serious disease, in which case there is no gestational limit to the lawful termination of pregnancy.

The proportion of pregnancy terminations carried out because of genetic disease in the fetus is very small – being vastly outnumbered by essentially "social" terminations of pregnancy.

Public support for the continued legal availability of abortion is strong, even for abortion on social grounds (at least for the first 3 months of pregnancy), and few health professionals would wish to return to the days of backstreet abortion. Womens' control over their reproductive lives both depends upon and has contributed to their increased social status; any attempt to reverse this would be strongly opposed.

Angus Clarke

References

Advisory Committee on Genetic Testing. Code of Practice and Guidance on Human Genetic Testing Services Supplied Direct To The Public. London: Department of Health, 1997.

Advisory Committee on Genetic Testing. Report on Genetic Testing for Late-Onset Disorders. London: Department of Health, 1998.

Department of Health and Social Security. Report of the Committee of Enquiry into Human Fertilization and Embryology. Chair: Dame Mary Warnock, London: Her Majesty's Stationery Office, 1984.

House of Commons Committee on Science and Technology. Third Report. Human Genetics: The Science and Its Consequences. London: Her Majesty's Stationery Office, 1995.

Human Fertilisation and Embryology Authority. Code of Practice, Fourth Edition. London: Human Fertilisation and Embryology Authority, 1998.

Human Genetics Advisory Commission and Human Fertilisation and Embryology Authority. Cloning: Issues in Reproduction, Science and Medicine. A Consultation Document. London: Human Genetics Advisory Commission and Human Fertilisation and Embryology Authority, 1998.

United States

The United States has no national laws on medical privacy and no national laws specifically on genetic privacy. Over 100 bills on genetic privacy have been introduced in Congress, without success. It is difficult to write a privacy bill that does not also hinder research. The terms "genetic information" and "genetic test" are difficult to define.

At least 39 states have laws governing genetic information and health insurance (most exclude life insurance, because this is a commercial product not usually considered a human right and is not ordinarily linked to home mortgages, as it is in the United Kingdom). These laws vary widely. At least two states – New Hampshire and New Jersey – have included family history under the heading of genetic information that insurance companies are not permitted to request. Most states, however, limit genetic information and genetic testing to molecular or chromosomal tests. Many specifically exclude HIV tests or cholesterol tests from their definitions of genetic testing. Some states forbid insurance companies to require tests (with some exceptions for tests that are proven to be actuarially relevant – these are rare, so far); others forbid access to results of tests that people have already taken; most allow people to present test results that are in their own interests. Some states allow access to test results but require provision of insurance regardless of results, and also forbid companies to charge higher premiums. Some states, such as New Mexico, require provision of insurance but allow companies to raise premiums. In sum, there are no universal provisions.

In view of the overall structure of health insurance (see the US section in Chapter 10), these state privacy laws can protect few people – perhaps 4–5% of those insured (Reilly, 1998). Most people have their insurance through self-insured employers, Medicare (a federal program for people over 65), or Medicaid (a federal program for low-income people of all ages), which are not subject to state regulation.

The federal Americans with Disabilities Act (1994) requires that people with disabilities be hired as long as they can do the job with reasonable physical and technical support to be provided by the employer. The Equal Employment Opportunity Commission has ruled that presymptomatic status for a genetic disorder (e.g., having the Huntington mutation, in the absence of symptoms) is a disability covered under the ADA. A genetic condition that is socially perceived as being a disability, even if it causes no other problems to the individual, would also be covered under the ADA, just as HIV infection (in the absence of AIDS) is covered, because most people perceive HIV as a disability. The ADA prohibits employers from testing for non-work-related conditions before hiring. They may, however, give physical examinations after hiring.

There are no federal laws specifically on genetics and employment. So far, at least seven bills have failed in Congress and passage seems unlikely. About 13 states have such laws, which follow patterns similar to the more prevalent laws on genetics and insurance. President Clinton's 1999 Executive decree prohibiting genetic discrimination in employment affects only employees of the Federal Government and duplicates protections already existent in the ADA.

Abortion is regulated by the 50 states, but states are limited by the US Supreme Court *Roe* v. *Wade* decision in 1973. The Court struck down state laws restricting abortion, in the name of protection of a right to privacy and constitutional requirements of equal protection under state and federal laws. This court ruling permits abortion for any reason in the first and second trimesters. States may regulate the practice of abortion in the interest of woman's physical and mental health, but not abrogate the liberty of women to seek abortions or of physicians to help them. States may also prohibit abortions in the third trimester, which most (but not all) do.

Since 1973, *Roe* v. *Wade* has been under relentless attack by the "right to life" movement in the US. However, in several decisions, the Supreme Court has refused to overturn it, even though the original concept of "right to privacy" has been replaced in these decisions by the principle of "a liberty interest" of the woman. The burden of proof is on the states to show why any infringement on this interest is justified. No state has specifically attempted to legislate against abortions for genetic indications, although two states attempted unsuccessfully to prohibit abortions for sex selection. Although conservative American presidents (Ronald Reagan and George Bush) have supported a Human Life Amendment to the US Constitution that would make abortion illegal for any reason except rape, incest, and to save the mother's life, the fact is that they did not take the initiatives necessary to pass this Amendment at the state or federal levels. There are no data on the incidence of abortion for genetic reasons, but it could not be more than 1–2% of the total number. Total numbers of abortions have declined in recent years, partly because of increased use of contraception, and partly because fewer physicians perform them.

There are no federal or state laws governing IVF, donor insemination, egg or embryo donation, preimplantation diagnosis, or sex selection (whether pre or post conception). These activities have evolved mainly in the private sphere, with no government support, because "right to life" groups opposed public funding of experimentation regarding embryos. There is no federal law banning human cloning, despite recommendations from the National Bioethics Advisory Commission (NBAC) in 1998 for a 5-year legal moratorium. NBAC based its recommendation on the fact that cloning was "unsafe", and allowed for a re-examination of the issue in the light of future scientific developments. Barring some announcement that human cloning is actually being attempted, it is unlikely that the US will ever have such a law. It is difficult to write a law that would not also ban therapeutically useful research on cloning tissues. Nevertheless, about five states have banned cloning. About one-third of states allow surrogacy, including use for-profit agencies.

Dorothy C. Wertz and John C. Fletcher

Reference

Reilly, P.R. Laws to regulate the use of genetic information. In: Rothstein, M.A. (ed.), *Genetic Secrets*. New Haven: Yale University Press, 1997, pp. 377–391.

NORTHERN/WESTERN EUROPE

Denmark

Genetic testing and counseling are not specifically regulated by law, but are generally considered the responsibility of the medical profession. Medical doctors are responsible for their actions by an act of parliament regulating the medical profession. The practice of genetic testing and data banking must adhere to other acts of parliament, such as the law on patients' rights (2000), where the principle of informed consent and confidentiality are regulated. It should be noted that Denmark, as a full member of the European Union, is obliged to implement all EU directives. The developmental pattern in testing and counseling is determined by economic and medical factors modified by individual demand. The recommendations of the National Board of Health are not legally binding in a strict sense of the word. Hospitals are owned by the regional governments, which determine their priorities according to economic and other criteria. Often dissemination of new genetic services is the result of local initiatives rather than a national decision.

Registries

Denmark is famous for its well-developed registries of the population. This tradition goes back to 1645, when all births and deaths were entered into church records. In 1769, the first national census was taken.

In 1924, all local authorities created offices to register the population, and since 1968 these records have been kept on file in the national Central Person Registry (CPR). This registration of the population means that authorities know the composition of the population by age, sex, birthplace, residence, citizenship, religious affiliation, marital and employment status etc. Each individual has a unique CPR-number that will identify this individual. Public registries use this individual identification code, and the establishment of computerized registries has created powerful means of coordinating data from various registries.

The long tradition for registering the population goes together with a strong democratic tradition, and only in rare instances, such as during the German occupation in 1940–45, has widespread worry been voiced that the registries would be abused. The development of the human genome project has also created some worry that the Danish registries are not sufficiently protected against abuse of sensitive individual data. Public as well as private registries are regulated by law to protect individuals from undue surveillance and new legislation is being considered to improve the security of registries, e.g., in connection with the rise of genetic data banks.

Lene Koch and Søren Nørby

France

In 1994, the French Government passed a set of bioethical laws (Laws no. 94-653 & 654, 29 July 1994) with the aim of better framing the application of genetics to medicine and to the social context as a whole.

According to these laws, the laboratories involved have to be 'accredited' by the Ministry of Health. Laboratories that carry out genetic (cytogenetic, molecular genetic) analyses must obtain a special authorization from the Ministry of Health (Article R 162-6-2 of Decret 95-559 of 6 May 1995). Genetic counseling is compulsory before any prenatal diagnosis.

The collection of biological samples for research purposes must be declared and authorized by the Ministry (Law no. 96-542, 28 May 1996). French law (Law 94-654;

29 July 1994, Titre VI, art. 145-15) stipulates that blood samples for genetic testing can only be taken for medical or research purposes after obtaining the person's consent. This consent should be written and signed, except under specific conditions. These conditions would certainly apply to a Huntington patient who is no longer mentally able, such as the patient in the vignette presented in the survey. In this case, however, if the legal guardian does not agree to the test, it is likely that the test would not be legal in France.

The research. As far as we know, no special written consent is required, however, before information about a donor can be made accessible to his/her relatives.

Genetic testing for other than research or medical reasons is at present forbidden by French law (Art. 16-10. Law 94-653 of 29 July 1994). Genetic testing cannot be required by an employer.

Claire Julian-Reynier

The Netherlands

There are many laws in The Netherlands that affect genetics services. The Hospital Provisions Act (1971) requires a license to provide certain services, including clinical genetics. Only the seven foundations for clinical genetics have obtained such a license so far. Another law, the Data Protection Act (1989), regulates privacy protection for registers containing personal data. All clinical genetics centers have regulations to conform with this act.

The Population Screening Act (1996) requires the central government to approve certain screening programmes before they are implemented. Under this new law, prenatal cytogenetic diagnosis for advanced maternal age requires a license. According to the Ministry of Health, another activity of clinical genetics – offering genetic testing to relatives of patients and consultants – should also require a license, but it was ultimately decided that this was not necessary at present.

Abortion in The Netherlands is legal under the Abortion Act (1984). The number of legal abortions among women living in The Netherlands increased slightly from 19,568 in 1991 to 20,811 in 1994. Still, the abortion rate in this country is the lowest in the world. The number of prenatal cytogenetic and molecular diagnoses in the same period increased from 10,126 in 1991 to 11,802 in 1994. The resulting numbers of terminations of pregnancies were 252 and 329, respectively. This is between 1.3% and 1.6% of all legal abortions. The indication for 72% of all prenatal diagnoses is advanced maternal age. About 47% of all pregnant women in this age category participate in this type of prenatal testing.

Preimplantation genetic diagnosis is allowed for severe, untreatable disorders. The first baby resulting from this procedure was born in 1997.

Leo P. Ten Kate

Sweden

Abortion and Preimplantation Investigations

There is no upper limit of gestational age for abortion in Sweden. Abortion may thus be performed until the 28th week of gestation; after that the fetus is by definition a viable baby. However, pregnancies may be interrupted after the 28th week if the mother's life is threatened or the fetus has a disorder not compatible with extra-uterine life, such as anencephaly.

Abortion is available on request in Sweden until the 18th week of gestation. Permission from the "Abortion Council" at the National Board of Health and Welfare to perform an abortion is necessary after the eighteenth week, but in practice abortion is almost never denied before 21 weeks of gestation.

The majority of abortions after prenatal diagnosis are performed after the screening ultrasound investigation has revealed a severe malformation of a fetus in early/mid second trimester. Such malformations are found in about one out of 200 pregnancies. Fewer than 100 pregnancy terminations performed annually are due to disorders indicated by prenatal diagnosis, compared to about 15,000 abortions for social reasons and about 100,000 live births.

Sperm donation is allowed when the man in a couple is infertile. The law states that a child born after sperm donation must be informed that the biological father is a person other than the social father. Also, the sperm donor may not be anonymous to "his child". Egg donation is so far not allowed in Sweden, but a new law permitting it is in process. Surrogate motherhood is forbidden. Preimplantation research is permitted up to 8-cell cleavage. Embryos may be preserved for three months.

Neonatal Screening and Registration

Neonatal genetic screening is performed for PKU, galactosemia, hypothyroidism, and the adrenogenital syndrome.

The National Board of Health and Welfare (Socialstyrelsen) administers six registries to which reporting is compulsory. They are the Medical Birth Registry, the Swedish Registry of Congenital Malformations, the Central Cytogenetic Registry (only unbalanced autosomal aberrations detected before one year of age), the Swedish Cancer Registry, the Cause of Death Registry, and the Prenatal Malformation Registry (newly introducted). Besides these central registries, there are numerous national or regional registries administrated by groups of doctors for research and clinical follow-up, such as the Huntington Disease Registry, the Paediatric Cardiac Malformation Registry, and the Colon Polyposis Registry.

Filing data is required by law, and the files are confidential. The patient must give informed consent for anybody who wants access to the information. For genetic disorders this is sometimes problematic, but oral or written informed consent to use the information in the files for genetic counselling for the family is usually obtained without difficulty. The general Health and Medical Services Act of 1982 is relevant to clinical genetics. In the present context, the most important piece of legislation is the 1990 government proposition on the application of gene technology to human beings, which covers research and experimentation on fertilized eggs. Prenatal diagnosis is regulated, as well as the use of gene technology in mass screening. Sperm donation has long been permitted in Sweden. In a recent report, the majority of the National Swedish Medical Council said no to egg donation, but a minority argued for permitting it.[1] The majority argument is that treating egg donation and sperm donation differently presupposes that there is a morally relevant difference between them concerning the risks for the woman, the child, or society. Those who argue that egg donation should not be allowed emphasize these differences and the risks. The minority argues that if no such differences can be demonstrated, either both sperm donation and egg donation should be accepted or both should be forbidden.

A proposal for a new law on the protection of privacy of genetic information from third parties in Sweden is currently being developed by the Ministry of Health. The problems concern in particular the possibility that employers and insurance companies will require genetic information as a precondition for work or for a life insurance contract. The proposal states that in Sweden, as in Norway, such uses of genetic information should be unlawful.

In addition to the laws regulating abortion, there is also a law permitting research on fertilized eggs up to 14 days after the fusion of the ovum and the sperm. This research is particularly important to help infertile couples by improving methods such as *in vitro* fertilization and preimplantation diagnosis, particularly when there are hereditary diseases on the male or female side.

DNA or RNA may not be used in mass screening for health purposes without permission from the National Board of Health and Welfare. Permission may be given only if the screening has a clearly stated medical purpose, if the genetic information gathered is safely protected, and if informed consent is obtained from the participants.

Ulf Kristoffersson and Gören Hermeren

Reference

1. Assisterad befruktning. Synpunkter på; vissa frågor i samband med befruktning utanför kroppen, Stockholm: Statens medicinsk-etiska råd, 1995. The minority included one of the present writers (GH).

SOUTHERN EUROPE

Greece

Following to the Hippocratic Oath, the Greek Penal Code (section 371) punishes infringement of medical confidentiality with a pecuniary penalty and up to one year imprisonment. Physicians are not obliged to testify in courts about what has been confided to them during the practice of their profession. This is reinforced by the 1939 Code of Practice of the Medical Profession and the Code of Medical Ethics (1955).

According to the Code of Medical Ethics, a physician cannot impose a treatment or perform an operation that might be inappropriate or experimental, if this may offend the personal freedom or free will of mentally competent individuals. Clinical drug trials on humans are regulated, through a ministerial decree issued in 1984, to comply with the Helsinki Declaration[1]. Although the first ethics committee for medical research was established by the Institute of Child Health in 1965, it was not until 1978 that a Ministerial decree imposed the establishment of hospital ethics committees.

Abortion is permitted, under Law 1609/1986, when performed by a gynecologist with the woman's consent. The physician should be assisted by an anesthesiologist in a health care establishment. Nevertheless, one of the following conditions should be met: (1) the duration of pregnancy should not exceed 12 weeks, (2) there is evidence of serious fetal abnormality (up to 24 weeks), (3) there is unavoidable risk to the life, physical or mental health of the pregnant woman, (4) conception is the result of rape, seduction of a minor, incest or abuse of a woman incapable of resistance (up to 19 weeks). Although the Greek Orthodox Church condemned abortion, more than 200,000 abortions were performed every year, even before enactment of the law.

Liberalisation of the abortion law in 1977 proved to be an excellent opportunity for development of prenatal diagnosis centres. This law allowed termination of pregnancy for medical reasons up to week 24. From the beginning, the Ministry of Health supported the establishment of prenatal diagnosis centres, by providing funds for the purchase of necessary equipment and hiring of personnel. Chorionic villous sampling (CVS) for prenatal diagnosis has been performed since 1983. A law on the practice of genetics, passed in 1980 (N. 1036), specified that the state must disseminate and apply current knowledge in genetic issues and matters of family planning. Health professional education, public information and practical applications are under the control of the Ministry of Health.

There is no explicit legislation governing the newer reproductive techniques. However, *Law No. 2071 of 15 July 1992* on the Modernisation and Organisation of the National Health System, in Section 59 provides for the establishment and operation of units of artificial fertilisation. According to this Law, a presidential decree will determine the terms and conditions for the establishment and operation of units of human artificial fertilisation and will regulate in detail all relevant ethical, deontological, legal and economic issues. These units must operate in specifically organised public or private hospitals or specifically organised private clinics. Furthermore, *Section 1471 of the Greek Civil Code* forbids a husband from disrupting fatherhood if he gave his consent to the artificial procreation.

Homologous insemination as well as donor insemination has been performed for over twenty years in the private sector. Also, there exist private banks for the storage of semen. *In vitro* fertilisation was introduced in 1984 and the first twin children were born in 1986. Ever since, *in vitro* fertilisation has been performed to a growing extent in the private sector, under no state control, and has been introduced in at least two university hospitals as well. The attitude of physicians as well as of lay people towards artificial procreation is positive. The Greek Orthodox Church has not taken an official position, but is positive towards tissue and organ transplantation.

Overall, one should take into consideration the general and constitutional principles governing human rights and respect of human personality, although there are no particular regulations or court precedents determining how the general principles apply or when a fetus becomes a person in the legal sense. There are also some specific legal provisions such as the safeguarding of the normal course of gestation (Penal Law), as well as the safeguarding of all possible rights of the person who will be born alive (Civil Law). Section 36 of Civil Law considers the embryo or fetus a person, provided it will be born alive. The embryo or fetus can be an heir (Section 1711 of Civil Law) or a legatee (Section 1936). Also it has the right to ask, provided it is born alive, for compensation for grief, distress or pain, or injury to psychological and bodily health or wholeness, caused by a third person during gestation.

There is no specific legislation covering the collection and use of fetal tissue. Law No. 1383/1983 on the "Removal and transplantation of human tissues and organs from a live or dead person" does not state whether fetal tissues are excluded from its provisions, while blood, sperm, ova, testicles, ovaries, and embryo transfer are explicitly excluded. However, from the examination of the law as a whole, it can be seen that its provisions do not refer to fetal tissues. With regard to the use of fetal tissue for research, transplantation etc., it is clear that fetal tissue, such as blood, skin etc. is used for fetal diagnosis and therapy (e.g., thalassemia). In this context therapeutic research is being undertaken. On the other hand, fetal tissues are not used for transplantation or any other purpose. (Commission of the European Communities, EC Working Group on Human Embryos and Research.)

The abortion law actively encourages the development of genetic services and access to them. There are no laws affecting population screening or genetic registers.

Christos S. Bartsocas and Ariadne Malamitsi-Puchner

Reference

1. Metaxotou, C. Diffusion of prenatal screening in Greece. In: Reid, M. (ed.), *The Diffusion of Four Prenatal Screening Tests Across Europe*. London: King's Fund, 1991, pp. 86–90.

Italy

Italy faces a very peculiar cultural situation. A strong tradition of Catholic thinking, which tends to identify morality with religion, opposes lay thinking. Italy has abstained from any bioethical reflection for a long time, considering it synonymous with religious ethics. This tradition has not given life to a fruitful discussion, but has created a stalemate in bioethics and biolegislation, thus preventing comprehensive regulations and public policies. Italy, however, also has an extensive legal culture, which guarantees the rights of the individual and which has provided examples of advanced legislation, such as Act no. 194 on the voluntary interruption of pregnancy, issued in 1978 and confirmed by a people's referendum in 1981. This law makes a distinction between the first trimester, within which termination is permitted on the basis of mother's free choice, and the second trimester, when termination is allowed (i) when the pregnancy or childbirth entails a serious threat to the woman's life; (ii) when serious abnormalities or malformations of the fetus constitute a serious threat to the physical or mental health of the mother; or (iii) when the pregnancy is the result of rape.

In other areas where Catholic and non-Catholic ethics conflict, such as assisted reproduction, no agreement has been reached in parliament on a draft of a proposed law, leaving Italy without any regulation on this delicate matter. While awaiting a legislative solution, the Code of Medical Professional Ethics (http://www.fnomceo.it/) has issued rules to guide physicians in their everyday practice. In the section devoted to "Sexuality and reproduction," this Code introduces limitations on the use of assisted reproduction techniques: these are allowed only for stable heterosexual couples when the woman is pre-menopausal, and are explicitly forbidden for surrogate mothers or when the husband is deceased. Thanks to the absence of law in this field, private practice has thrived and grown considerably, with the virtual absence of any form of control whatsoever.

Quite recently (January 2001), the Ministry of Health has tried to regulate research on cloning for therapeutic purposes, by setting up an expert commission. The commission has not been able to produce a commonly agreed-upon document. In fact, the jointly written proposal was later rejected by some of the Catholic experts.

As far as genetic testing is concerned, in 1998 Italy developed a comprehensive document entitled "Guidelines for Genetic Testing," which was officially approved in

1999. This document was developed by a group of experts appointed by the Presidency of the Ministers' Council-Scientific Committee for the Risks Deriving from the Use of Biological Agents, and was co-ordinated by the Istituto Superiore di Sanitá (National Institute of Health). The document contains very general guidelines concerning the quality of genetic tests as well as their management, in terms of provision, counseling, informed consent, confidentiality, and access. Particular sections are devoted to prenatal diagnosis and tumor risk assessment. Working groups on best practice guidelines for genetic tests for specific disorders also operate within the Italian Society of Human Genetics (available in Italian only at http://sigu.uniroma2.it/sign/double.htm). A doment on genetic tests has also been prepared by the National Committee on Bioethics, reaffirming, with some ambiguities, the basic principles of the individual's autonomous choice and of clinical benefits of test results (available in Italian only at http://www.palazzochigi.it/bioethica/Temi_problemi/bioetici.htm).

No advisory committee specifically guiding government policies and regulations on genetic testing is operative in Italy. Confidentiality of genetic information is, instead, regulated by law, namely the "Protection of individuals and other subjects with regard to the processing of personal data" Act No. 675, 1996. It states that any processing (i.e., "any operation ... concerning the collection, recording, organization, keeping, elaboration, modification, selection, retrieval, comparison, utilization, interconnection, blocking, communication, dissemination, erasure and destruction") of sensitive data, including genetic data, should be authorized by a Data Protection Authority. The law and the subsequent acts are available at http://www.privacy.it/, both in Italian and English.

In 1995, the Professional Code of Physicians introduced the concept of informed consent as a requirement for any treatment (Codice Deontologia Medica, Art. 3, 1995). From an ethical point of view, it guarantees the individual's self-determination. From the point of view of medical professionals, promulgation of who were quite unprepared, this Code is a complex event bringing changes they are often unwilling to accept.

Marina Frontali and Gioia Jacopini

Spain

Spain has modern laws on issues related to reproduction and genetics that nowadays permit the practice of genetic counseling in a context of freedom.

Abortion

Voluntary termination of pregnancy was approved in 1985, in article 417 bis of the Código Penal (Penal Code).[1,2] Termination is allowed in three circumstances: "If the pregnancy may endanger the physical or psychological health of the mother" (no limit on gestational age), "in case of pregnancy as consecutive to a reported rape" (until the 12th week of gestation) and "presumed severe physical or mental handicap for the fetus" (until the 22nd week). For this "eugenic abortion" a certificate from two qualified physicians is required. Application of prenatal diagnosis followed by abortion has made possible the reduction of birth rates for Down syndrome and neural tube defects. Although self-determination of abortion is an option not legally recognized for the mother, in practice the "risk to the psychological health of the mother" is used as a way to obtain elective termination apart from legal or medical causes.

Genetic Manipulation
The new Spanish "Código Penal" approved in 1995 includes some articles related to genetic manipulation, confidentiality, and reproduction (Código Penal Español, 1996).[3] Genetic manipulation is regulated by articles 159–162, punishing with prison the performance of procedures that are not oriented to cure or treat a disease and that are imprudent. Human cloning is also forbidden and fertilization of human ova for purposes other than human procreation is not allowed.

Confidentiality
Confidentiality is also protected. The use or disclosure of personal, familial or clinical data without an individual's consent is forbidden.

Assisted Reproduction Laws
Since 1988, several laws and decrees have been promulgated to regulate every aspect of present and future artificial reproduction. These laws and regulations are as follows.

Law on techniques of assisted reproduction (35/1988).[4] This law regulates assisted reproduction using artificial insemination with either husband or donor sperm, *in vitro* fertilization (IVF), and intratubal transference of gametes (GIFT). It allows donation of gametes and pre-embryos and regulates the legal consequences for the children (filiation, anonymous paternity, access to all information about parents except identity). A 1996 law[5] establishes a National Register of Donors of gametes and pre-embryos and sets forth legal conditions and mandatory clinical protocols for donors as well as recipients. It is forbidden to fertilize human eggs for purposes other than human procreation. It defines the pre-embryo status and the application of these reproductive techniques to prevent or treat inherited diseases. Genetic manipulation for race, for sex selection or to change non-pathological characteristics is forbidden, as is human cloning.

Real Decreto 415/1997 creates the National Committee on Human Assisted Reproduction.[6]

Carmen Ayuso, Carmen Ramos, and Isabel Lorda

References

1. La indicación eugenésica. C. Ayuso, C. Ramos, T. Pampols, A. Ribes, J. Diaz-Recasens, y Carmen Echevarría. En: La reforma del delito de aborto. Estudio del nuevo artículo 417 bis del código penal. Análisis jurídico y médico. Tomo IX JL Diez Ripollés. Colección "Comentarios a la legislación penal" Revista de Derecho Privado (ed.), Editoriales de Derecho Reunidas. EDERSA, 1989.
2. Delito de Aborto. Código Penal Español (1995) Artículo 417 bis. Manipulación genética y reproducción. Arto. 159–162. Aborto. Artco: 144–146 y 417 bis. Violación del secreto profesional. Artco: 197–199. Ley Orgánica del Código Penal Español. 24 Mayo, 1996.
3. Manipulación Genética. Código Penal Español, Título V, libro II, artículos 1995: pp. 159–162.
4. Técnicas de reproducción asistida. Ley 35/1988 de 22/11/1988 (88/27108). Protocolos estudio donantes y usuarios, técnicas de reproducción humana asistida y registro nacional de donantes de gametos y preembriones. Real decreto 412/1996 de 01/03/1996.
5. Requisitos de homologación y autorización de Centros y Servicios sanitarios para técnicas de reproducción asistida. Real Decreto 413/1996 de 01/01/1996 (96/06645).
6. Comisión Nacional de Reproducción asisitida. Real Decreto 415/1997 de 21/03/1997 (97/06157).

EASTERN EUROPE

Czech Republic

The Health act was enacted in 1966 and is still effective today, with only a few small amendments. This act states that everyone has the right to health care; however, the act restricted patients from making decisions about their health care, and medical professionals were given this responsibility. The physician was obliged to inform the patient and family appropriately in order to win their trust and cooperation during examinations and treatments. In the past, health care consumers accepted their subordinate position, but consumers' desire for full information is increasing. The Health Act of 1966 created an atmosphere of medical camouflage, concealment, or "white lies," and even today many physicians have retained their paternalistic attitudes toward patients and their families. The truth about cancer diagnoses is sometimes not revealed to the patient because this information is considered highly detrimental to the patient's mental health.[1] The option of a second opinion concerning the treatment strategy is usually not offered to the patient.

Patients' rights were described briefly in the Ethics Code in 1992, but this is not a legal document. The Ethics Code of the Czech Association of Physicians recommends a patient–physician relationship without a paternalistic attitude, but also accepts the possibility of retaining information in the case of an unfavorable prognosis. In the future it will be necessary to create amendments to medical legislation and to establish ethical codes and guidelines. The privatization of health services, the patient's right to choose a health care provider, and the right to seek justice in cases of treatment that results in injury will definitely bring more awareness of ethical issues to physicians.

In 1957, abortions were legalized, and until 1986, special abortion committees evaluated each application for an abortion individually. According to the new law passed in 1986, all women have the right to terminate their pregnancies up to the twelfth week.[1] This procedure is inexpensive and paid for by the patient. The pregnancy can be terminated until the 24th week if there is a severe defect in the fetus's development, high risk (more than 10%) of a severe genetic disease, or exposure to teratogenic agents during pregnancy. There are some situations when pregnancy can be terminated even after the 24th week (lethal or very severe complications of fetal development or the woman's life is in danger). The rights of women to make free decisions about their pregnancies are protected by law; however, not much publicity is given to this issue. There are some "pro-life" movements, but their policy is to educate people about the risks of abortion, to increase responsibility in people's decisions and behavior and to help pregnant women in troubled situations. Religion does not play as important a role in the abortion issue as in Western European countries or the United States. The status of the fetus is not specified in law and is not discussed among the public or among medical professionals. Primary consideration is given to the physical and psychological health of women, rather than the status of the fetus.

Laws concerning gamete donation do not exist, but in 1982 a memorandum was created which stated that sperm could be donated only for medical reasons and used only if both husband and wife agreed. If the risk of severe genetic disease is greater than 10% (autosomal dominant, recessive, polygenic, or chromosomal), then donated sperm may be used for artificial insemination. The Central Ethical Committee decided that donation of eggs should be considered equal to sperm, and the same regulations apply.

Jana Zidóvska and Lenka Foretova

Reference

1. Stolinova, *J. Citizen, physician and the law* [Czech], Avicenum, Prague 1990.

Hungary

Abortion is permitted for social and medical reasons before the twelfth week of gestation. Among medical indications are those of genetic-teratogenic nature if the risk is 10% or more for severe, untreatable fetal defects and genetic disorders that are not diagnosable prenatally. About 5% of induced abortions are performed for medical indications, however; most are carried out for 'social' reasons. So-called elective abortions are allowed before the twentieth week when there is a prenatally diagnosed severe defect or disorder or a 50% risk for a severe X-linked disorder. Abortions may be performed at any gestational age when a lethal disorder, incompatible with postnatal life, is diagnosed prenatally.

Andrew E. Czeizel and Laszlo Timár

Poland

Legal regulations affecting the practice of medical genetics only concern abortion laws. The abortion law implemented in 1952 was not restrictive; pregnancy could be terminated until the 12th week of gestation if social reasons existed. In practice this meant abortion on demand. In 1993, following fierce debates in the parliament, a new, more restrictive law was introduced.

According to this new law, a prenatal test (invasive) could be performed when:

- "the conceived child belongs to a family at increased risk"
- "there is a suspicion that the fetus is affected with an incurable genetic disease"
- "there is a suspicion of a severe disease in the fetus."

Pregnancy may be terminated if "prenatal test(s) – certified by two physicians (other than the one performing termination) is indicative of a severe and irreversible condition in the fetus" (Dziennik Ustaw No. 78, 1993).

On August 30, 1996, a new bill was passed in the Polish parliament and was implemented in January, 1997 (Dziennik Ustaw No. 139, 1997). The following are the articles relevant to prenatal diagnosis:

- "State and local administration authorities are obliged to guarantee free access to information and prenatal diagnosis, particularly in cases of increased risk, suspicion of a genetic defect or incurable disease of the fetus;"

- "Termination of pregnancy may be carried out by a physician if prenatal test(s) or other medical data indicate that there is high risk of a severe and irreversible disorder or incurable life-threatening disease in the fetus;"
- "Termination is permissible as long as the fetus is incapable of survival outside the mother's body."

As can be seen, the law of 1997 is more liberal than that of 1993. Its implementation, however is, not always easy. A good example of this difficulty occurred in a department of medical genetics in a large university town, which recently had to stop performing prenatal tests because gynecologists at the university hospital refused to collaborate, on account of their opposition to termination in cases of positive prenatal tests.

One of the most important and controversial problems in Poland is the attitude towards induced abortion. Table 1 represents attitudes of 150 persons involved in medical genetics as compared to those from a representative sample (1209 persons) of the general population (CBOS, 1996). As Table 1 shows, the group of medical geneticists is slightly more permissive for social indications for abortion and for abortion of affected fetuses. There is complete agreement between the two groups when maternal health or maternal life are in danger.

There is no legislation in Poland relevant to *in vitro* fertilization (IVF), artificial insemination by a donor (AID), egg donation, surrogate motherhood, pre-implantation diagnosis with embryo selection, or sterilization. In the absence of legislation permitting these procedures, IVF and AID are performed in a few centres. Legal regulation of these above named procedures would help considerably in the provision of genetics services.

Jacek Zaremba and Elzbieta Fidzianska

Reference

CBOS. Center of Studying Public Opinion: Attitudes toward legal regulations of abortion and possible change of law, Report of the Survey, Warsaw, April 1996, p. 6.

Table 1. Percent Agreeing that Abortion should be Legal*

Indication for abortion	Medical geneticists (n = 150)	General population (n = 1,209)
On demand	16	13
Social	60	50
Maternal health in danger	77	78
Maternal life in danger	89	88
Fetus affected	85	65
High genetic risk of fetus being affected	66	n.a.

*Representative sample; data of the Center for Studying Public Opinion (CBOS), March, 1996.
n.a. question not asked.

Russia

The people of Russia, including both laypeople and medical professionals, are not, as yet, well accustomed to medical genetic services and modern procedures. Ethical and legal aspects of these services are now *in statu nascendi*. Up to now, only a few ethical and legal items appear in Russian Federation Basic Legislation on National Health Care and Ministry of Health decrees regulating medical genetic services. However, due to current progress in reproductive and genetic technologies, the legal and ethical implications are being widely discussed in Russian society and by professionals and authorities. The Russian Federal Law, regulating the handling of non-human genetic material and activities in genetic engineering in general is already in place. This document considers mainly the problem of biological safety of genetic technologies for personnel, population and environment. Draft bills concerning human reproductive rights and the bioethical regulation of applications of modern biomedical technologies are now at the stage of professional and social review. The contents of these drafts include family planning, *in vitro* fertilization, surrogate motherhood, adoption, prenatal diagnosis, abortion, gene therapy, cloning, etc. Since these are still drafts, it seems premature to discuss them in detail.

Vladimir Ivanov and Vera Leonidovna Izhevskaya

NEAR EAST

Israel

Abortion for genetic reasons or after the detection of severe malformations is legal in Israel. The law establishes that the abortion is to be performed in a public hospital and authorization given by a commission of three professionals, including a social worker. One of the three must be a woman. After receiving genetic counseling, a family that decides to interrupt the pregnancy fills out a comprehensive form, and the case is presented to the commission.

For the last several years, requests for pregnancy terminations that were out of the ordinary, including interruptions after 23 weeks gestation, have been dealt with by a special committee that functions in a limited number of hospitals.

Surrogate motherhood is still under discussion in Israel, and individual cases are dealt with on a case-by-case basis.

Zully Gelman-Kohan

Turkey

There is no body able to regulate services conducting genetic testing or applications of biotechnology.

Since 1983, abortion for social reasons has been legal until the 10 week of gestation with consent of the couple, not only the mother. Among respondents, only 30% agreed that a woman's abortion decision should be her own. On the other hand, there is no time limitation on medical abortions. These facts are not widely known or openly discussed among geneticists in Turkey.

In vitro fertilization is marketed for infertility, and there are regulations prepared by the Ministry of Health. IVF can only be used for married couples. Gamete donation or embryo transfer to a surrogate mother is prohibited. Regulations put some restrictions on preservation and destruction of extra embryos. On the other hand, there are no regulations for preimplantation diagnosis.

After the Turkish Medical Association's and National Genetics Society's declarations, the Higher Council of Health prohibited sperm separation for preconceptional sex selection in the absence of X-linked diseases. However, it may still be going on illegally. There is always the possibility of illegal procedures such as usage of sperm from a donor or use of imported sperm. In the very near future preimplantation genetics will take place in Turkey. There is a great need for a law to regulate it.

Isik Bökesoy and Berna Arda

LATIN AMERICA

Argentina

Argentine laws do not explicitly provide for medical procedures involved in human reproduction. Neither do they refer to genetics. Only very recently have public hospitals developed programs for family planning. "Genetic" abortion is not authorized under current laws. The codes that are in force at present do not consider the interruption of pregnancy as a crime if the patient is an insane woman who was raped.

There are no laws on embryo cryopreservation, egg donation or surrogate motherhood, although some sectors of Congress (the House of Representatives and the Senate) are currently studying several bills that will regulate these aspects of health care. There are multidisciplinary groups discussing these bills with legislators of different political parties.

Argentina's population is mainly of European and Catholic origin. It is the result of many immigration waves, mainly from Spain and Italy. The Jewish population, descendants of Central European immigrants, has a very important say in all aspects of our national life.

The laws that regulate the procedures involving human reproduction should take into account these traditions, in order not to affect feelings that express a true national culture. There is agreement to respect the embryo and the fetus at any gestational age.

Brazil

The following legal documents regulate aspects of reproduction and genetics in Brazil: (a) The Brazilian Constitution, enacted on October 27, 1988; (b) The Penal Code, promulgated on December 7, 1940; (c) The Code of Medical Ethics, established by the Federal Council of Medicine on January 8, 1988; (d) The Resolution on Assisted Reproduction, enacted by the same council on November 11, 1992; (e) The Biosecurity Law, promulgated by the Congress on November 24, 1994; (f) The Law of Family Planning, published on January 12, 1996; (g) The Patents Law, enacted by the Congress on May 14, 1996; and (h) The Resolution on Research Involving Human Subjects, approved by the National Health Council on October 10, 1996.

These documents establish the following: (a) Abortion is prohibited, except when the mother's life is in danger or in cases of rape (article 128 of the Penal Code). (b) It is forbidden not to use all the means of diagnosis and treatment available in favor of patients (article 57, Code of Medical Ethics). (c) It is forbidden to abbreviate the lives of patients, even if requested by them or by their legal guardians (article 66, same code). (d) Individual (voluntary) or group induction of surgical sterilization is prohibited (Family Planning Law). (e) In relation to assisted reproduction, selection of sex is forbidden; the number of oocytes and pre-embryos transferred to the uterus of a woman having IVF cannot be higher than four, and if multiple gestation occurs it is prohibited to perform selective elimination of excess embryos; also prohibited is the commercial donation of eggs, or paid surrogate motherhood; the maximum time of *in vitro* pre-embryo development is set at 14 days; the destination of unused cryopreserved embryos should be established in advance by the couple involved; and a temporary donor of a uterus should be biologically related to the genetic donor of the embryo up to a second degree relationship. (f) The manipulation of human germinal cells is forbidden (Biosecurity Law). (g) No patent can be applied to the whole or parts of living beings, except genetically manipulated microorganisms (Patents Law). (h) Research on humans is strictly regulated in accordance with international standards (National Health Council Resolution).

Since 1993, the Senate has considered liberalization of abortion (a measure introduced by Senator Eva Blay), including its legalization in cases of affected embryos. Attempts have also been made for a complete reevaluation of the Civil and Penal Codes, so far without success.

Specific Examples of the Laws' Implementation

Several instances of situations that involved the application of these laws and of ethical principles have been documented in the journal Bioética, published by the Federal Council of Medicine since 1993. They included: (a) a patient with the autosomal recessive form of muscular atrophy known as Werdnig-Hoffmann disease. After repeated interventions with pulmonary problems, her parents requested the interruption of medical care. The request was not accepted (Kipper and Hossne, 1993); (b) a judicial mandate for the operation on a patient with Down syndrome who had duodenal atresia, independent of parental approval; (c) the authorization, by judge Miguel Kfouri Neto, of the abortion of an anencephalic fetus (Barbosa de Deus and Dallari, 1993); (d) the denial of a request for sterilization of a patient with severe mental retardation, and of separation of Siamese twins (França, 1993); (e) the prohibition of requests for pregnancy tests for the admission of women to employment (França, 1994).

Francisco M. Salzano and Livinia Schüler-Faccini

References

Barbosa de Deus, B. and Dallari, S.G. Bioética e direito. Bioética 1993; 1: 91–95.
França, G.V. Pareceres. Bioética 1993; 1: 195–198.
França, G.V. Pareceres e resoluções. Bioética 1994; 2: 101–103.
Kipper, D.J. and Hossne, W.S. Caso clínico. Bioética 1993; 1: 85–89.

Chile

The most recent official text of the Chilean Sanitary Code (Decree No. 967 of July 1992) does not contain specific regulations to control genetics services. The Law No. 18173, included in the Sanitary Code, regulates organ transplantation, but it does not restrict gamete donation. Article 199 of the Sanitary Code clearly states that "It is forbidden to execute any action whose objective will be to produce an abortion". This sentence complies with the Constitution of 1980.[1]

According to this legislation, in Chile it is forbidden to perform induced abortions and research involving the human embryo. Preimplantation diagnosis and surrogate motherhood have not yet been legally defined.

Ricardo Cruz-Coke

Reference

1. Republica de Chile. Codigo Sanitario, Edit Jurídica, Santiago, 1993.

Mexico

Abortion on the basis of a fetal defect is illegal in most Mexican states except for Quintana Roo and Véracruz, and very few public or private hospitals are involved in prenatal diagnosis programs. On the other hand, there are no laws regarding gamete donation, research on human embryos, or preimplantation embryo selection.

Rubén Lisker, Salvador Armendares, Alessandra Carnevale, and Osvaldo Mutchinik

ASIA

Thailand

There is no abortion law in Thailand. The decision depends on the doctor's judgement. There is no time limit for abortion. Referral is very rare.

Recently, the government has issued a law supporting abortion of fetuses with medical or genetic problems. There is no law about gamete donation at the present time, since this is quite a new technology here and there are very few services. There are national and university ethical committees to monitor research on embryos.

Valairat Dhamcharee, Apwat Mutirangura, and Orasri Romayanan

CHAPTER 12

CHINA: VIEWS OF A BIOETHICIST

Ren-Zong Qiu

This chapter consists of two parts. The first part describes a brief history of the development of genetics in China, within the Chinese cultural context, and the laws that affect genetic practices. The second part addresses ethical issues in genetic practices, including access to and distribution of genetics services, values and presumptions underlying the use of genetic knowledge, voluntary versus mandatory screening and testing, directive versus non-directive counseling, privacy and confidentiality in genetics services, and eugenics versus healthy birth. Hopefully, the paper will help to explain in part the differences between the responses from Chinese geneticists and those of their Western colleagues.

BACKGROUND

First we will provide some basic facts about China. The area of China is 9.60 million square km. The population of China is 1.2 billion, about 22% of the world's population, with 7% of the world's arable land. At present, annual births are still 40 million, and the net growth of the population is 14 million per year. Each year, one quarter of the increase in national income has to be spent on the costs of the population increase. Each year 20 million people are added to the labor force. Currently there are 190 million unemployed and underemployed people in rural areas. The GNP per capita was about US$ 800 in 2000. 300,000 hectares of arable land disappear each year, and there is soil erosion in one-third of the arable land. 85% of fuel is coal, which worsens environmental pollution. The reasons for controlling population are the following: in the next few years, the average increase in GNP will be 8–9%, the population will reach 1.3 billion, the natural growth rate will be below 1.00%, and women's total fertility rate will be below 2.00. These basic facts will help readers to appreciate the challenges and the pressures facing the Chinese people.

A Brief History of Genetics in China

Although Chinese had observed in earlier times that some human traits are inherited, the science of genetics could not develop in the Chinese cultural context with its unique cosmology, which is based on a conceptual framework of *Yin* and *Yang*. One Chinese adage says: "Dragon gives birth to dragon, phoenix gives birth to phoenix, and a thief has a son who is a burglar." Traits that are inherited and those that are not inherited cannot be distinguished by daily experience. During five thousand years of her history, China has developed her own unique culture, the basis of which consists of three main teachings, which are Confucianism, Taoism, and Buddhism. In Chinese culture, there has developed a unique cosmology, according to which the universe consists of *qi*, a vital energy or a physico-psychological entity, and all things or beings in the universe are nothing but the different patterns of *qi* (vital energy). *Qi* exists in two forms: *yin* and *yang*. The *yin-yang* dualistic theory emerged from the worship of sex and reproduction. When ancient Chinese observed that all human creatures come from the woman's vagina after sex, they tried to make an analogy of the universe (macro-world) with human beings (micro-world). They inferred from the observation that there is an intercourse or interaction between *yang* (male) and *yin* (female) in the universe and all things in it come from "a mysterious gate." This belief is reflected in the sentences of *Dao De Jing*, such as "the mystery of mysteries is the gate of all wonders", and "the gate of the mysterious female is the root of the world." For the Chinese, the universe is an organism which is capable of creation, reproduction, and transformation. The concept of *yin-yang* interaction might explain some facts about human inheritance of traits, but scientific knowledge of genetics could not and did not develop from this metaphysical concept.

Before 1949, knowledge of genetics in China was basically introduced from the West. Many Chinese geneticists had studied in Britain, the US, and other Western countries in the 1920s, 1930s, and 1940s. Some had studied and conducted research in laboratories directed by American geneticist Thomas Hunt Morgan (1886–1945), who discovered X-linked disorders. Although generally apolitical, Morgan was a prominent critic of the eugenicists, arguing that behavioral differences between social groups probably stemmed from environment rather than heredity. After their return to China, most of Morgan's students taught in universities, where they carried out research. In 1949, research in genetics was minimal. Although research groups within the Academy of Sciences had undertaken related research, there was no institute specializing in genetics. Although university students could study the subject, no department offered it as a major. During the period of time between July and August 1948, at the meetings of the Academy of Agricultural Sciences and Academy of Sciences of the Soviet Union, the Morganist approach was labeled as a reactionary idealist pseudo-science and the Michurinist approach (I.V. Michurin was a Soviet fruit horticulturist) was labeled as progressive and materialist. Michurin believed that environmental shocks could make the hereditary constitution receptive to change by "assimilating" external conditions. Thus winter grains could be converted into Spring grains by planting them in a different season. On August 26, 1948, the Academy of Sciences of the Soviet Union adopted a resolution in which the Academy closed down the department of cell genetics and determined to eliminate all influences of the Morganist approach. In China at that time, only a few geneticists disagreed with the Morganist approach. The overwhelming majority of Chinese geneticists remained unconvinced and resented the Soviet approach to genetics and its wholesale transfer to China. On June 29, 1952 an article entitled "Struggle to Persist in the Michurinist Direction in Biological Science" was published in *People's Daily*, the official organ of the Chinese Communist Party. Echoing the position that then prevailed in the Soviet Union, the article argued that "Michurinist biological science is a great achievement of the conscious, thorough application of Marxism and Leninism to biological science ...[and] represents a fundamental revolution in biology." It totally repudiated the Morganist school of genetics as "reactionary" and "idealist." The article called for "using Michurinist biology to thoroughly transform various disciplines of biology," and specifically pointed out that "not only the old genetics, but also other biological disciplines, including the old ecology, cytology, embryology, and microbiology... must be thoroughly transformed." After this article was published, departments in colleges and universities and other institutions involved in biological research and study made plans to implement these directives. Beginning in the Fall of 1952, the universities basically ceased offering courses on Morganist genetics, and research that was directed by Morganist theories was forced to stop. Even middle school textbooks were revised according to the guiding ideology of Michurinist science. Academic journals published only articles supporting the views of the Lysenkoist school. Lysenko (1898–1976), an agronomist, believed in the inheritance of acquired characteristics, an idea that he got from Michurin. He rejected Mendel's theories. Lysenko believed that mathematics could not be applied to living beings, and eliminated all control groups and statistical analyses from his experiments. He was politically astute, and appeased the Soviet government for many years by promising high yields in agriculture and by fabricating results. His influence waned with the fall of Kruschev and he was dismissed from his posts by the Soviet Government in 1965. One difference from the situation in the Soviet Union was that Chinese geneticists remained in their institutions, engaged in teaching or research in other areas. This situation persisted until the 1956 Qingdao Conference, although different voices could be heard from time to time. For example, in December 1953, an article that presented Soviet criticism of Lysenko's views was published in *Science Bulletin*, a journal under the auspices of the Chinese Academy of Sciences, and in 1955 a senior Chinese botanist, Hu Xiansu (1894–1968), criticized some of Lysenko's views in his book, *A Short Course on Plant Taxonomy* (Li *et al.*, 1995).

The Qingdao Conference on Genetics was held in August 1956, cosponsored by the Chinese Academy of Sciences and the Ministry of Higher Education. About 130 people participated in this conference, and geneticists of all major approaches, including those who favored the Morganist approach as well as those who favored the Michurinist approach, took part. A total of 56 people spoke, some individuals as often as seven or eight times. This was the first time, after several years of dispute, that advocates of both schools sat down together and jointly participated in academic discussions. The Morganists boldly stated their views with perfect assurance, and Lysenkoists, while maintaining their own opinions, also criticized some of Lysenko's errors. Guang-Yuan Yu, then Director of the Science Section, Ministry of Propaganda, Chinese Communist Party, a famous Chinese economist and philosopher of science (Yu, 1995), elaborated on the importance of distinguishing between academic and political issues, announced the wholesale removal of various political labels imposed on the Morganist school of genetics, and advocated that different academic opinions be resolved through free discussion and scientific practice. Problems which could not be resolved immediately could wait. Conclusions should not be rushed, and one should learn from the strong points of the opposition (Li *et al.*, 1995).

In the early 1960s, there were articles published in *Chinese Medical Journal, Chinese Journal of Internal Medicine* and *Chinese Journal of Obstetrics & Gynecology* describing the results of cell genetics in China, such as detecting congenital abnormalities with chromosome and sex chromatin, the abnormalities of sex differentiation, and the karyotype changes in malignant cancers. A Chinese geneticist, Shou-Yuan Zhao of Fudan University, pointed out in 1983 that attention should be paid to the following aspects of the practical application of medical genetics: genetic screening, prenatal diagnosis, and genetic counseling, and that in the future the development of medical ethics can help to "heighten the quality of the population, and improve the genetic quality of the people." The urgent tasks facing medical genetics would be heightening intelligence, increasing physical strength and life expectancy of all Chinese people, and decreasing the morbidity of all kinds of genetic diseases. He argued that the long-term goal of medical genetics is to make human beings better individuals mentally and physically by modifying prenatal or postnatal individual development. He also claimed that medical genetics can help to cure incurable diseases and benefit mankind (Zhao, 1983, pp. 9–10). In the late 1980s, some labs equipped with advanced appliances were set up under the sponsorship of the National Commission of Family Planning and Ministry of Public Health, and their results began to be applied in clinics. In 1984, the journal *Genetics and Diseases* was established; in 1992, its name was changed to *Chinese Journal of Medical Genetics*. In 1992, National Number 863 Programs began to sponsor projects on gene therapy for cancers. In October 1992, the First International Conference on Gene Therapy was held in Beijing. In 1994, a National Conference on Genetic Diagnosis, Therapy and Prevention was organized by the Editorial Board of the *Chinese Medical Journal* and the Chinese Academy of Medical Sciences (Wu, 1993, 1995). Currently there are more than 400 medical geneticists in China who are working in internal medicine, gynecology and obstetrics, pediatrics, psychiatry, or other departments in general hospitals, departments of medical genetics in some medical schools, institutes of maternal and child care, and institutes of science and technology in family planning in large cities (Qiu, 1995).

Wilson Lo, a Chinese geneticist who had been educated in the medical school of Johns Hopkins University, wrote that the following figures show the increase of genetic diseases in China:

About 1.3% of newborns in China suffer a congenital abnormality, among which 80% are due to genetic diseases. Congenital abnormality is the main factor causing the deaths of 30–40% of newborns.

There are about 50 million persons with disabilities in China; about 20–30% of these disabilities are caused by genetic diseases.

Among children between 7 and 14 years old in China, there are about 1 million whose IQ is below 50; 50% of this retardation is due to genetic diseases.

There has been some change in the spectrum of diseases in big cities. In the pediatric wards of general hospitals in these cities, 20% of in-patients are victims of genetic diseases (Luo, 1995, p. 44).

The National Natural Science Foundation of China, the country's main funding agency for basic research, launched a national human genome project. It named Professor Bo-Qin Qiang, a Member of the Chinese Academy of Sciences, Vice-President of the Chinese Academy of Social Sciences/Peking Union Medical College, and Professor Chu Chen, Shanghai 2nd Medical University as its coordinators in

1993 as part of its contribution to international efforts to sequence the human genome. The Chinese Human Genome Project mainly aims at: (i) development and improvement of techniques for genome research; (ii) collection and storage of genetic samples; and (iii) analysis of some particular disease-related loci in Chinese genomes. In addition to the Han nationality, which is the largest gene pool in the world, China has more than 55 ethnic groups with a total population of 91.2 million (1990), or 8.04% of the whole population of China. This project is also designed to study genetic diversity of the different ethnic groups and to find out the frequencies of main alleles at specified loci, and differences between them. The scientific goals of the genetic investigation of ethnic groups are to determine the types and incidences of genetic diseases and to preserve the genetic materials of diverse groups. Cancer-related genes in several cancers, such as esophageal cancer and liver cancer, have been localized or cloned. Progress has also been made in several chromosomal regions, such as the Duchenne muscular dystrophy region, the region around the fragile X, and chromosome 17 (Qiang, 1996). Since 1996, the Centers for the Human Genome Project, Chinese Academy of Science, Northern Center for Human Genome Project in Beijing and Southern Center for Human Genome Project in Shanghai were successively established. In 1999, Chinese scientists joined an international Human Genome Project and shared 1% of the whole work in mapping and sequencing of the human genome in collaboration with scientists from the US, Britain, France, Germany, and Japan.

Cultural Context in Which Genetic Practices Developed in China

The practice of medical genetics, including geneticists' treatment and clients' responses, have been greatly affected by cultural factors and the moral ethos in China.

Filial Piety, Sexism, and Pro-reproductionism

In Chinese culture, Confucianism is dominant. Confucianism is mainly an ethics or philosophy of morality that teaches us how to be human. We were all born as humans biologically, but not morally. So we have to learn how to be human morally. Confucian ethics are not only elaborated in many texts and their interpretations, but are also embodied in social conventions and institutions. There are a number of principles in Confucian ethics that regulate interpersonal relationships, such as: filial piety in the child–parent relationship, fraternity in the sister–brother relationship, trustworthiness in the friends' relationship, and kindheartedness and benevolence in the physician-patient relationship. Among them, filial piety is very important. The second greatest Confucianist, Mencius, said "There are three vices that violate the principle of filial piety, and the biggest is being without offspring" (*Men Zi*). One of the requirements of filial piety is to extend the ancestors' life. But for Chinese, only boys, not girls, can do this. From the principle of filial piety follow the values of respecting the elderly, and also the ethos of sexism and emphasis on reproduction. The social and moral status of women has always been lower than men's. Before 1949, women in the lower classes had no names of their own; the "name" used to be three Chinese characters, which meant whose daughter the woman was and to whom she was married. Women used to be treated as machines of reproduction or as sex tools. When a dynasty was in decline, some woman was always selected to be blamed for it. Infertile couples without a child experienced great psychological or familial pressure, and childlessness was also blamed on the woman. Producing no children could be a reason for divorce by the husband, and her status in the family or the community would be greatly lowered. This pro-reproductive and sexist view explains why many Chinese with genetic diseases are not willing to be tested or screened, or to receive follow-up care by geneticists.

Status of the Fetus and Healthy Birth

Under Confucian teachings, the Chinese hold a hierarchical view of the value of life. The human fetus has a value, but it has never been treated as a person. In approximately the third century B.C., the third greatest Confucian, Xin Kuang said: "Birth is the beginning of a person, and death is the end of a person. If one has a good birth and a good death, then he fulfills the *tao* of mankind" (*Xun Zi*). Since then, this view has become a conventional Chinese view. Traditionally, the mourning period after a person's death continued for seven weeks, but a dead fetus was never mourned. When an abortion, spontaneous or induced, took place, the Chinese never said that a person died. As for the phrase "good birth and good death," it can be interpreted both morally and medically. Morally speaking, a person should do good in his life time. Medically speaking, a person should have a healthy birth and a peaceful death. Modern Chinese interpret the word "eugenics" as "good birth" or "healthy birth" without any knowledge of the tragic history of eugenics in Nazi Germany. However, when the idea of "good birth" is combined with collectivist social and political philosophy, the idea of a eugenics program easily emerges in the mind of some geneticists (Qiu, 1993, 1994, 1997).

Collective-oriented Social and Political Philosophy

Confucianism is an ethics that teaches people how to be human. For Confucians, the demarcation between a human and an animal lies in whether you care for others or you care only for yourself. The basic concept of Confucianism is *ren*, which means "loving people" (*The Analects of Confucius*). The golden rules of Confucianism are: "You should not do to others what you don't want to do to yourself," and "After establishing yourself, you should help others to establish themselves; after you develop yourself, you should help others to develop" (*op. cit.*). In contrast to Western individual-oriented social and political philosophy, the Chinese focus has always been on the collective – family, community, society, country or nation. One reason is that the Chinese hold a different concept of the "person" from some Westerners. For the Chinese, a person is not as independent as some Westerners presume. No person can survive without support from others; so a person is a relational person or a person in relation. Metaphorically, a person is an atom in the West, but a person is a drop of sea water in China. For some Westerners, if an individual person does not exist, the whole world becomes meaningless. But for the Chinese a drop of water evaporates, but the sea will still exist. So the collective is more important than the individual in China. If there is any conflict of interests between the individual and the collective, the former should be subordinated to the latter. This is why the majority of Chinese respondents in the survey always gave priority to family, community, society, country or nation over individuals in their responses to many questions (Qiu, 1993, 1994).

Health as the Balance of Yin *and* Yang *and the Status of the Disabled*

Presently in China, traditional medicine still survives with a unique theoretical framework in which the influences of Confucianism, Taoism, and Buddhism are mixed. Health is the balance of *yin* and *yang*. Excess or deficiency of *yin* or *yang* will cause illness. Genetic diseases, too, can be explained by the imbalance between *yin* and *yang*. This imbalance may be caused by wrongdoing by the parents or even ancestors of those with genetic diseases, because the wrongdoing may result in the imbalance between *yin* and *yang*. So traditional Chinese thought that any genetic disease a newborn suffers was caused by *que de* (lacking virtues) of his/her parents or ancestors. For Chinese Buddhists, genetic disease is a retribution for wrongdoing in the newborn's previous life. Historically there have been negative or contradictory attitudes toward the disabled. On one hand, Confucianism's principle of *ren* and its ideal society, Great Harmony, in which all persons with disabilities should be taken care of, are helpful to the disabled. However, the concepts of *que de* and retribution have led to much discrimination against the disabled, and they have had great difficulty in getting support from society. The lives of persons with disabilities in China can be miserable, due both to these cultural reasons and to an underdeveloped economy. This can explain why the majority of Chinese respondents to the international survey gave a positive answer to the question of aborting fetuses with disorders or defects and always emphasized the value of delivering a healthy baby.

Medicine as the Art of Humaneness (yi ben ren shu)

In traditional medicine in general, and in Confucian medicine in particular, medicine is taken as the art of humaneness (*yi ben ren shu*). Here *ren* (humaneness) means loving people, caring for people, or doing good to people. For a traditional or Confucian physician, the responsibility is the same as a prime minister's – practicing *ren*. A Chinese adage says: "If you did not become a prime minister, then become a doctor." So the essential core of the relationship between physician and patient is doing good – beneficence. Doing good is the main standard for evaluating whether an action is obligatory or not. In Chinese culture, the concept of "right" is very weak. There is not even a word in the Chinese language which is equivalent to the English word "right." Westerners always ask "Is it right?" In contrast, Chinese always ask "Is it good?" These different ways of ethical thinking explain why the responses from the majority of Chinese respondents to the survey indicate that they are not very sensitive to the patient's or client's rights, (such as reproductive rights) but they are very sensitive to whether the action will bring benefit or harm to the patient–client, family and society. In Chinese medicine, bioethical principles of beneficence, respect, justice, and solidarity are widely accepted, but the principle of beneficence is given priority in most cases.

Doctors who practice medicine in China have the capacity to decide the life or death of a patient, so they bear a very heavy responsibility (Qiu, 1988). Traditional China is a patriarchal society in which paternalism is very strong in social life. For example, a magistrate used to be called "an official like parents," and paternalism is also strong in Chinese traditional medicine as well. In the encounters between physician and patient, medical decisions are made by the physician, and the doctor's opinion is directive. Although there has been a great change in modern China, and paternalism in medicine has been considerably weakened, it still prevails. This is why the majority of Chinese respondents to the survey tended to support "directive" counseling in the questions regarding directive versus non-directive counseling.

Confucian ethics as well as traditional medical ethics are mainly "virtue ethics." Chinese ethicists and doctors tend to think that a doctor should have virtues and cultivate them and that this is more important than norms or ethical standards. Confucius said: "If a man has no *ren, li* (norms) are useless to him." (*The Analects of Confucius*). So in its two thousand year history there has been no ethical code in Chinese medicine. Even now, the Chinese Medical Association has no ethical code for its members. In December, 1988, China's Minister of Health drew up and promulgated a brief ethical code for medical personnel. This code instructs them to "heal the wounds, rescue the dying, practice socialist humanitarianism;" to "respect the patient's responsibility and his or her rights, treat all equally without discrimination;" and to "respect the patient's privacy and to keep his or her confidentiality" (*Chinese Hospital Management*, 1988).

Contextualism and Balance of Values

Confucianism is not principlism. The moral judgment a Confucian makes is not a deduction from a principle. One reason is that for Confucians, a person is not an independent entity. On the contrary, a person exists only in an interdependent relationship with other persons, especially with other members of her/his family or community. This concept of person implies that when making a moral judgment and behaving according to this judgment, the context in which the moral agent exists has to be taken into account. When we take the context into account, we have to balance values that may be in conflict and all possible consequences of various options. In a sense, Confucian ethics is mainly an axiological ethics with a contextual orientation.

According to Confucian rules of etiquette in giving and receiving, men and women should not touch each other (*nan nu shou shou bu qin*). But if your sister-in-law is drowning, you should give a hand to her (*Meng Zi*). In the ethics survey, one Chinese respondent provides an example. A couple had no child for many years after their marriage. The family suggested that the husband's brother have two children, so the couple could adopt one of them, but the wife disagreed. She preferred to have sex with her brother-in-law, and had her own biological child. This is a case that may be called "surrogate fatherhood." The similar case called "borrowing a wife" exists too. In the latter, the husband of a couple without a child can have sex with another man's wife ("surrogate motherhood"), and have his own biological child. In both cases there is a balance of values in the family; a woman has a responsibility to uphold the principle of chastity, and yet she also has the responsibility to extend the ancestors' life. When they are in conflict, priority is usually put on the latter over the former. However, which value is given priority depends on one's preferences, values, and beliefs. This is why so many responses from Chinese geneticists vary. Although the majority gave priority to family, community, or society, there were still some respondents who gave priority to individuals.

Laws Affecting Genetic Practices

Since the founding of the People's Republic, many laws have been promulgated that are helpful to genetic services.

In the Chapter of Fundamental Rights and Duties, Constitution of People's Republic of China (PRC) promulgated on December 4, 1982, it is stipulated that "Citizens of the PRC are all equal before the law" (Article 33); "Citizens' personal freedom in the PRC must not be infringed upon" (Article 37); "Citizens' personal dignity in the PRC must not be infringed upon" (Article 38); "Citizens of the PRC have the right and duty to work" (Article 42); "Citizens of the PRC have the right of access to material support from state and society when they are old, diseased or disabled" (Article 45); "Citizens of the PRC have the right and duty to education" (Article 46); "Women of the PRC have equal political, economic, cultural, social, and familial rights with men" (Article 48). It is also stipulated that "A couple have a responsibility for family planning. Parents have the duty to rear their minor children, and children have the duty to support their parents. The violation of freedom to marry and the abuse of the elderly, women and children are prohibited" (Article 49), "When Citizens of the PRC exercise their freedom and rights, they may not do harm to the interests of the state, society, community, and others' legal freedom and rights" (Article 51). *(A Collection of Laws of People's Republic of China*, 1992 Edition, pp. 7–9.)

In the General Principles of Civil Law of the PRC promulgated on April 12, 1986 it is stipulated that minors, mental patients and other incompetents should have a guardian as their proxy in civil activities (Article 12–19). In this law there is a special section on "Personal Rights," including "Citizens enjoy rights to life and health" (Article 98); "Citizens enjoy the right to portrait; their portrait [photograph or likeness] may not be used for the purpose of profits without their own consent" (Article 100); "Citizens enjoy the right to fame, and citizens' personal dignity is protected by law" (Article 101); "The handicapped's legal rights are protected by law" (Article 104); "Women enjoy equal civil rights with men" (Article 105) (*op. cit.*, pp. 190–191, 200–201). Personal rights are interpreted by Chinese legal experts as including right to protection of body, health, life, name, portrait, fame, freedom and autonomy, privacy, marriage, and so on.

In the Law on Safeguarding Women's Rights and Interests promulgated on April 3, 1992 it is stipulated that "Women enjoy equal political, economic, cultural, social and familial rights with men. The state protects women's special interests according to law, and will gradually improve the system of social security for women. Discrimination against women, abuse and injury or killing of women are prohibited" (Article 2), the state guarantees women's equal political rights, and rights to culture and education, employment, to property, and the right to marriage and family with a man (Chapters 2–7) (*op. cit.*, pp. 69–73).

In the Law on Safeguarding the Handicapped promulgated on December 28, 1990, it is stipulated that "The handicapped enjoy equal political, economic, cultural, social, and familial rights with other citizens. The handicapped's civil rights and personal dignity are protected by law. Discrimination against and stigmatization, bullying or humiliation of the handicapped are prohibited" (Article 2); "The state should take measures to give the handicapped special support, to lessen or eliminate the impact of disability and external barriers, and to guarantee the handicapped the opportunity to exercise their rights" (Article 3) (*op. cit.*, p. 76).

In the Marriage Law promulgated on September 10, 1980, it is stipulated that "Marriage should be prohibited in the following cases: (1) Directly-related members of one's immediate family and indirectly-related members within three generations; (2) Those who have contracted leprosy without cure or other diseases which are considered as medically inappropriate for marrying" (Article 6); "Husband and wife are equal in the family" (Article 9); "Husband and wife have the right to keep his/her own surname and given name" (Article 10); "Husband and wife have the freedom to participate in production, work, study, and social activities; one side should not restrict or intervene against the freedom of the other side" (Article 11); "Husband and wife have a responsibility for family planning" (Article 12); "Husband and wife have equal rights to dispose of their common property" (Article 13); "Parents have the duty to rear and educate their children; children have the duty to support their parents. Infanticide by drowning or other acts to injure or kill infants are prohibited" (Article 15); "Children can take their father's surname or their mother's surname" (Article 16); "Children born out of wedlock enjoy equal rights with children born in wedlock. No one may discriminate against or do harm to them" (Article 19) (*op. cit.*, p. 373).

The controversial Law on Maternal and Infant Health Care, which was promulgated on October 27, 1994, will be discussed in the second part of this chapter.

ETHICAL ISSUES IN GENETIC PRACTICES

Access to and Distribution of Genetic Services

Genetic services have been provided to clients for more than ten years in China. However, genetic services are still not universally available, accessible, and affordable. Most clinics for prenatal monitoring and diagnosis mainly use genetic counseling, physical examination, and ultrasonography. Cytogenetics and biochemistry facilities exist only in municipal hospitals or family planning centers in large cities. More advanced molecular genetic techniques are only available in a few university hospitals or institutes for genetic research. In the survey, 15% of Chinese respondents estimated that genetic services in China are accessible to only 1% of those who need them, 11% estimated 5%, 24% estimated 10%, 12% estimated 20%, and the remaining 38% estimated variously; 74% thought that the highest quality genetic services are currently accessible only to those who can pay by themselves, and 26% thought that such services are accessible to all who need them regardless of the ability to pay. It is difficult to estimate accurately the accessibility of genetic services, but there is no doubt that there is a wide gap between demand and supply, between those who enjoy public health services free of charge or can pay for them and those who are unable to pay, and between those who live in big cities and those who live in remote or mountainous rural areas. There are about 60 million people who live in poverty and lack food, clothing, and basic health care. Others are vexed by skyrocketing medical costs. In view of the current disease spectrum and overpopulation in China, genetic services are not high on the agenda of priority, and adequate resources are not available for genetic services. This will not change in the near future. However, there is no discrimination with regard to sex, class, race, or ethnic group.

For Chinese, genetics is an art of *ren* too. So the state and society have a responsibility to provide genetic services which are accessible, quality-guaranteed, effective, and affordable to all people. However, in China or other developing countries with few resources this can only be achieved gradually. The state should have a program to develop genetic services step-by-step, but steadily, insistently, and progressively. The ethical issues in access to and distribution of genetic services include: (i) Just allocation of resources. In present circumstances, how many resources is it just to allocate to genetic services, in view of competing claims for funding from other health care services? (ii) Just distribution of genetic services. How can the genetic services now available be justly distributed in different areas so as to make them accessible to those who need them? (iii) Just payment system for genetic services. Should genetic services be paid for by public health care programs like other ordinary health care services? Or should they be paid for by patients or clients themselves or by third parties? Because China's health care system is now being reformed, public health care programs are near bankruptcy, and genetic services are at stake.

Values and Presumptions Underlying the Use of Genetic Knowledge

The values that can be promoted by the use of genetic knowledge include: treating genetic diseases, preventing genetic diseases, reducing the occurrence of non-genetic diseases where genetics may contribute to susceptibility, enhancing human traits, promoting individual and family happiness, saving resources for society, respecting the well-being of future generations. Any action using genetic knowledge is predicated on the presumption of dichotomies such as nature versus nurture, or biological determinism versus social determinism. With the development of human genetics and the success of the human genome mapping project, more weight seems to be put on biological determinism, or genetic determinism, which claims that all human diseases, traits and behaviors are determined by genes, and leaves no room for environment and an individual's free will. For some Chinese legislators or officials, human intelligence seems to be determined merely by genetic factors, and they want to prevent the births of children with congenital defects and to improve the quality of the population through genetics. However, they are ignorant of naturally occurring new mutations in the population. Genetic determinism also leads to the attempted medical solution of social problems. If all human behaviors are predetermined by genes, then not only risky behaviors, but also unethical and illegal actions can be explained by abnormal genes and can be corrected by gene therapy, but not by health education, moral education, or correctional institutions. Genetic determinism will also raise some ethical and legal issues about whether offenders should be held responsible or accountable for their unethical or illegal actions, because they have no choice or free will. Genetic determinism will produce over-expectations about genetic knowledge and over-ambitious genetic programs, and this in turn may lead to genetics becoming discredited as eugenics has been.

As for the enhancement of human traits, it raises equally serious issues. First, which kind of human traits should be enhanced? This will be a problem, since there will be no agreed-upon answer among human beings. Second, if we reach an agreement on what kind of human traits should be enhanced, it may be unavoidable that people with the opposite and "undesirable" traits that are not chosen for enhancement will be stigmatized or discriminated against. Third, there is no guarantee that the enhancement of human traits will not lead to eugenic practices such as those in Nazi Germany. Although the enhancement of human traits has not become a problem in

China, because gene therapy is just at a beginning stage, but some Chinese legislators or officials tend to be genetic determinists and want to use genetic knowledge to reduce the population with disability, including the mentally retarded, in the name of the state's interest or society's interest. These officials believe this is a main long-term goal of genetics; this belief led them to take eugenic programs for granted. In echoing them, many Chinese respondents gave similar answers to the question on the goals of genetics.

Voluntary vs. Mandatory Screening and Testing

Genetic screening and testing in general should be voluntary, according to the principle of respect for persons. However, "voluntary" means informed choice. If people are not informed, voluntariness is meaningless. If we believe that both genetic screening and testing will benefit people and their families, that genetic services will become part of universal health care, and that genetic knowledge will become part of compulsory public education, geneticists and institutions concerned have the responsibility to provide genetic information and education to people. However, if people choose not to have genetic screening and testing after being informed and educated, we have to respect their choices. But there are exceptions to voluntariness: (i) screening of newborns can be compulsory when treatment is available for those affected by genetic disorders; (ii) testing the sex of the fetus, unrelated to genetic diseases, should be prohibited (Fletcher and Wertz, 1991). Although the Chinese Ministry of Public Health promulgated a prohibition on sex selection by genetic testing in January, 1989, this practice is still continuing illegally. It is an important factor in sex ratio imbalance.

Genetics is an art of *ren*. In order to do good to people, the Chinese think that we should not absolutely exclude any compulsory genetic screening or testing if we recognize that we have an obligation to the health and life of the next generation. If I have a dominant genetic disease, then my child will have a 50% chance of having the disease too. I would feel I have the obligation to be screened or tested. Otherwise, my child who developed the disease could sue me for wrongful life in the courts, when he/she learned the truth. So this obligation may form a moral basis for compulsory screening of newborns, which would be in the best interests of myself, my child, and my family. With the spread of genetic knowledge and the change of ethos leading people to a greater sense of responsibility for the health and life of future generations, either compulsory screening for some diseases will be generally accepted, or it will become unnecessary because all people will have it voluntarily. This is why the majority of Chinese respondents to the survey favored genetic screening for newborns, but rejected prenatal sex selection, because it will do harm to women.

The potential benefits of genetic screening and testing must outweigh the potential harms, risks, or costs, according to the principle of *ren* or beneficence. If genetic screening involves a particular group or several groups, the benefits and costs should be distributed fairly within or between groups, according to the principle of justice. Any unfair distribution in which the benefits are disproportionally given to some people or some groups and the costs are disproportionally borne by others is unjust.

Directive vs. Non-Directive Counseling

Genetic counseling is provided only in the hospitals or institutes sponsored by the Ministry of Public Health of the central government or by bureaus for public health of municipal/provincial governments. Those who come to clinics to seek genetic counseling are mainly couples who worry about whether their future baby will be genetically normal, or who have a relative who is genetically diseased, or who have had a defective child before; or those who have read something on inherited diseases in journals or books and worry about their future child or already-born child. Genetic counseling is directed to potential parents, couples, and individuals, anyone who wants to seek counseling. There is no limit on seeking genetic information about children, even when treatment is impossible or unavailable.

Chinese genetic counselors seem in-between providing non-directive advice – simply providing information – and serving social interests in public health. They cannot and should not make decisions for clients based on the counselor's own values. However, because of ignorance and frustration, in some cases clients not only need information, but also want counselors to tell them what they should do. So it is justifiable for genetic counselors to play a more active role, what may be called a guiding role, in counseling and in clients' decision-making. One reason is that in many cases counselors and clients share the same values in making decisions. However, this does not mean that the decisions should be made by counselors. After the possible consequences of the various options have been disclosed to the client, and the counselor's suggestion ("If I were you, I would ...") has been made to the client, the final decision should still be made by the client. If one thinks of genetics as an art of *ren*, genetic counseling should not only provide information passively, but should also actively guide the client to make her/his own decision in her/his best interest. The underlying presumption in genetic services, as well as in other medical services, is that priority should be given to beneficence. Most Chinese geneticists work in government medical schools, hospitals or institutes, and they feel they have the responsibility to do their work within the framework of government policies. The Chinese government now has a basic policy, i.e., reducing the quantity of the population and raising its quality. Many geneticists, in counseling, tend to suggest that a client abort a fetus, for example, with Down syndrome, and think this is in the client's best interest as well as in the interest of society. However, if the client insists on carrying this fetus to term, the counselor cannot oppose this decision. This is why many Chinese respondents to the survey favored directive counseling in their responses.

There are some difficulties in genetic counseling. Genetic knowledge is based on the language of atom and molecule, but traditional Chinese are familiar with the language of *yin* and *yang*. This situation will be improved with the extension of compulsory education. In addition to the language barrier, there is a cultural barrier. A few clients do not believe what the counselor says, or they think that "good fortune" will help them. Parents may worry about the future marriage of their child, so they are reluctant to be followed up. For traditional Chinese, the marriage and reproduction of their child are very important values.

The information obtained in genetic counseling has never been reported to public health authorities, because obligatory reporting is only applied to some communicable diseases, such as cholera, typhus, diphtheria, smallpox, and sexually transmissible diseases, including HIV, according to the Law on the Control of Communicable Diseases. Whether genetic information is disclosed to the spouse or any member of the family depends on the nature of the information, the consent of the client, and, more importantly, the consequences of the disclosure. If the client's genetic disease may affect any member of the family, that member should have a share in the information. If the disease may affect her/his future children, the spouse should share in the information. Generally speaking, consent for any disclosure of the information should be obtained from the client, except in the case where, even though the client insists on not disclosing the information, keeping this information confidential would harm a third party and the disclosure could not harm the client. Genetic counselors should balance the positive and negative consequences caused by disclosure versus non-disclosure of the information. Underlying all these practices is the belief that genetics is an art of doing good to people.

Privacy and Confidentiality in Genetic Services

According to the proposition that genetics is an art of *ren*, all clinically relevant genetic information after screening or testing should be fully disclosed to patients and family members, except two kinds of cases: (i) XY genotype in a female, and (ii) an accidential finding that a woman's husband is not the biological father of her child. In both cases, the family will be happier without the disclosure. The disclosure of genetic information to patients and family members should be followed by counseling and education. If the patient refuses to disclose proven risks of harm to family members, the obligation of preventing harm to others will override the obligation of keeping confidentiality, according to the principle of beneficence or/and nonmaleficence.

Chinese would claim that efforts should be made against genetic discrimination in insurance and employment and in the delivery of genetic services, and against abuse of eugenics, according to the principle of *ren*. The genetic screening of employees for increased occupational risks should rarely be necessary, and efforts should be made by employers to improve occupational health conditions and to reduce occupational risks. However, when the danger cannot be significantly reduced by the employer, it becomes acceptable to screen employees and move those susceptible to safer jobs, but any action of discharge by the employer should be legally prohibited. The state should be prevented from taking any action to protect the collective "gene pool," or to engage in "genetic cleansing," or positive eugenics. However, so-called "negative eugenics," i.e., reduction and prevention of genetic diseases through genetic programs, can be ethically accepted, but these programs should be implemented with clients' informed consent.

Eugenics vs. Healthy Birth

The term "Eugenics" has different meanings today than at the time of Darwin and his cousin Galton. Now eugenics usually means either a practice used in Nazi Germany for the purpose of ethnic cleansing on the basis of racism by misuse of medical genetics, or use of compulsory sterilization in the US and other countries for the purpose of reducing the population with physical, mental, or social abnormalities. The difference between eugenics and medical genetics is that eugenics is a program imposed by the state, regardless of the preferences of the individuals concerned, whereas the prevention of genetic diseases is a voluntary choice by individuals (Wertz *et al.*, 1995). In other words, informed consent is a criterion for distinguishing prevention from eugenics. Actually, apart from rejecting informed consent, eugenics as a program can be massively implemented only on the basis of racism.

The Law on Maternal and Infant Health Care, which was promulgated on October 17, 1994, consists of 7 chapters and 39 articles. There are 4 articles which may have caused it to fall under the suspicion of eugenics. Others deal with the responsibility of the government at various levels to provide pre-marital and perinatal care (19 articles), procedures and qualifications (7 articles), incentives (1 article), ethical requirements (3 articles), legal liability (3 articles), definition (1 article), time of implementation (1 article). There are three groups of articles which will be discussed below.

(1) *Ethical requirements*: The following Articles could be taken as being in accord with conventional ethical principles: Article 19 (consent or proxy consent for any termination of pregnancy or application for a sterilization operation), Article 32 (prohibiting sex identification of a fetus for non-medical reasons), and Article 34 (keeping confidentiality).

(2) *Legal liability*: The prescribed punishments are imposed only on those who practice without qualification, or do harm to clients, or commit fraud. There is no punishment imposed on the clients. In the authoritative "Answers to Questions about the Law on Maternal and Infant Health Care," it was stated that those who will be punished are institutions and persons who practice in violation of the Law, for example having no certificate or committing fraud, not the clients (Answers, 1995, p. 330).

(3) *Suspicious ("eugenic") articles*: Article 18 prescribes that the physician shall give advice to married people for a termination of pregnancy if some disorder or malformation is detected in prenatal diagnosis. But what the physician provides is "medical advice." If a termination of pregnancy is made, consent has to be obtained from the couple or a guardian if the client is incompetent. And for Chinese, a fetus is never considered as a human being from a Confucian perspective. Article 9 prescribes that physicians should provide medical advice to the couple, but this is still advice, not a directive. Article 10 is the only article that can be called eugenic:

"Physicians shall ... explain and give medical advice to both the male and female who have been diagnosed with certain genetic diseases of a serious nature which are considered to be inappropriate for childbearing from a medical point of view; the two may be married after [only if – in English version] both sides agree to take long-term contraceptive measures or to have a ligation operation for sterility."

The Chinese word "*yousheng*," which occurred very frequently in the responses of Chinese geneticists in the survey, has dual meanings. It may refer to the Nazi practice during the Second World War, but may also refer to healthy birth. In today's Chinese language, "*yousheng*" routinely means "healthy birth," which is always used with "*youyu*," which means "rearing healthy children." In Chinese documents and in the media it is more often used to refer to healthy birth. The translation of Chinese "*yousheng*" into English as "eugenics" has caused misunderstanding and confusion. There is no racist tradition in China, and the Chinese have been the victims of slaughters by Western imperialists and Japanese militarists from the middle of the

19th century to the middle of this century. But no Chinese think that they are superior to any Western peoples or to the Japanese, or that the Han people are superior to Chinese minorities. Chinese have made many grave mistakes in different times, but all those had nothing to do with racism. So the difference between Chinese "*yousheng*" and the German Nazis' eugenics lies in: (1) Underlying German Nazi eugenics was racism, but for the Chinese "*yousheng*" is a hope to reduce the number of defective births. When Chinese say "quality of population," it does not mean certain racial or ethnic groups. In the Chinese language, "population" is "human being + mouth," so it is the sum of individuals. And the Chinese consist of more than 50 ethnic groups. (2) Nazi eugenics was imposed without any informed consent, but in the Chinese Law informed consent is a principle which is to be observed, although the language in some articles is not duly explicit and should be improved.

There have been two approaches to drafting Chinese laws. One approach is to focus on reducing the population with physical deformities or mental retardation by limiting marriage and reproduction by compulsory means. One representative of this approach is Gansu's Law, which is called Regulation on Prohibiting Reproduction of Dull-witted, Idiots or Blockheads, promulgated by the Fifth Session of the Standing Committee of Gansu People's Congress on November 23, 1988 (Lei *et al.*, 1991). The definition of dull-witted, idiots or blockheads is:

1. Congenitally caused by familial inheritance, inbreeding or parents under external influences;
2. Mental retardation of middle or severe degree with IQ below 49;
3. Behavioral disorders in language, memory, orientation, thinking etc.

The law prescribes:

Prohibiting the reproduction of dull-witted, idiots, or blockheads.
Dull-witted, idiots, or blockheads may be married only after sterilization. If a couple are both dull-witted, idiots, or blockheads, only one has to be sterilized; if only one of the couple is dull-witted, idiot, or blockhead, only he or she shall be sterilized (Lei *et al.*, 1991).

Gansu's Law is an eugenic law similar to those in the US before the Second World War. And some Chinese scholars who are involved in medical ethics and health law played a negative role in describing people with disabilities, which provided a negative (pessimistic) background for drafting the Law. First, they called physically deformed or mentally retarded people "inferior births," which is a term discriminating against the disabled (Zheng, 1987; He, 1989; Wu, 1990; Chen, 1995). This term was also used in some government documents and ministers' speeches. Second, they defined those parents with severe genetic diseases, severe schizophrenia, inbreeding and old age as "parents without reproductive value" (He, 1985, 1989; Wu, 1990) which is also a discriminatory term. Third, they favored an ideology in which priority was given to the social good or state interest regardless of individual interests (Du, 1985; Zheng, 1987; Chen, 1992; Zhan, 1996). There is some eugenic thought in their minds.

The other approach in drafting laws is to focus on improving pre-marital, prenatal, and perinatal health care with the purpose of reducing the misery and enhancing the happiness of families and individuals by providing information, education, counseling, and services to prevent severe congenital or genetic diseases, while respecting the decisions made by families and individuals. The reduction of the number of disabled persons is only an indirect consequence of these individual and familial decisions. Informed consent is a crucial principle in this approach. Representative of this approach is the Minutes of the First National Workshop on Ethical and Legal Issues in Limiting Procreation (Minutes, 1996). The main points of the Minutes are:

1. *Medico-genetic perspective*. In the etiology of mental retardation, 82% results from non-genetic congenital or environmental factors. Participants agreed that the sterilization of the mentally retarded, if their retardation is caused by genetic factors, may reduce a small part of the mentally retarded population, but the focus should be on developing pre-conception, prenatal and perinatal care, maternal and child care and programs of community development so as to effectively prevent mental retardation and reduce its incidence.

2. *Ethical perspective*. Participants agreed that the ethical principles of beneficence, respect, justice and solidarity should be applied to the evaluation of the action taken on limiting procreation by the severely mentally retarded. The sterilization of severely mentally retarded people may be in their own best interests. They should enjoy the same rights as others. Their families, communities and society ought not to discriminate against them, and ought to protect and not infringe upon their deserved rights. The workshop called for the whole society to identify itself with physically and mentally handicapped compatriots, and to do the best to provide social support for them and guarantee them quality of their lives. If procreation would do harm to them, limiting their procreation can be ethically justified. Because the severely mentally retarded are incompetent, they have no responsibility for the various consequences of the condition of mental retardation. So limiting their right to procreation or sterilization is not a punishment for them, but a contraceptive measure to reduce their misfortune and their families' misfortune. Participants agreed that because the incompetence of the severely mentally retarded renders them unable to make rational judgments on what is in their best interests, one has to appeal to the decision of a guardian or proxy who has no conflicting financial or emotional interests.

3. *Legal perspective*. If a law of compulsory sterilization were drawn up, it would infringe upon some civil rights laid down in the Constitution and other laws, such as the right to inviolability of the person and the incompetent's right to guardianship. If a law on voluntary sterilization were drawn up, such infringements would be avoided.

In Chen *et al.* (1989) the author emphasized that women have the right to decide on the termination of pregnancy or on sterilization (pp. 261–262), and in Chen (1992) the author claimed that it is necessary to obtain consent for sterilization and termination of pregnancy (p. 205). Feng (1994) pointed out that the proponents of eugenics tend to ignore the difference between genotype and phenotype, and to give priority to the nation over individuals (pp. 246–247). She argued that it is ethically justifiable to sterilize severely mentally retarded persons only if it is in their best interests; sterilization should be voluntary, not compulsory. Consent or proxy consent should be obtained, both to sterilize severely mentally retarded persons and to abort fetuses with a genetic defect (pp. 250–253). For more detailed arguments please see Qiu (1996). Many officials do not know the natural or "new" mutation rates of human genetic disorders nor the limits on reducing the disabled population. At any time, 3–5% of the population will have disabilities, and the society or country always has the responsibility of caring for them.

Any action taken in the drafting and implementation of Law should be evaluated in an ethical framework that consists of basic principles of bioethics: nonmaleficence/beneficence, respect, justice and solidarity. Four applications of these principles, accompanied by illustrative cases, appear below.

1. Welfare of individuals and family vs. reduction of the disabled population
The purpose of the Law should be to improve human existence through prevention of genetic diseases. This should be done for the welfare of individuals and families. The reduction of the disabled population is an indirect consequence of prevention.

2. Voluntary vs. Compulsory Services: Informed consent versus Imposition by the State
Anybody has the right to self-determination in marriage and reproduction, with only minimum exceptions, e.g., inbreeding, incest and rape. This is a principle accepted by the international community. Under China's unique circumstances of overpopulation, it is justifiable to limit the right to reproduction for a certain period. But it is a limit beyond which compulsion cannot be justified. Because the purpose of the Law is the welfare of individuals and families, the logical consequence is that any decision on marriage or reproduction should be made by the client her/himself. Any medical advice on the postponement or withdrawal of a marriage application, sterilization or termination of pregnancy should be voluntary and based on informed consent from the client, not imposed by the physician or by the state. There is no guarantee that the use of the Law by physicians will avoid a eugenic purpose.

3. Medical advice: Fact vs. Value
There is the wording "medical advice" in the Law. But medical advice is not only the statement of medical facts, but also involves a value statement. So any medical advice may be a mix between medical facts and values held by the physician. However, the physician's values are not the same as the client's values. But many physicians, geneticists, and officials always confuse facts with values and the physician's value with the client's value.

Case 1: Conflict of values: Love vs. healthy offspring A musician Y and an engineer W went to a clinic to undergo pre-marital examination. The physician told them that blood types were not compatible, and that the incidence of newborn haemolysis is 3%, and suggested that they consider this possibility and consult their parents. They said that they would be married. But just before the wedding Y changed her mind: "After deliberate consideration, I will not marry W in order to avoid his family being without offspring, because he is a single son of his parents." The author of this report remarked that Y was wise, because she believed in science and overcame her passion through her reason, and prevented the possible newborn haemolysis (Zhan, 1996, p. 30). There were many options facing Y. The choice she made is only one among them. Which option she chose depended on the priority she gave to the different values.

Case 2: Conflict of values: Love vs. having offspring L and Y wanted to marry. After pre-marital examination L was found infertile. After learning the truth, Y wanted to break off with her. But she did not believe the diagnosis because she had always been healthy. She insisted on marrying Y, and Y could not refuse. After marriage she was proved to be infertile. But Y and his mother were anxious to have a baby. The marriage was unhappy. Ten years later they divorced. The author of the report made a comment to the effect that if they had believed in science, the marriage and divorce would not have happened (Zhan, 1996, pp. 30–31). However, it is not an issue of science, but of values. If Y had not given priority to having an offspring over their mutual love, the story would have had a happy end.

4. Confidentiality and Privacy

In implementation of the Law, keeping confidentiality and privacy for clients is very important.

Case 5: Breach of confidentiality A girl L and X wanted to marry. After the pre-marital examination, "hymen was broken, had sex behavior" was written in the form by the physician and seen by X. Then X refused to marry L. The author of the report said nothing about preserving confidentiality and privacy or about the physician's duties. Instead, he remarked that L should be happy not to marry a man who gave priority to a woman's hymen over herself (Zhan, 1996, p. 30).

Now the authorities have a regulation prohibiting mention of a woman's hymen in the pre-marital examination form or on any other occasion.

Recently the Chinese "*yousheng*" (have a healthy baby) endeavor has instituted two "engineerings," or public health interventions. One is to popularize iodized salt in the areas with high a prevalence of cretinism. The other is to encourage fertile women in the areas with a high incidence of neural tube defects to take folic acid. These two "engineerings" show that "*yousheng*" is now on the right track.

China is in transition. During this period there are many opportunities and challenges in medical genetics. The dialogue, communication and collaboration between Chinese geneticists and their foreign colleagues will help the healthy development of medical genetics and its application in China.

REFERENCES

A Collection of Laws of People's Republic of China, 1992 Edition, Beijing: The Law Publishing House.

Answers to Questions about the Law on Maternal and Infant Health Care, *Maternal and Child Care of China*, 1995; 10(6): 9, 330–331.

Bankowski, Z. and Capron, A.M. (eds). Genetics, Ethics and Human Values: Human Genome Mapping, Genetic Screening and Gene Therapy. *Proceedings of the XXIVth CIOMS Round Table Conference*, 22–27 July 1990, Geneva: CIOMS, 1991.

Chen, M.G. *et al.* (eds). *An Introduction to Medical Law*. Shanghai: Science & Technology Press, 1989, pp. 248–263.

Chen, M.G. *et al.* (eds). *Health Law*. Shanghai Medical University, 1992, pp. 200–205.

China's Ministry of Health: Ethical Code for Medical Personnel, Chinese Hospital Management 9 (1988), no. 3, 5.

Du, Z.Z. *An Outline of Medical Ethics*. Jianxi: People's Press, 1985, pp. 220–223.
Feng, Jianmei, *Modern Medicine and Law*. Nanjing: University Press, 1994, pp. 245–260.
Fletcher, J.C. and Wertz, D.C. An International Code of Ethics in Medical Genetics: before the human genome is mapped. In: Bankowski, Z. and Capron, A.M. (eds), Genetics, Ethics and Human Values: Human Genome Mapping, Genetic Screening and Gene Therapy, *Proceedings of the XXIVth CIOMS Round Table Conference*, 22–27 July 1990. Geneva: CIOMS, 1991, pp. 97–116.
He, L. *et al.* (eds). *Modern Medical Ethics*. Zhejiang: Education Press, 1989, pp. 139–146.
He, Z.X. (ed.) *An Introduction to Medical Ethics*. Jiangsu: Science & Technology Press, 1985, pp. 142–143.
Law on Maternal and Infant Care of the People's Republic of China, printed by the Ministry of Public Health of the People's Republic of China, 1994.
Lei, Z.H. *et al.* (eds). *Handbook of Family Planning Administration*. Printed by Gansu Commission on Family Planning, 1991, pp. 131–132.
Li, P. *et al*. The Qingdao Conference of 1956 on Genetics: the historical background and fundamental experiences. In: Danian, F. and Cohen, R. (eds), *Chinese Studies in the History and Philosophy of Science and Technology*. Dordrecht: Kluwer Academic Publishers, 1996, pp. 41–54.
Luo, H.Y. (Lo, Wilson). Medical Genetics. In: Fang, Q. (editor-in-chief), *Contemporary Internal Medicine*. Beijing: The People's Military Medicine Press, 1995, pp. 43–70.
Minutes of the First National Workshop on Ethical and Legal Issues in Limiting Procreation, November 11–14, 1991, Beijing, China. In: Qiu, R.Z. (ed.), *Reproductive Health and Ethics*. Beijing: Beijing Medical University & Peking Union Medical College Joint Press, 1996, pp. 316–319.
Qiang, B.Q. The Chinese Human Genome Initiative, an unpublished article.
Qiu, R.Z. Medicine – the art of humaneness: on ethics of traditional Chinese medicine. *Journal of Medicine and Philosophy* 1988; 13: 277–300.
Qiu, R.-Z. The fiduciary relationship between professionals and clients: a Chinese perspective. In: Pellegrino, E. *et al.* (eds), *Ethics, Trust and the Professions: Philosophical and Cultural Aspects*. Washington DC: Georgetown University Press, 1991, pp. 247–262.
Qiu, R.-.Z, Morality in flux: medical ethics in PRC. *Kennedy Institute of Ethics Journal* 1991; 1: 16–27.
Qiu, R.-Z. What bioethics has to offer the developing countries. *Bioethics* 1993; 7: 108–125.
Qiu, R.Z. Chinese medical ethics and euthanasia. *Cambridge Quarterly of Healthcare Ethics* 1993; 2: 69–76.
Qiu, R.Z. Medical ethics and Chinese culture. In: Pellegrino, E. (ed), *Transcultural Dimensions in Medical Ethics*. 1994, pp. 155–174.
Qiu, R.Z. Genetic counseling in China. *Proceedings of the Third Session of International Bioethics Committee of UNESCO*, vol. II, September 1995, pp. 25–26.
Qiu, R.Z. Ethical issues in compulsory sterilization of seriously mentally retarded. In: Qiu, R.Z. (ed.), *Reproductive Health and Ethics*. Beijing: Beijing Medical University & Peking Union Medical College Joint Press, 1996, pp. 319–327.
Wertz, D.C., Fletcher, J.C., and Berg, K. *World Health Organization: Guidelines on Ethical Issues in Medical Genetics and the Provision of Genetic Services*. Geneva: WHO, 1995, p. 3.
Wu, M. Some remarks on gene therapy. *Chinese Medical Journal* 1993; 73(7): 389–392.
Wu, M. Genetics and medicine. *Chinese Medical Journal* 1995; 75(10): 1–3.
Wu, X.Z. *et al.* (eds). *Modern Clinical Medical Ethics*. Tianjin: People' Press, 1990, pp. 354–357.
Wu, X.Z. *et al.* (eds). *Medical Morality: Theory and Practice*. Tianjin: Academy of Social Science, 1990, pp. 733–741.

Yu, G. Speeches at the Qingdao Genetics Conference of 1956. In: Danian, F. and Cohen, R. (eds), *Chinese Studies in the History and Philosophy of Science and Technology*. Dordrecht: Kluwer Academic Publishers, 1996, pp. 27–40.

Zhan, Z. The first step of sexual love. *Happy Families* 1996; 9: 30–31.

Zhao, S.Y. Human diseases and genetics. In: Du, C.S. and Liu, Z.T. (eds), *Medical Genetics*. Beijing: People's Health Care Press, 1983, pp. 1–10.

Zheng, R.S. Healthy Birth Should Be Taken as Moral Concept and Norm of Human Beings. In: Zhu, Xuefu, Moral Consideration on Eugenics, Proceedings of the Fourth Conference on Medical Morality. *Journal of Medicine and Philosophy* 1987; pp. 43–47.

CHAPTER 13

CHINA: VIEWS OF A MEDICAL GENETICIST

Xin Mao

INTRODUCTION

China is a country of great contrast. It has the world's largest population, yet a relative dearth of natural resources per capita. It has a very long history and rich culture, but an uneven educational system; there are 300 million illiterate and semi-illiterate people. It has the fastest growing economy due to a decade of economic reforms, but the socialist party has remained as the ruling party of the political system. China covers a vast territory, yet the level of development varies greatly from region to region.

CULTURAL AND SOCIAL BACKGROUND

The Chinese View of Life, Health, and Medicine

China has the world's oldest continuous civilization. The evolution of its social systems can be divided into the following periods. The Neolithic Age (7000–1700 BC) was the beginning of Chinese civilization, and the Yellow Emperor is thought to be the Chinese ancestor. In the Bronze Age (1500 BC–AD 220), China emerged as a united nation with a central government, standard writing system and a diversity of religious teachings. In the Iron Age (AD 220–1910), Chinese empires (feudalism) reached a peak and then collapsed. Confucianism, Taoism and Buddhism became the dominant religious teachings. In the Pre-modern Age (1910–49), China was a mosaic of feudalism and bureaucratic capitalism. Modern China (1949–present) is a socialist nation. Marxism and Mao Zedong's socialism are mainstream ideologies. This modern period can be subdivided into Mao Zedong's utopian socialism (1949–77), Deng Xue Ping's pragmatic socialism (1978–94), and post-Deng Xue Ping's national socialism (1994–present).

Historically, Chinese culture was mainly shaped by Confucianism, Taoism and Buddhism. Since 1949, it has also been influenced by Marxism and Mao Zedong's ideology. More recently, Chinese culture appears to be integrated by the concept of pragmatism. However, Chinese moral attitudes towards human life, health, the status of woman and of the fetus are mainly and profoundly affected by the longstanding values of the traditional religious teachings rather than by Marxism or Christianity.

As early as the third century BC, Chinese have viewed human life as beginning with birth and ending with death. Traditionally, the mourning period after a person's death lasted for seven weeks, but a dead fetus or child was never mourned. This is because Chinese emphasized the continuity between the fetus before and after birth, but they did not consider the fetus a human being or person. When a baby is born, it is always considered to be one year old; however, most Chinese do not feel a child has died or has been killed when a spontaneous or induced abortion occurs.

What about Chinese views on health and medicine? Taoists regarded life as the coordination of Qi (vital force or energy) and good health as a harmony or balance of Yin Qi (negative force) and Yang Qi (positive force); Yin comes from mother and Earth and Yang from father and Heaven. If Yin and Yang were imbalanced, illness would occur. This philosophical idea not only dominated Chinese religious beliefs on health but also constituted the key theories of traditional Chinese medicine (TCM). For example, in TCM Yin and Yang are equivalent to physiology/pathology; and five elements, Jin (gold), Mu (wood), Shui (water), Huo (fire) and Tou (earth), are thought to be mechanisms of physiopathology: Zang-Fun (organ), Channel and Collateral are thought to be anatomy; Six Weathers, Fong (wind), Han (cold), Shu (summer), Shi (damp), Zao (dry), Huo (internal heat), and Seven Emotions, Xi (happy), Nu (anger), You (anxious), Si (thinking), Bei (sad), Kong (panic), Jing (scary) are thought to be etiology. Herbs and acupuncture are used as treatments of diseases; and Qigone (meditation) and Taoji (breath and physical exercises) are used as prevention of diseases. Even from the viewpoint of modern cancer genetics, the idea of Yin and Yang still seems to be accepted. For instance, it has been widely stated that the initiation and progression of human malignant diseases are likely caused by loss of control of cell proliferation and differentiation *in vivo* due to loss of function of tumour suppressor genes (Yin) and activation of oncogenes (Yang), or a genetic imbalance (imbalance of Yin and Yang) in normal cells.

In the actual practice of maintaining moral, mental, and physical health, Confucianism taught people to know the importance of loyalty, filial piety, humanity and brothership. In contrast, Taoism preached keeping good health and being immortal by means of proper breathing, exercise, diet, sex, and taking medicine. Buddhism emphasized that to reach Nirvana, the highest goal of people in the world, and to save all forms of life, a person had to get rid of karmas and return to the emptiness of the original mind. He or she would then not experience reincarnation.

All these teachings have developed a doctrine or "psycho therapy" that might alleviate Chinese from suffering both physically and mentally. This doctrine includes two elements. The first one is "nonaction" (Wu Wei), which does not simply imply "do nothing," but rather do nothing unnatural or supernatural. The second element is "unattachment", which means that one can pursue and seek a goal, but one should not be attached to that goal emotionally. However, this doctrine also upheld paternalism in China.

Ideologically, Marxism advocated a society of Utopia or a communist "Nirvana" without class, individualism, egoism and ethnic discrimination, but instead a society of absolute equality. Politically, it favored a powerful centralized government and one party – the Communist Party – ruling the social system. Economically, it stood for "equal poor and rich", namely, equal allocation of social wealth and resources as well as centralized economic planning. It had a holistic social philosophy, which meant that each individual is seen as a component of the whole society. Thus, an individual's interests should be properly subordinate to the interests of the whole society or nation. This added more weight to the existing paternalism in China. However, Marxism did not have special philosophical views of human life and health. Chinese pragmatism or "Socialism with Chinese Characteristics" emerged following the government's implementation of the policies of economic reform and openness. It belongs to a mosaic of modified Marxism and a market economy. It favours "getting rich as soon as you can," but it also stands for solving various social issues with some sort of "Chinese method."

The Status of Chinese Women

Although remnants of matriarchal society still exist in one of the 55 ethnic minorities present in China today, patriarchy has dominated Chinese society for thousands of years. Historically, Chinese social and moral attitudes towards women were described as "elite men and lowly women"; in other words, men were respected and women were not. Chinese women were regarded as dependent on their husbands and families economically and politically. Their duties were taking care of their husbands, parents, and husband's parents, and bearing and raising children. Since 1949, China has undergone a significant shift from a traditional society to a developing socialist society in which women's social status has changed radically. For example, women's rights are now protected by the Constitution of the People's Republic of China and relevant laws ("The outline of Chinese women's development 1995–2000," Renmin Ribao, 9 August 1995). Officially, women's organizations of national and local governments have been established which address such issues as equal education, employment, payment, benefits, health care, etc. Chinese women's organizations emphasize the improvement of all aspects of women's lives, and they advocate family planning, genetics services and legal abortion. A Chinese political slogan saying "Women can hold half of heaven" indicates the belief that women are equal to men, at least theoretically. As Wen shows, however, a large proportion of families, especially those in rural areas, still have a preference for male children, which is leading to a serious sex ratio imbalance in China.[1]

National Policy: Family Planning

The population "explosion" is one of the most important issues worldwide. Wu documents an increasing rate of growth of the Chinese population.[2] China is also confronted with certain unalterable national conditions, such as reduced arable land, inadequate natural resources, and unbalanced economic development among different regions of the country. In order to reach a balance between population growth and national economic growth, the Chinese government has made it a fundamental national policy to practice family planning, control population growth, and improve the quality of China's human resources. In article 25 of the Constitution of the People's Republic of China, it is stipulated, "The state promotes the practice of family planning in order that population growth may be in keeping with socio-economic development." Article 49 of the Constitution and article 12 of the Marriage Law state

"Both husband and wife have the duty to practice family planning." The key points of the current family planning policy are promoting deferred marriage and childbearing, fewer but healthier births, preventing genetic defects, advocating the practice of "one couple, one child," and encouraging birth spacing for those who have practical difficulties with having only one child. This "Fundamental State Policy" has been strictly and successfully implemented in the urban areas, but it has been difficult to enforce in the countryside. In minority areas, it is more flexible.[2] This policy has certainly had an impact on genetic research and services in this country.

GENETICS

Prevalences and Incidences

So far, detailed national data on the prevalence of genetic disorders have not been available in China. In the middle of the 1980s, however, population-based epidemiological investigations of the prevalence of genetic disorders and birth defects in Sichuan province, with a population of 120 million, were carried out.[3,4,5] As the International Clearinghouse for Birth Defects Monitoring Systems reported, the overall prevalence of birth defects in China was very similar to those in the developed countries, while the prevalence of some types of birth defects was highest in China.[11] Given the prevalence of genetic conditions in the Chinese population, it can be predicted that there are an enormous number of people with genetic disorders and birth defects, and the absolute number of patients will increase with the growth of population.[6–10]

Genetic Research and Services

Advancement and Achievements

Human cytogenetic techniques were first introduced to China in the early 1960s.[12] In 1975, chorionic villi sampling (CVS) for prenatal diagnosis of genetic disorders was established. Subsequently, cytogenetic and biochemical analyses of cells from amniocentesis and cordocentesis, fetoscopy-based analysis, as well as sonographic techniques were applied to prenatal diagnosis. Since the 1980s, molecular genetic techniques, such as Southern and dot-blot hybridization, polymerase chain reaction (PCR), allele specific oligonucleotide (ASO), denaturing gradient gel electrophoresis (DGGE), single strand conformation polymorphisms (SSCP), DNA sequencing and fluorescence *in situ* hybridization (FISH), have been used in genetic research. In 1988, *in vitro* fertilization, embryo transfer and gamete intrafallopian transfer were available in several hospitals.[13,14] In 1992, the techniques of enrichment of fetal nucleated erythrocytes from maternal blood for prenatal diagnosis and sex determination in the first trimester were introduced to China.[6] In 1993, China launched its Human Genome Initiative, which had several goals. These goals included (1) development and improvement of the techniques for genome research, (2) establishment of banks for collection and storage of genetic samples, (3) analysis of several disease-associated loci in Chinese genomes, (4) investigation of genetic diversity and the frequencies of major alleles of specified loci in 56 Chinese ethnic groups, (5) determination of the types and prevalence of genetic conditions in Chinese ethnic groups, and (6) preservation of genetic resources.[15,16] In the same year, the largest intervention trial in the world, aimed at reducing the high rate of neural tube defects by allowing hundreds of thousands of women in three provinces to take small doses of folic acid daily before and during the first weeks of pregnancy, was started.[17] Gene therapy for patients with hemophilia B has been initiated in Shanghai.[18] There is also a national plan to launch an ambitious campaign to establish national networks to monitor and intervene in nine birth defects and 14 genetic disorders in the general population.

Allocation of Genetic Resources

Genetics services, including genetic counseling and prenatal diagnosis, have been provided for more than 10 years in China. However, the basic tools used for prenatal monitoring and diagnosis by most clinics are pedigree analysis, physical examinations, and ultrasonography. Cytogenetics and biochemistry facilities are mainly available in municipal hospitals and family planning centres. More advanced molecular genetics techniques are available only in a few university hospitals and institutes for genetic research. There are few comprehensive reports in the literature on genetics services in university hospitals. For instance, from 1980 to 1988, 4628 patients (578.5 cases per year) were counseled in the genetics clinic of Peking Union Medical College. Altogether 966 patients (120.75 per year) were diagnosed as having genetic disorders. Among them were 436 patients (54.5 per year) with chromosome aberrations, 161 (20.1 per year) with inherited metabolic disorders, 111 (13.9 per year) with constitutional bone disorders, 103 (12.9 per year) with genetic neuromuscular disorders, 109 (13.6 per year) with dysmorphic syndromes and 46 (5.8 per year) with different kinds of genetic dermatoses and oculopathies.[19]

Due to the absence of a centralized registration system, it is difficult to estimate an accurate number of genetics services being provided in China. There has been an extremely wide gap between need and availability of such services. In the Chinese population of 1.2 billion, the absolute number of persons with genetic disorders is enormous. Until now, only civil servants and employees of state-owned enterprises, institutes or universities were covered by social health insurance plan, which provided them with free medical care. Currently most genetics services are not covered by free medical care. Medical and life insurance systems are still underdeveloped in China. Therefore, almost all of the people who need genetics services have to pay for them.

In China, 91.96% of the population are Chinese (Han) while the 55 ethnic minorities, who live in the most remote and poorest inland and boundary areas, constitute only 8.04% of the total population. Currently, relatively better genetics services are available in several metropolitan cities (Beijing, Shanghai, Harbin, Changsha, Chengdu, Wuhan and Guangzhou). Like other developing countries, genetics services, therefore, are definitely not distributed equally among the 56 ethnic groups, the cities and the countryside, or the economically developed coastal areas and poor inland areas. Whatever the situation will be in the future, unequal distribution of genetics services cannot be radically changed in the short term, as China is still one of the poorest countries in the world. At this time, the top priorities of the government are attempting to provide food, shelter, transportation, education, and basic health care for its rapidly growing population.

Cytogenetics
To our best knowledge, in 1987, the average income of Chinese workers in factories or universities was approximately US $30 per month, while the cost of cytogenetic testing was ¥8–30 (approximately US$ 1.60–6.00). As previously mentioned, genetics services are not free in China; patients have to pay for such services out-of-pocket. In 1987, the population prevalence of chromosomal disorders in Sichuan province was 31.5 per 100,000, which implied that there would be 346,500 persons with chromosomal disorders in China, with its population of 1.1 billion at that time. If all of these cases were diagnosed cytogenetically, diagnosis would cost US$ 54,400–2,079,000, which is equivalent to the monthly incomes of 18,480–69,300 workers. Therefore, cytogenetics services are very expensive for each patient or family in China, and essentially unavailable for the average Chinese worker.

Genetic Screening and Testing: Cost–Benefit

Newborn Screening and Testing Children
Before 1950, the infant mortality rate in China was 200 per 1000 births. In 1989, it fell to 13.4/1000 in urban areas and 34.3/1000 in rural areas. In 1993, it dropped further to 10.6/1000 in the Shanghai area. The major causes of death have shifted from infectious and parasitic diseases to malignant tumors and congenital and genetic disorders.[43] Newborn screening for hereditary metabolic disorders including phenylketonuria (PKU), hypothyroidism, and galactosemia, is routinely carried out in six screening centers in China. From 1982 to 1994, the total number of screenings were as follows: 550,229 tests (33 cases positive) for PKU, 176,170 (24 positive) for hypothyroidism, 51,793 (1 positive) for congenital adrenocortico-hypertrophy, 329,250 (none positive) for galactosemia and 280,000 (none positive) for histidinemia.[6] Regarding the costs and benefits of newborn screening, one report said that the 10-year cost of screening, diagnosis, medical care and dietary therapy for each PKU case identified would be ¥20,000. If PKU infants were not diagnosed and treated, the estimated cost of living for one untreated, mentally retarded individual with a mean life span of 40 years would also be ¥20,000. Income loss, special education costs, and inability to contribute to society would cause a total loss of at least ¥100,000. The long-term estimated benefit due to early screening, diagnosis and treatment minus the cost of screening and treatment is about ¥100,000. Thus, the ratio of benefit to cost would be about 5 : 1.[23]

Social Views
Neonatal intensive care units (NICU) and advanced surgical facilities of pediatrics are available in several metropolitan hospitals. Most physicians, however, make a social cost-benefit analysis before using these facilities to treat very low birthweight infants and infants with genetic disorders or birth defects, even if the treatment will benefit these infants. This is primarily due to the "one couple one child" family planning policy. Almost all Chinese parents wish to have a healthy child, and many low birthweight babies will be unhealthy even if they survive. Second, there are many full-term infants with acute disease who need NICU facilities to save their lives. Third, Chinese

physicians cannot persuade parents of infants with very low birthweight or with genetic conditions to consent to using these facilities to treat their children, as these treatments are usually too expensive for the parents. Therefore, social and economic factors seem to limit, to some extent, the autonomy of parents of seriously ill newborns in decisions about medical care in China.

The majority of respondents would support parents' decisions for nontreatment of their newborns with severe genetic diseases and birth defects who will die without treatment. 15.8% said they had counseled such parents who refused an operation for their newborns with genetic diseases. These data indicate that most physicians and genetics services providers in China consider the costs and benefits to society, even in the cases where the treatment will benefit those children with genetic diseases and birth defects.

No Opposition To Genetics Services?

Is there any opposition to genetics services in China? In our survey of genetic services providers, 85.8% of respondents thought no opposition existed while 14.2% said some opposition existed. Examples of comments among those who wrote that some opposition exists are "Because superstition still exists, science is not believed by some people who refuse to have genetic services", "People without genetic training judge things based on their own interest. They have bad impacts on genetic services", "There is no organization that opposes genetic services, but some traditional ideas such as 'more offspring, more happiness' conflict with genetic services", "Feudalistic remnant. The activity is found among people with feudalism," "None. But people traditionally do not want to let others know that there is a genetic trait that runs in the family."

Although genetics research in China has not been conducted for very long, it has made remarkable progress compared with other developing nations in the world. This is because the social and cultural climate supports its development, and almost no public opposition exists.

DISABILITIES IN CHINA

The Status of Disabled People

Based on the 1987 national sampling survey data on disabled people, it was estimated that in China there were 51.64 million disabled people (4.9%) in the population at that time, which is approximately equal to the whole population of the United Kingdom. Among people with disabilities, there were 8.17 million children, or 2.7% of all children. The majority (64.9 %) of these disabilities were due to postnatal diseases and injuries, while the remaining (35.1%) were due to genetic disorders and *birth defects*, which indicates great potential for the prevention and treatment of disabling conditions.[24,25] In China, the rights of disabled people have been protected constitutionally. There are official bureaus of civil administration, and semi-officially the Chinese Association of Disabled People is responsible for taking care of disabled people. Efforts to prevent disabling conditions among children have been made in China.[25] These efforts are usually categorized on three levels: the primary level includes genetic counseling, immunization programs, improved prenatal, perinatal and postnatal health-care, and legislation; the secondary level includes prenatal diagnosis and newborn screening; and the tertiary level includes special education and comprehensive rehabilitation services.[25]

Moral Attitudes Towards People with Disabilities

Although the Chinese government has made some achievements in the protection of people with disabilities, most of the Chinese still regard disabilities as a severe burden for both family and society. Majorities of respondents would not support disabled parents' decisions to have disabled children. 40% of the respondents explained that their reason for not supporting such parents' decisions was that they regarded disabilities as a burden on family and society, and 10.7% gave eugenic reasons. Some of the respondents commented, "The survival of these children is a very heavy burden to both the individuals and society"; "Is it necessary to allow children with genetic disease to live in the world miserably and bring suffering to the society and family economically?"; "For these kinds of children, living is worse than death. Don't bring financial loss to the family and the society."

LAWS AND REGULATIONS

Abortion

Induced or artificial abortion, including first-, mid- and late- trimester termination of pregnancy, is legally available on request and widely utilized in China. In the recently promulgated Law on Maternal and Infant Health Care, the indications for using medical termination of pregnancy have been stipulated as follows (Article 8, The Law on Maternal and Infant Health Care, Renmin Ribao, 28 October 1994):

(1) a fetus has a severe genetic disease,
(2) a fetus has a severe malformation or defect,
(3) continuation of pregnancy will endanger the pregnant woman's life or seriously damage her health.

What are the abortion rates in Chinese women? What are the reasons for these abortions? What percent of women choose abortion after a positive prenatal diagnosis, and what proportion of abortions are done for genetic reasons in the context of the total number of abortions provided? As Wang, et al. show[26] the abortion rates (per 100 pregnancies and per 100 live births) in 12,000 women in eight provinces of China in a period of 4 years (1983–87) were higher in urban areas (48.6 and 101.5) than in rural areas (36.5 and 61.9). The main reasons for abortion for women in urban areas, were contraceptive failure and "voluntarily observing government requirement for population control" while for those in rural areas it was "voluntarily observing government requirement for population control." No genetic reasons, however, were listed. The national data on the percentage of women who choose abortion after a positive prenatal diagnosis or the proportion of abortions done for genetic reasons in the context of the total number of abortions provided have been unavailable. One report from Beijing said physicians terminated all of the 7 pregnancies at risk for Tay–Sachs disease, GM1 gangliosidosis, Gaucher disease, metachromic leukodystrophy, Niemann–Pick disease, mucopolysaccharidosis IH and mucolipidosis prenatally diagnosed by amniocentesis in the second trimester in their practices.[27] Another report from Shanghai said that 42 pregnant women were diagnosed as at high risk for delivering children with alpha- and beta-thalassemias, and none of these pregnancies were terminated.[28] Therefore, whether the pregnancy was terminated or not depended mainly on the doctor's view of the severity of the genetic disease involved.

What are Chinese genetics services providers' views on abortion in this country? In our survey, 89% of the respondents said China legally permitted abortion on request, or for non-medical (social) indications; 95% agreed with the current legal stance on abortion in China. 78% said the reason that they agreed with abortion being available for social indications was population control and family planning. 42% of respondents said they had had experience with late abortion (after 24 weeks). 52% of respondents who offered a referral for late abortion did so because they felt disability was a burden, while 17% said they did so because of eugenic considerations. In practice, there have not been any legal restrictions on late abortion, but there are technical limitations in some village and primary clinics in China.

REPRODUCTIVE OPTIONS

In 1988, few genetics services providers in China mentioned that embryo or egg donation might cause ethical and/or legal problems.[13] Some of these reproductive options are new subjects for most of the genetics services providers in this country, and the likely ethical and legal issues involved in these options have not been investigated. Some special cases reported in China may also raise ethical issues. For example, from 1984 to 1990, 34 homotransplantations of testes, an alternative treatment for male infertility, were successfully performed. From the viewpoint of physiology, implantation of testes enables a male with infertility to restore his reproductive function naturally while artificial insemination does not. Most of the donors of the testes were the patients' fathers or brothers.[29] As some of the patients have married and had children, ethical issues associated with testes transplantation have now emerged. In our survey, two respondents reported two cases of "surrogate fatherhood." The first one was a case of a mentally retarded man who was married and advised not to have a child at counseling. His brother was mentally normal, and his parents wanted his normal brother to have two children so that the mentally retarded brother could adopt one. However, the mentally retarded man's wife wanted to have her own child and indicated that she could have a child with her brother-in-law. The second case was that of a man of marrigeable age who was infertile due to undescended testes. He requested implantation of one of his father's testes. His father agreed. He is now married and has had a child with the implanted testes of his father.

So far, there are no national laws or regulations on gamete donation or research involving the human embryo in China. This is because there has not been much demand for donated gametes and some up-to-date techniques for research on the human embryo may not be available in China. In contrast, there are laws and regulations on prenatal diagnosis. For instance, the indication for prenatal diagnosis has been stipulated in the Law on Maternal and Infant Health Care as follows: "The pregnant woman should have prenatal diagnosis as long as doctors suspect the fetus has an abnormality." It should be pointed out that although these laws and regulations should help promote the provision of genetics services in this country, the major incentives for development of genetics services are national economic growth, reasonable allocation of resources, and developing the public's full awareness of genetics and its social significance.

EUGENICS IN CHINA

Eugenics, meaning bearing and raising healthy children, has been popular in China since the 1970s. For example, the recently promulgated Law on Maternal and Infant Health Care was entitled "Eugenics Law" before being passed. There are many national, provincial, and municipal eugenics organizations which are funded in part by the government, and function to integrate human and medical genetics, pediatrics, gynecology-obstetrics, urological surgery, demography and sociology. These organizations publish books and journals, and organize conferences on every aspect of "well bearing and well raising" in maternal and infant health care. There has also been mass propaganda on "well bearing and well raising": eugenics accompanied by family planning and birth control. In our survey, 1/5 of Chinese respondents explained that their choices in answering some of the questions were based on eugenic principles. They commented, "it is the geneticists' or doctors' duty to improve the population quality." Some respondents regarded eugenics as indistinguishable from genetics services. Moreover, there has been no opposition to eugenics in this country so far. Why did the Chinese respondents seem conceptually to favor eugenics? The following may offer explanations:

(1) pressures of the population explosion, with a large number of patients with genetic disorders and birth defects, scarcity of resources and lack of funding;
(2) traditional preference for healthy children and great expectations for children by Chinese parents;
(3) little awareness of the Western history of eugenics and unfamiliarity with the cultural norms of international science;
(4) absences of abuses of eugenics in Chinese history (but an imbalanced sex ratio will provide them a history!);
(5) lack of ethical training in genetics, leading to conceptual confusion of genetics services and eugenics;
(6) increasing nationalism, due to heavy suffering of the Chinese people caused by foreign invaders in the past century.

The Chinese eugenics program, which is a part of the Chinese government's family planning program, seems to be negative eugenics and similar to eugenic programs of major western nations. However, most Chinese geneticists believe that its purposes are absolutely different from those in the USA and Germany in the late nineteenth and early twentieth centuries. The Chinese program has not had any tendency towards racial discrimination or ethnic cleansing and will never allow genocide, because the program is mainly accepted among the Han majority and may not be enforced among other ethnic minorities. China's Law on Maternal and Infant Health Care and relevant programs, as one Chinese geneticist said, only facilitate practices common "for decades … in western countries", and there is no similarity between what is practiced in China and "Hitler's concept of eugenics".[33] From a Western ethicist's viewpoint, the difference can only be demonstrated if there is voluntariness on the part of the Chinese public. There is little consensus between Chinese and Western scientists regarding the degree of the Chinese public's voluntary acceptance.

In recent years, there has been criticism of Chinese eugenics, especially the Law on Maternal and Infant Health Care, by a number of Western geneticists, journalists and politicians ("Scientists attack China over selective breeding", *The Times* Monday June 5 1995).[34,35,36] Some of them even intended to take sanctions against the Genetic Society of China.[33] This criticism seems to have very little impact on the Chinese situation. Some other Western geneticists said "We can understand China's practical situation and the purposes of its eugenic or family planning program, but it is unnecessary to have a national law to punish people with genetic diseases who do not want to be sterilized as they are not criminal. A guideline on such issues may be better" (personal communication). However, China has a very long history of feudalism, and 25% of its population are illiterate and semi-illiterate people. Guidelines will be easy to ignore for the Chinese public if they are not legislated. While China is shifting from a totalitarian government to a nation ruled by laws, the legislative construction seems essential. As it stands, the Law on Maternal and Infant Health Care is similar to a medical "standard of care" or guideline; it has no penalties for nonobservance and is not enforced.

ETHICS

Chinese Medical Ethics

Traditional Chinese medical ethics, based on the practice of Chinese medicine and shaped by traditional Chinese humanitarianism, can be traced back to two treatises: "On the Absolute Sincerity of Great Physicians" written by Sim Simiao (581–682 AD) of the Tang dynasty, which was similar to the Hippocratic Oath, and "Five Admonitions and Ten Maxims for Physicians" advocated by Chen Shi-gong (1555–1636 AD) of the Ming dynasty. Their ideas can be summarized as follows[38]:

(1) The life of a human being, including every part of the human body, is supreme and sacred. Physicians must rescue every life without any preconditions, and seek to extend life.
(2) Physicians should be very conscientious in their service. All patients should be treated with respect, concern, sympathy and equality.
(3) Physicians should not aim to obtain profit for themselves. It is immoral to seek after money, or official position, or to accept payment in the form of sex from patients, even prostitutes.
(4) Physicians should devote themselves to medicine, be diligent in their work and study, and not fear for their personal safety. They should constantly strive to improve their medical skills while treating patients carefully and efficiently.

(5) Abortion is immoral; even to touch the body of female patients for diagnosis and treatment is considered immoral behavior.

In 1937, "The Creed of Doctors" was published by the Chinese Association of Doctors as the motto for modern medical practitioners in the Republic of China. Since 1949, Mao Zedong's articles entitled "Serve the People" and "In Memory of Comrade Norman Bethune" were used as the political and moral guide for medical professionals. In 1988, a brief ethical code was promulgated by the Ministry of Public Health of China, which called for all medical professionals to "heal the wounds, rescue the dying, practice socialist humanitarianism," to "respect the patient's personality and his or her right, treat all equally without discrimination", and to "respect the patient's privacy and keep his or her confidence."[39]

Compared with western bioethics, which is focused on what western medical scientists consider the serious issues associated with great medical advances, and on the principles of respect for persons, beneficence, nonmaleficence, proportionality and justice,[40] Chinese medical ethics appears to be more dependent on the moral self-discipline of medical professionals in practice. This is partly due to the fact that the medical litigation system is still undeveloped in China.[39]

Sex Selection by Prenatal Diagnosis

The majority of Chinese respondents oppose sex selection in the absence of an X-linked disorder, by either prenatal diagnosis or preconceptional means. This is because an unbalanced sex ratio exists in the Chinese population, and it is becoming a severe social problem. The Law on Maternal and Infant Health Care stipulates, "sex identification of a fetus by technical means shall be strictly banned."

Expertise

High quality genetics services depend on well-trained providers. The extreme scarcity of well-trained genetic service providers in China has been a problem. This scarcity is due to the fact that genetics education in medical schools only began about 10 years ago, and the major course of medical genetics is only a 24-hour elective course in the medical school curriculum. Currently, most genetics services providers have received very limited training in genetics, and the number of such professionals is declining. One reason is that health care reform in China has driven all hospitals into a "medical free market", meaning that in order to be able to pay staff salaries, stipends and bonuses, each hospital has to make profits from the medical services it provides. Genetics services, however, make much less money compared to other medical specialties in a hospital. Therefore, nobody is willing to practice genetics without reasonable payment.

Another reason the number of genetics professionals is declining is unfair treatment in promotion to professional positions. One Chinese respondent to the international survey commented, "In the early 1980s, there were many people engaged in research and clinical work in genetics and eugenics in Liaoning province. However, in 1987, when appointments with professional titles started, unlike internal medicine, surgery, pediatrics and obstetrics/gynecology, there was no such specialty of genetics and eugenics with a professional title. Many professional personnel could not be promoted to the corresponding positions and therefore left this professional group, which had serious impact on the development of genetics and eugenics." This situation is not uncommon in other provinces of China. Because of these problems in the health care system in this country, it seems that it will take a very long time to reform the provision of genetics services.

Public Education

Public education in genetics and relevant issues appears to be the most important and effective way to reduce genetic disorders and birth defects in the general population. The Law on Maternal and Infant Health Care stipulates that the state undertake research and education in the field (Article 5), and it also specifies that the services to be provided for premarital health care include education about sex, reproduction and genetic diseases (Article 7). In our survey, 50% of the Chinese respondents thought that public education in genetics should be the top priority of the government health budget. In China, however, genetics has not become a core curriculum even in medical schools. Public education in genetics is hampered by vast illiteracy, and lack of funds and expertise in China. It is hoped that rapid growth of the national economy will eventually enhance public education about genetics and relevant issues.

Unlike developed countries, there is very limited national funding for life sciences research. The total investment in China's HGP by the National Natural Sciences Foundation was only ¥3 million (approximately US$ 361,446).[17,18] It appears that there have not been any funds to support investigation into ethical, legal and social issues. Furthermore, there has not been any sort of engagement of the public, including politicians, lawyers, social workers and clients, in discussing these issues, although there has been mass propaganda on family planning and birth control in China.

Most Chinese geneticists have had very little or no ethics training in their careers. One report said that Chinese medical students' professional behavior was significantly different among those who received top scores in medical ethics courses and those with low scores or those without such training.[42] Therefore, the present situation needs improvement. To draw up ethical guidelines and implement them in genetics practice and research in China will be a challenging task. The best approach to this task is good communication between western and Chinese geneticists, with the conditions of mutual respect and appreciation for each other's cultural values, and social and political system. The task will require great effort and hard work from Chinese pioneers in this field.

Acknowledgements

I am grateful to all of our Chinese colleagues who participated in and supported this survey. I am also grateful to Drs. Dorothy C. Wertz and John C. Fletcher for their collaboration. I would like to thank Professor Renzong Qiu for providing East China data, and Dr. Denise Sheer for allowing me to use her facilities to complete this work.

REFERENCES

1. Wen, X.G. Effect of son preference and population policy on sex ratios at birth in two provinces of China. *Journal of Biosocial Science* 1993; 25: 509–521.
2. Wu, J.P. Population and family planning in China. Verhandelingen Koninklijke Academie Voor Geneeskund Van Belgie 1994; 56(5): 383–402.
3. Xiao, K.Z. *et al.* Consecutive three-year birth defects monitoring in Sichuan province. *Heredity and Disease* 1998; 5(2): 65–68.
4. Zhang, S.Z. *et al.* The prevalence of chromosome diseases in the general population of Sichuan, China. *Clinical Genetics* 1991; 39: 81–88.
5. Liu, X.Z. *et al.* Epidemiological and genetic studies of congenital profound deafness in the general population of Sichuan, China. *American Journal of Medical Genetics* 1994; 53: 192–195.
6. Zhou, Z.L. Recent advances of perinatal medicine in China. *Chinese Medical Journal (English)* 1995; 108(5): 387–389.
7. Zuo, Q.H. *et al.* Epidemiological studies of etiology of mental retardation in children aged 0–14 years old in China. *National Medical Journal of China* 1994; 74 (3): 134–137.
8. Zeng, Y.T. and Huang, S.Z. Disorders of haemoglobin in China. *Journal of Medical Genetics* 1987; 24: 578–583.
9. Zhang, Y.Q. and Cao, X.Q. Experience in neonatal screening for congenital hypothyroidism. *Chinese Medical Journal (English)* 1993; 106(3): 216–219.
10. Xiao, K.Z. *et al.* Epidemiology of neural tube defects in China. *National Medical Journal of China* 1989; 69(1): 189–191.
11. The 1992 Annual Report of the International Clearinghouse Birth Defects Monitoring Systems.
12. Wu, M. The human genome project – a challenge and an opportunity. HUGO Pacific Genome Newsletter 1996; 2(1): 2.
13. Zhang, L.Z. *et al.* Pregnancies following *in vitro* fertilization and embryo transfer and following gamete intrafallopian transfer. *Chinese Medical Journal (English)* 1988; 101(5): 303–304.
14. Xu, L.L. *et al.* Successful pregnancy and birth after transfer of donated embryo fertilized *in vitro*. *Chinese Medical Journal (English)* 1988; 101(12): 876–878.
15. Qian, B.Q. *et al.* The Human Genome Project in China. *HUGO Pacific Genome Newsletter* 1996; 2(1): 2–4.
16. Maddox, J. and Swinbanks, D. Faith in academic excellence, in: Science in China. *Nature* 1995; 378: 547.
17. Mervis, J. China's unique environment favors large intervention trials, in: Science in China: A Great Leap Forward. *Science* 1995; 270: 1149–1151.
18. Kinoshita, J. Agriculture finds a niche; drug researchers seek help, in: *Science in China*: A Great Leap Forward. *Science* 1995; 270: 1147–1149.
19. Ye, L.Z. *et al.* Analysis of 4628 cases in the genetic counselling clinic of PUMC hospital. *Proceedings of CAMS and PUMS* 1989; 4(3): 126–130.
20. Li, L.Y. *et al.* Chromosome analyses of 2,319 cases in genetic counseling clinic. *Chinese Medical Journal (English)* 1986; 99(7): 527–534.
21. Wang, S.G. *et al.* Cytogenetic study of 1633 cases. *Chinese Medical Journal (English)* 1988; 101(4): 231–236.
22. Sun, Y.Y. 8 cases of abnormal karyotypes first reported in the world identified in periphery blood chromosome analysis of 2018 cases. Yearbook, Chinese Academy of Medical Sciences and Peking Union Medical College 1994, p. 117.
23. Liu, S.R. and Zuo, Q.H. Newborn screening for phenylketonuria in eleven districts. *Chinese Medical Journal (English)* 1986; 99(2): 113–118.

24. Guo, M. and Meng, J.X. Demography of people with disabilities in China. International *Journal of Rehabilitation Research* 1993; 16: 299–301.
25. Chen, J.H. and Simeonsson, R.J. Prevention of childhood disability in the People's Republic of China. *Child: Care, Health and Development* 1993; 19: 71–88.
26. Wang, Y. *et al*. Induced abortion in eight provinces of China. *Asia-Pacific Journal of Public Health* 1991; 5(1): 32–40.
27. Shi, H.P. *et al*. Diagnosis and prenatal diagnosis of lysosomal storage diseases. *Chinese Medical Journal (English)* 1988; 101(6): 383–387.
28. Zeng, Y.T. *et al*. Prenatal diagnosis of thalassemia: experiences at the Shanghai Children's Hospital. *Hemoglobin* 1998; 12(5&6): 796–800.
29. Zhan, B.Y. *et al*. Advances of homotransplantation of testes in China. *Chinese Medical Journal (English)* 1993; 106(7): 486–488.
33. Maddox, J. and Swinbanks, D. When is prenatal diagnosis "eugenics?," In: Science in China. *Nature* 1995; 378: 549.
34. Editorial of The Lancet. Western eyes on China's eugenics law. *The Lancet* 1995; 346: 131.
35. Opinion of Nature. China's misconception of eugenics. *Nature* 1994; 367: 1–2.
36. Dickson, D. Concern grows over China's plans to reduce number of "inferior births". *Nature* 1994; 367: 3.
38. Wu, Z.H. Conflicts between Chinese traditional ethics and bioethics. *Cambridge Quarterly of Healthcare Ethics* 1994; 3: 367–371.
39. Qiu, R.Z. Chinese medical ethics and euthanasia. *Cambridge Quarterly of Healthcare Ethics* 1993; 2: 69–76.
40. Wertz, D.C., Fletcher, J.C., and Berg, K. Guidelines on ethical issues in medical genetics and provision of genetics services. Hereditary Diseases Programme, Division of Noncommunicable Diseases, WHO, Geneva, 1995.
41. Malkin, D. and Knoppers, B.M. Genetic predisposition to cancer – issues to consider. *Seminars in Cancer Biology* 1996; 7: 49–53.
42. Lu, Y.H., Meng, X.Y., and Liu, X. Professional behavior of medical school graduates: an analysis. *Medical Education* 1994; 28: 296–303.
43. Ministry of Public Health of China: Health statistics information in China 1949–88. Beijing: People's Publishing House, 1989.

CHAPTER 14

GERMANY

Irmgard Nippert and Gerhard Wolff

PART I. GENETIC SERVICES

A. Demography

The Federal Republic of Germany (FRG) is located in central Europe. It covers a total area of 356,910 km^2. Following unification on October 3, 1990, it includes the formerly separate Federal Republic of Germany (West Germany), the German Democratic Republic (East Germany) and Berlin. In 1997 it had a population of 82 million, and a density of 230 people/km^2. The age structure is characterized by an increasing proportion of elderly people. In 1997, 16% of the population was 0–14 years old, 68% 15–64 years and 16% 65 years and over. The population growth rate in 1997 was 0.0%, the birth rate 8.98 births/1,000 population and the death rate 10.82/1,000 population.

The infant mortality rate in 1997 was 5.3. Life expectancy in 1996 was: men, 73.7 years (West Germany) and 71.2 years (East Germany); women, 80.0 years (West Germany) and 78.5 years (East Germany). The fertility rate is 1.24 (1997).

In 1996, Germany had 7.4 million residents of foreign nationality and, in 1997, there was a net migration rate of 1.87 migrant(s)/1,000 population. Over 90% of the population is ethnically German.

The ethnic divisions are German 91.5%, Turkish 2.4%, Italian 0.7%, Greek 0.4%, Polish 0.4% and other 4.6%, made up largely of people fleeing the war in the former Yugoslavia.

Of the population, 38% are Protestant, 34% are Roman Catholic and 28% are unaffiliated or belong to other religions. More than 99% of the population of 15 and over can read and write (1991); 38.7 million are in the labor force, 41% work in industry, 3% in agriculture, and 56% in other sectors (1995).

It is estimated today that it will take 10–15 years for the eastern states to match living standards in western Germany. Western Germany, with three times the per capita output of the eastern states, has an advanced market economy and is a world leader in exports.

B. Health Insurance

The German health care system is a blend of government-mandated financing by employers and employees combined with private provision of care by physicians, and controlled hospital expenditures and administration by not-for-profit insurance organizations. These insurance organizations, known as "sickness funds," establish and collect the contributions of employers and employees.

About 89% of the population is covered by the sickness funds and about 9% by private health insurance, 0.2% are not insured and the remainder have other insurance coverage. The sickness funds provide full coverage for all medically necessary services, including ambulatory and in-patient care, prescribed drugs, medical appliances, dental care and early detection of and screening for genetic diseases.

At the moment the major challenges facing Germany's health care system are rising costs; in 1995, 11.1% of the GDP was spent on health care.

The German health care system clearly favors ambulatory care by private practitioners contracting with sickness funds. This means that the economic resources of university or hospital-based genetic centers and the income generated by providing ambulatory genetic services to sickness fund patients are threatened if a private practitioner (for example, a medical genetic specialist in conjunction with an obstetrician) sets up a private practice and a laboratory near a university or hospital center. As a consequence, the university/hospital genetic center may lose patients and income.

Research money does not sustain routine genetic services. On the contrary, income stemming from reimbursement by the sickness funds is often used to finance research, the development of molecular genetic diagnostics, and the implementation of novel services.

Most, but not all, primary health care physicians are private practitioners. Private practice is mostly carried out in self-owned private practices, but there are exceptions, for instance, at German universities. Here, the directors of clinical departments are often allowed to charge for the medical services they provide.

Physicians may increase their income by increasing the numbers of diagnostic tests or treatments offered. This is a strong incentive to offer tests (fear of litigation may also be a strong incentive to offer prenatal tests). The advantage of this kind of funding is that each patient has access to most genetic tests he or she wishes. A disadvantage is that in this provider-driven system, the patient may be offered tests prematurely and/or without adequate information and counseling. This situation is not likely to last. In the future, ambulatory care will be held to a budget by implementing upper limits to a practitioner's earnings. Predictive and confirmatory genetic testing might be limited to situations of immediate medical relevance.

C. Historical, Social, and Legal Background of Medical Genetics

In post-war Germany, human genetics was totally discredited by its use in the service of the Nazi state, when prominent German human geneticists actively participated in spreading Nazi race ideology, declaring Jews to be foreign genetic material that had to be removed from the German people. A eugenic law made sterilization compulsory for a variety of illnesses thought to be genetic in origin. Thousands of children born with severe birth defects as well as mentally retarded adults were put to death in so-called mercy killings in the process of the "euthanasia" program (Burleigh, 1994).

Thus, after World War II, human genetics as a scientific discipline not only had a very doubtful reputation, but was academically in a very fragmented state. It was scarcely represented in medical school curricula. As late as 1959, only four departments of human genetics existed in medical schools in the FRG. One must bear in mind this dreadful heritage when examining the post-war development of human genetics in Germany and when looking at the public debate on controversial ethical and moral problems generated by the "new genetics". It is evident in the public debate that the past still evokes special fears and mistrusts.

In the early 1970s, with the advent of new cytogenetic and biochemical prenatal genetic testing techniques, a joint 7-year multicenter collaborative study on the diagnosis of genetic defects by university-based geneticists and obstetricians and the Deutsche Forschungsgemeinschaft led to the establishment of 41 genetic service centers all over the FRG and West Berlin. Training and manpower were funded from research sources. Thus in the 1970s, the structure was made up of genetic diagnostic services as well as genetic counseling services situated predominantly at specialized university-based centrers. This implementation and diffusion of genetic services was accompanied by the following legal and structural changes.

(1) Since 1975, the costs of prenatal genetic testing and genetic counseling have been covered by the FRG sickness funds.

(2) In 1976, Section 218 of the Civil Penal Code in the FRG legalized the interruption of pregnancies up to the 22nd week of gestation if the fetus was found to be "defective" and the pregnant woman wished to terminate the pregnancy.

All terminations required medical approval. The law in East Germany allowed abortion on request until the twelfth week of pregnancy, and permitted abortion without limit of time if the mother's physical or psychological health would be endangered or the infant would be severely disabled. The western and eastern laws were reconciled in 1995. There is no longer a special reference in the law to the fetus's condition, and no special time frame for a medically indicated abortion is required.

Under the new law, the only indications for abortion are maternal. If a fetus with a disability is aborted, it is because the birth would damage the mother's psychological or social health (maternal health indication), not because of the "quality of life" of the infant. This indication is usually described as "the anticipated burden" (Belastung) of the child on the mother.

(3) Direct state government financial support (mainly of the state universities) has allowed expansion of (prenatal) genetic testing and genetic counseling services. Cytogenetic laboratories and genetic centers were installed (mainly in medical schools) as follows: Bavaria 1975, North-Rhine-Westphalia 1976, Schleswig-Holstein 1977, Baden-Würtemberg 1978, West Berlin, Lower Saxony 1979, Bremen 1979.

D. Certification, Education, and Recognition as a Speciality

In 1978, medical genetics was acknowledged as a "medical special education" (Zusatzbezeichnung Medizinische Genetik) by the certification boards of the states (Landesärztekammer).

In May, 1992, the general assembly of German physicians (Deutscher Ärztetag, Köln 1992) agreed to acknowledge human genetics as a new speciality "Facharzt für Humangenetik". The difference between recognition in 1978 and 1992 is very important.

It took 14 years for human genetics to grow from an area that is recognized as a "Zusatzbezeichnung" to a speciality, which is official recognition that this area of medical practice (human genetics) has gained sufficient importance within the health care delivery system to merit being provided by specially trained physicians. Recognition as a speciality is first by the national board (Bundesärztetag) and subsequently by the state board of physicians (Landesärztekammer).

In 1988, the Society of Human Genetics was established, and by 1995, membership was 713. In 1988 a Board of Medical Genetics was founded, and by the end of 1996 membership was 441. The Board issues professional guidelines, provides quality control and since 1988 has published its own journal *Medizinische Genetik* in conjunction with the German Society for Human Genetics.

At present, genetics is taught in the undergraduate medical curriculum at two points. During preclinical education, which extends over two years, genetics is taught in conjunction with basic biological training. However, the amount of time devoted and the content chosen is not formally specified. Universities are free to decide on the number of courses, intensity and time requirements for medical students. During clinical training, human genetics is taught and includes an obligatory written examination. The German Society of Human Genetics actively lobbies to strengthen the role of human genetics in the medical curriculum. There are some indications that medical curriculum changes to be decided upon by the Federal Government in the future will put more emphasis on human genetics.

At the postgraduate level there are detailed established requirements for education and training in medical genetics: (1) Specialists in human genetics, "Facharzt für Humangenetik", with special training ("Fachkunde") in cytogenetic diagnostics or special training in molecular laboratory diagnostics. The training period covers 5 years of postgraduate training, including 2 years for the "Fachkunde" (subspeciality). (2) Special education in medical genetics "Zusatzbezeichnung Medizinische Genetik" (2 years of training in clinical genetics and genetic counseling after 4 years of clinical practice or recognition in a different medical speciality).

For non-medical postgraduate training there is a 3-year training program with special priorities that can be chosen either in molecular genetics, biochemical genetics, cancer genetics or teratology and mutation genetics. The Society for Human Genetics is responsible for this postgraduate program, "Fachhumangenetiker/in."

E. Current Status of Services

In 1995, there were 105 facilities specialised in genetic counseling and diagnosis, 50.5% of which were in private practice. In 1996, there were over 440 medical and non-medical specialists working in genetic service provision in private practice and secondary specialist care. No data are available on their relative numbers. University-based genetic counseling centers exist in each medical school in Germany. These combine genetic counseling with special diagnostic and research laboratories and provide access to specialist obstetric and pediatric care at the university level. There was a 32.4% increase in services over 3 years and a 53.5% increase in private practitioners performing these services. In 1993, two-thirds of all invasive prenatal genetic diagnostic procedures reimbursed by sickness funds in West Germany were performed by private practitioners. In contrast, genetic counseling is still mostly provided in the specialist care sector.

Approximately 90% of the female population is covered by sickness funds, which will reimburse the physician for genetic counseling only if the physician has qualified in the speciality of human genetics or has the Zusatzbezeichnung (special education in medical genetics).

However, private practitioners are becoming more involved, so that in 1991, 27.4% of genetic counseling services were reimbursed to private practitioners, whereas by 1993, 47.8% of all genetic counseling services reimbursed by sickness funds were rendered by private practitioners. Genetic counseling services are not growing as quickly as amniocentesis, chorionic villus sampling or isolation of DNA.

F. Access to Services

Genetic services, especially prenatal genetic services, are expanding, and there are no economic hindrances to access to genetic services in Germany. The sickness funds will reimburse for all genetic services a physician deems necessary. In theory, there are no obstacles to accessing genetic services due to age, income, race, health status or long-term need because of the comprehensive coverage by the sickness funds. In practice, however, the most serious obstacle to accessing genetic services is lack of professional education in genetics by physicians practising in primary or secondary care.

The better a patient is educated (including university or college degree), the higher the income and the more urban the area, the better access he or she has to genetic information. This was clearly shown by two utilization studies in prenatal genetic diagnosis services (Nippert, 1984; Nippert and Horst, 1994). We estimate that groups relatively disadvantaged in their access to genetic information and services may be those with low socio-economic status, those living in rural areas in northern or eastern Germany, and immigrants not fluent in German, who may have a high risk for diseases rare in central Europe, such as hemoglobinopathies, and an increased genetic risk due to consanguineous marriage.

G. Goals of Medical Genetics

There are no defined measurable outcomes for the effectiveness of genetic services generally recognized and used in Germany. Nevertheless, there is general agreement that reducing the prevalence of genetic diseases that cannot be cured is not a primary goal of genetic service provision. The 1996 position paper of the German Society for Human Genetics clearly articulates this view. Instead, it sees the primary goal as informing people about their individual genetic risks and letting *them* decide how to act upon this information. This view is shared by 93.2% of all medical geneticists in Germany according to the Wertz–Fletcher International Survey conducted in 1994/1995 (Nippert, Wolff, Ethische und Soziale Aspekte der Humangenetik,

DFG funded study 1992–1996). Therefore, implementing screening programs in prenatal care or for carrier status in adults on a population basis is not actively promoted and is not to be found on the policy agendas of the professional and scientific associations in Germany.

H. Expansion of Prenatal Diagnosis and Genetic Testing

Genetic testing, especially prenatal diagnosis (PD), has undergone rapid expansion in Germany. The German sickness funds in West Germany made the following reimbursements: 1991, 42,745 amniocenteses and chorionic villus sampling services with 722,250 live births; 1992, 49,233 amniocenteses and chorionic villus sampling services with 720,794 live births and 70,280 mothers aged 35 years, and 1993, 56,594 amniocenteses and chorionic villus sampling services with 717,915 live births but with no data for mothers aged 35 years and older.

In 1993, the average age of mothers giving birth was 30.2 years; the average age of mothers at first birth was 27.6 years; 10.5% of all mothers were >35 years and 1.1% were >40 years of age. In relation to live births, 16.5% of all pregnancies were aborted in 1993; reported abortions due to fetal abnormalities were 1.1% of all abortions.

Not all women at elevated risk have PD.

In 1992, 49,233 amniocenteses and chorionic villus sampling procedures were reimbursed by the sickness funds in West Germany. In 1992, 70,280 mothers aged 35 years and older gave birth (live births) in West Germany. Even if we assume that 10% of these mothers were privately insured and that an unknown proportion had invasive PD, these data suggest that not every pregnant woman aged ⩾35 years will undergo invasive PD. Some may opt for non-invasive procedures and special ultrasonography; some may decline invasive or non-invasive procedures entirely. In Münster (Department of Human Genetics) in 1992, 18% (n = 248) of all pregnant women eligible for invasive PD, for reasons such as maternal age, previous child with chromosomal aberration, increased risk for genetic disease and others, after pretest counseling declined the offer of PD. Of all pregnant women who came to the genetic counseling center in 1992 prior to any PD tests, 19% opted for non-invasive procedures only. Of all women who underwent invasive PD in Münster in 1992, 3.5% were listed with the indication “maternal anxiety” (average age: 30.9 years).

Only a small fraction of pregnant women receive comprehensive pretest counseling in Germany. The reasons for this less than desirable practice are manifold. In most cases, the primary care obstetrician is responsible for the pretest counseling and referral of patients if deemed necessary. There are few financial incentives for the private practitioner to carry out pretest counseling. A substantial number of obstetricians in primary care are not specially trained to offer appropriate genetic counseling (either pre- or post-test counseling).

Counseling paid for by the sickness funds increased by only 23.1% in 3 years as compared to a 32.4% increase in invasive PD. And PD is not always accompanied by pre-test genetic counseling. Professional boards have recommended that women with an increased genetic risk undergoing invasive PD should have pretest genetic counseling to ensure informed consent, not scanty information from an obstetrician or a written leaflet.

I. Consumer Satisfaction

There have been no formal measures instituted to assess consumer satisfaction or to safeguard the rights of consumers to be informed.

In a study by Nippert and Horst in 1993 for the German Federal Parliament Office for Technology Assessment, consumer satisfaction in prenatal care was one major issue (Nippert and Horst, 1994). Data were reported on complaints by pregnant women about their primary care obstetricians' lack of adequate information and violations of informed consent. Women who reported negative experiences regarding test information, violation of informed consent and coercion to undergo tests (for instance, the triple test) were less likely to consider PD in their next pregnancy than women who had been appropriately counseled.

The introduction of the triple test in Germany was not regulated. The above-mentioned study found that informed consent was not obtained in many cases. Obstetricians often did not understand that this was a risk assessment test and not a diagnostic test. As a consequence, they could not explain a positive test result to the women and scared them unnecessarily. The laboratory data were unreliable because of the uncalibrated tests. The number of "false positives" was staggering.

In a sample of 1,157 women who underwent PD at a genetic counseling center in 1992, 90.7% reported satisfaction with the way PD was delivered, 85.3% would consider using PD in their next pregnancy, and 13.2% reported that the PD was not worth the stress they had experienced (Nippert and Horst, 1994).

Some consumers have successfully sued their primary care physicians or geneticists for not providing adequate information and test results. The first successful legal action for malpractice was brought in 1984 by a couple not offered PD and to whom a child with Down syndrome was born. More physicians and obstetricians became aware that they were vulnerable to such actions if they failed to take into account, or act upon, the knowledge of an increased risk of the birth of an infant with physical or mental abnormalities. These suits have increased physicians' willingness to offer PD.

J. Patient Organizations

There are more than 70 patient organizations. Some disability organizations voice concern about new eugenics created by the new genetics and are lobbying for more legal controls and restrictions, whereas other patient organizations, for instance the Huntington group, the hereditary ataxia group, the Society for the Prevention of Muscular Diseases, Society for Mucopolysaccharidoses, German Retinitis Pigmentosa Association, Cystic Fibrosis Association, Marfan group and others are actively collaborating with human geneticists to improve genetic counseling, research, distribution of information and public awareness. In 1996, an umbrella organisation (Selbsthilfegruppen und Humangenetiker e.V.) was founded, including both patient groups and human geneticists (Rüdel, 1995).

PART II. ETHICAL ISSUES

A. Directiveness

The major ethical issue faced by professionals and patients is directiveness in counseling after PD. Tables 38–44 show the percents of geneticists who would provide pessimistic, unbiased, or optimistic counseling for a variety of genetic disorders or other conditions. Although more German geneticists reported that they would be unbiased than did geneticists in most other continental European countries (with the possible exception of The Netherlands) the percent who would attempt to offer unbiased counseling falls far short of the goals of nondirectiveness promulgated by Fraser (1974) and by professional societies in English-speaking countries.

A comparison between West German and former East German respondents found that former East Germans were more directive than West Germans for 10 of 26 conditions listed (Cohen *et al.*, 1997). Former East Germans were also more pessimistic than West Germans for Down syndrome, sickle cell anaemia, cystic fibrosis, Huntington disease, familial hypercholesterolemia, XXY, fetal HIV infection, fetal rubella, Hurler syndrome, severe open spina bifida, and trisomy 13. As in some other socialist or former socialist nations in the international survey, these geneticists may have felt that children have "a right to be wanted," and that a child not wanted, for any reason (including genetic abnormalities) may be better off not being born. In socialist nations, there was also a societal interest in avoiding birth defects and a belief that children are accepted as human beings only after they have been integrated into the social system, and a belief that parents would presumably not take the trouble to socialize an unwanted child.

The new German abortion law encourages people to consider the mother's perspective on having a child with a disability. This, rather than the child's own "quality of life" is the determining factor in indications for abortion.

In order to find out how geneticists assess the importance and the consequences of the concept of nondirectiveness in genetic counseling, one of us (G.W.) added a lengthy series of questions to the international survey; these were asked only in Germany. Over 90% of the respondents did not agree that a medical doctor is sufficiently educated in regard to the doctor-patient relationship so that he/she does not need a special rule like nondirectiveness. There was a more than 75% consensus that

- nondirectiveness is a necessary ethical standard for genetic counseling (91.4%);
- nondirectiveness is a sound principle for acting in a responsible way with genetic knowledge (79.1%);
- the "golden rule" that the doctor should do nothing he or she would not wish for himself or his/her close relatives, is not sufficient for genetic counseling (82.4%);
- a medical doctor should not be active in genetic counseling without a special education and qualification in counseling techniques (81.8%);
- good genetic counseling includes addressing psychological and ethical questions without being asked to do so by the counselee (85.3%).

Nevertheless, a majority (72.2%) agreed that most of the genetic counselors are acting more directively than they are willing to admit. Nearly 1/3 (29.8%) agreed that the counselor also has to decide, together with the patient, which are the best options for the patient. Therefore, the attitudes surfacing in the answers seem to fulfil the purpose of demonstrating a responsible way of dealing with a technique (genetic testing) considered potentially harmful, rather than describing a current practice (genetic counseling). Moreover, several answers show that the evolution of the concept of nondirectiveness and its implications for genetic counseling (Wolff and Jung, 1995) have not been sufficiently reflected upon and acknowledged by human geneticists.

B. Personal Attitudes toward Abortion

Comparison between personal attitudes toward abortion and professional direction (pessimistic or optimistic) in counseling showed a significant ($p = 0.02$) relationship between personal attitudes and professional counseling among West Germans for 25 of 26 conditions listed in the questionnaire, and for 13 of 26 conditions among former East Germans (Cohen *et al.*, 1997). Thus, personal bias is influencing counseling.

C. Patient Autonomy

There is a growing trend around the world for people to demand respect for their autonomy. As in the US [see Statistical Appendix and chapter on patients in Germany believed that they had rights to information and services. Especially telling are their beliefs that withholding any requested service is a denial of patients' rights (66%) and that patients are entitled to any service they request and can pay for out-of-pocket (48%). Although these percents are somewhat lower than in the United States, they follow an emerging pattern of patient beliefs about entitlement.

D. Testing Children

Current professional guidelines (Stellungnahme zur genetischen Diagnostik bei Kindern und Jugendlichen, 1995) state that children should be tested only if there is a medical benefit to the child.

E. Disclosure to Relatives

Disclosure to relatives at risk, against a patient's wishes, is one of the most difficult problems in medical genetics.

Current professional recommendations in regard to this problem in Germany state that "if medical genetic information is not shared with family members, it is possible that information that some members would like to have and that is important and relevant to their health is withheld from them. In principle, this is an unresolvable conflict for a medical geneticist with this information. No matter what is done, important ethical principles are unavoidably compromized. Geneticists who press for the information to be relayed or who themselves relay the information violate patient autonomy and in some cases confidentiality, and by not relaying the information, they may violate their obligation to offer medical assistance. There are no general rules to deal with this dilemma; rather, each individual case must be separately considered with inclusion of as many as possible of the parties involved. With disorders that are neither treatable nor preventable, the right of self-determination concerning information takes precedence over the right to obtain information. If family members themselves have not requested information, it should not be given to them." (Positionspapier der Gesellschaft für Humangenetik e.V. (GfH). Med. Genetik 1996; 8: 125–131) see also: "Leitlinien genetischer Beratung (Principles of genetic counseling) of the Professional Association of Medical Genetics (Berufsverband Medizinische Genetik, BMG) Med Genetik 8/3 (1996) Sonderbeilage, S. 1–2).

F. Privacy

German privacy laws require written consent for third-party access to genetic information. The "Positionspapier" of the German Society of Human Genetics sets the following standard. "Genetic data must be particularly protected from third-party interest and inquiry, since these data can be considered the core of an individual's person. Genetic data should not be generally accessible, but only in specific instances when the affected individual has been completely informed of the usefulness and purpose of passing on the data and has given written permission to release them. In each individual case, before the physician releases his patient's data at the latter's request, the physician should weigh whether the data are absolutely relevant to the question at hand and to the person to be informed and whether the patient is aware of all of the facts that may have to be relayed and the significance of relaying the information. Without written exemption from professional confidentiality, inquiries about patient data must not be complied with. A medical geneticist is allowed to release data only if the patient expressly addresses a request for release of data to him, if the purpose of releasing the data is defined, and if the examination results and the genetic counseling content to be passed on are specified. Furthermore, the medical geneticist should establish that the patient is aware of all possible consequences of releasing the data. These obligations clearly go beyond the general concept of confidentiality in medical practice." (Positionspapier der Gesellschaft für Humangenetik e.V. (GfH). Med. Genetik 1996; 8: 125–131:)

G. Provider–Patient Differences

The surveys suggest that there are substantial differences in ethical views between providers and the people they serve. There is need for (1) more extensive patient surveys to validate the extent of differences; (2) consultations between patients and professional organizations to develop dialogue; (3) incorporation of patient views into counselor education; (4) increased public education and discussion of ethical issues.

PART III. CONCLUSION: LINGERING EFFECTS OF THE PAST

The experience of the Third Reich still affects German geneticists, although no one practicing today had professional experience under Nazism. Responses to a special set of questions appended to the German version of the international questionnaire showed that the majority agreed that, due to the history of crimes and human rights abuses through the eugenics programs under the Nazis, German geneticists have a special responsibility toward persons with genetic disorders (77%) and are more aware of ethical issues in human genetics than geneticists in other countries (69%). In addition, most respondents (64%) felt that ethical and social problems in genetics cannot be adequately discussed in public. When asked whether they feel that it is fair to blame today's German geneticists for past abuses of genetics under the Third Reich, 40% agreed, 35% disagreed, and 25% were neutral.

The greater nondirectiveness (at least in comparison with most other countries in continental Europe) of German geneticists may reflect lingering guilt over the Third Reich, even though German geneticists were not nondirective by North American or British standards. As in all questionnaire studies, there is some lingering doubt as to whether responses reflect actual practice. One could ask whether some responses, including nondirectiveness or respect for people with disabilities, reflected, not actual practice, but a belief that practices of the Third Reich must not return.

Even 30 years after the end of World War II, human genetics and psychiatry textbooks in Germany or elsewhere failed to acknowledge the past abuses of these sciences (Mueller-Hill, 1991). While privately distancing themselves from the crimes of their predecessors, German geneticists have only recently begun the process of dealing with this past in public writings, meetings, and discussions (Vogel, 1988, 1994). Sensitivity to ethical issues may connote not only care and conscience, but also guilt and shame. The experience of the Third Reich may have stifled open discussion of ethical issues. It is now time for the German genetics profession to emerge from this shadow and to develop its own direction and profile as a mature profession.

REFERENCES

Burleigh, M. *Death and Deliverance: "Euthanasia" in Germany c. 1900–1945*. Cambridge, UK: Cambridge University Press, 1994.

Cohen, P.E., Wertz, D.C., Nippert, I., and Wolff, G. Genetic counseling practices in Germany: a comparison between East German and West German geneticists. *Journal of Genetic Counseling* 1997; 6: 61–80.

Fraser, F.C. Genetic counseling. *American Journal of Human Genetics* 1974; 26: 636–659.

Mueller-Hill, B. Bioscience in totalitarian regimes: the lesson to be learned from Nazi Germany. In: Roy, D., Wynne, B. and Old, R. (eds), *Bioscience and Society*. New York: John Wiley & Sons, 1991, pp. 67–76.

Nippert, I. and Horst, J. Die Anwendungsproblematik der pränatalen Diagnose aus der Sicht von Beratenen und Beratern – unter besonderer Berücksichtigung der derzeitigen und zukünftig möglichen Nutzung der Genomanalyse. Gutachten im Auftrag des Büros für Technikfolgen-Abschätzung (TAB) beim Deutschen Bundestag, veröffentlicht als TAB-Hintergrundpapier Nr. 2, Bonn, Januar 1994.

Nippert, I. and Pawlowitzki, I.H. Who gets Amniocentesis? Utilization of prenatal cytogenetic diagnosis by Reg.-Bez. Münster residents. In: Eimeren, W. v., Engelbrecht, R. and Flagle, Ch.D. (eds), *3rd International Conference on System Science in Health Care*. Berlin-Heidelberg-New York: Springer-Verlag, 1984, pp. 612–614.

Positionspapier der Gesellschaft für Humangenetik e.V. (GfH), *Medizinische Genetik* 1996; 8: 125–131.

Rüdel, R. Selbsthilfegruppen und Humangenetiker beschließen Gründung eines Dachverbandes. *Medizinische Genetik* 1995; 7(1): 48.

Stellungnahme zur genetischen Diagnostik bei Kindern und Jugendlichen (GfH). *Medizinische Genetik* 1995; 7: 358–359.

Vogel, F. Human genetics and the responsibility of the doctor. In: Jonas, H., Vogel, F., Catenhusen, W.-M. and Reiter, J. (eds), *Ethics and Genetic Technology*. Gesellschaft Gensundheit and Forschung, e.V, Frankfurt, 1998, pp. 25–37 [German].

Vogel, W. On the self-understanding of genetic counseling. In: Neuer-Miebach, T. and Tarneden, R. (eds), *The Right to be Different: Inquiry into Prenatal Diagnosis and Human Genetic Counseling*. Marburg: Lebenshilfe-Verlag, 1994, pp. 53–60 [German].

Wolff, G. and Jung, C. Nondirectiveness and genetic counseling. *Journal of Genetic Counseling* 1995; 4: 3–26.

CHAPTER 15

HUNGARIAN TV VIEWERS' RESPONSES

Andrew E. Geizel and Laslo Timár

Our Hungarian colleagues took eight cases from the survey to the Hungarian public, in the "Health for All" TV program. This 30-minute program was shown once per month on TV Channel II (one of Hungary's three national channels) in the late afternoon on Thursdays, in general between 5 and 6 p.m. The estimated size of the viewing audience ("the public") was between 0.8 and 1 million. The cases constituted the introductory part of the "Health for All" program, and each received about 10 minutes. One of us (A.E.C.) wrote a screenplay for each case with the roles played by amateur actors and actresses. Afterward, the medical genetics and social implications of the issue were discussed. At the end of this discussion, viewers were asked to express their opinions by responding to fixed questions from the international survey. These questions were shown on the TV program, and viewers were informed that the text of the play and discussion, and the questions would also be published in a popular women's magazine. The magazine presented the questions in a "box" at the end of the text. Thus people had two possibilities to express their opinions: (1) reply to the questions and cut out the box from the text of the magazine and mail it to the Editor or (2) call the telephone numbers of our Family Planning Clinic, which were used for this purpose during the study period. One-third of the replies were mailed, two-thirds were received by telephone. Unfortunately, we have no knowledge about the sex of the respondents (however, the great majority of people who responded by telephone were women), nor did we learn the age, professional background, or other personal information of respondents. In addition to the telephone calls and mailed-in responses, we received more than 50 letters were sent to us with special personal opinions.

COMPARISON BETWEEN PROFESSIONALS' AND PUBLIC'S VIEWS

The following section describes the eight questions and compares the responses of the Hungarian geneticists and the public.

Case 1. Views on Confidentiality Versus Public Safety (see also Table 75)

Possible actions	Professionals	Public
1. Preserve the patient's confidentiality	2 (6%)	6 (1%)
2. Tell the transportation department if the patient refuses to do so	2 (6%)	202 (42%)
3. Warn the transportation department that passengers are at risk because of the medical status of a bus driver and ask them to have complete physical examinations for all their drivers, but *do not* tell them about your patient	24 (69%)	101 (21%)
4. Warn the transportation department that passengers are at risk because of the medical status of a bus driver and ask them to have complete physical examinations for all their drivers. If the department will not do so, tell them about your patient	7 (20%)	173 (36%)
Total	35 (100%)	482 (100%)

Obviously there was a significant difference ($\chi^2 = 48.01$; $p < 0.01$) between the opinions of professionals and the public. Professionals, mainly medical doctors, appreciated the patient's right to privacy or confidentiality (medical secrecy), while most of the public prefer the interest of the passengers (i.e., the society).

Case 2. Views on Prenatal Diagnosis for Maternal Anxiety (see also Table 25). If the woman was well-educated aged 25 with no first-hand knowledge of Down syndrome. I would or I would suggest.

	Perform the test (insurance should pay)	Perform the test only if she can pay out-of-pocket (insurance should *not* pay) n = 34 professionals, 296 public	Refer to someone who might perform the test	Refuse to perform the test and *not* refer
Professionals	3 (9%)	18 (53%)	3 (9%)	10 (30%)
Public	182 (62%)	108 (37%)	n/a	6 (2%)

Hungarian professionals and public had significantly different ($p < 0.001$) opinions. While 30% of professionals refused to perform the test in the second scenario, only 2% of the public thought they should refuse. In addition, the proportion of the public who wanted the doctor to perform the test and charge the insurance was significantly higher (6.6 times) than that of the professionals. These differences seem to indicate that people do not like the paternalism of medical doctors and they want them to provide more rights and options for patients and/or prospective parents.

Case 3. Views on Parenthood for a Blind Couple (see also Table 53)

	(n = 34 geneticists, 236 public)			p
	Agree	Neither agree nor disagree	Disagree	
Their decision may be the optimal one for them, in view of their situation and parenting goals				
Professionals	9 (27%)	11 (32%)	14 (41%)	0.001
Public	78 (33%)	6 (23%)	152 (64%)	
Their decision is unfair to the child				
Professionals	24 (69%)	6 (17%)	5 (14%)	0.001
Public	168 (71%)	0 (0%)	68 (29%)	
Their decision is unfair to society				
Professionals	15 (44%)	13 (38%)	6 (18%)	0.001
Public	82 (35%)	0 (0%)	154 (65%)	

Case 4. Views on Prenatal Paternity Testing (see also Table 24)

	Professionals	Public
1. Perform prenatal paternity testing without comment on her personal situation	2 (6%)	11 (7%)
2. Perform prenatal paternity testing with some comment on the possible negative consequences of her actions	12 (38%)	12 (8%)
3. Perform prenatal paternity testing only if she commits herself to long-term psychological counseling	2 (6%)	8 (5%)
4. Refer to someone who might perform prenatal paternity testing	7 (22%)	—
5. Refuse to perform prenatal paternity testing and *not* refer	9 (28%)	69 (46%)
6. Other	0 (0%)	51 (34%)
Total	32 (100%)	151 (100%)

For the public, point 4 was omitted and "and *not* refer" was also omitted in point 5. Instead of "other" as the sixth option, for the public we had a concrete alternative: "Perform prenatal paternity testing in the hope of saving the life of the fetus."

The difference was calculated only for points 1–3 and 5, and it was significant ($\chi^2_3 = 63.42$; $p < 0.01$) in the distribution of replies between the professionals and public. The difference is explained mainly by a significantly higher "refusal" reply to point 5 by the public (some women stressed that it is not good for a baby to be born in such a familial situation). Thus, people were more willing than professionals to refuse the test to protect the interests of the unborn child. One third of the public gave a positive reply to the new alternative, perform the procedures to protect the life of fetus.

Case 5. Views on Genetic Testing for Breast Cancer (see also Table 57)

	(*n* = 23 professionals, 420 public)			*p*
	Agree	Neither agree nor disagree	Disagree	
1. Tell her 42-year-old (healthy) sister that she is at risk for cancer, even if the patient will not tell her				
Professionals	12 (52%)	5 (22%)	6 (26%)	<0.001
Public	21 (5%)	0 (0%)	399 (95%)	
2. Encourage her 20-year-old daughter to be tested				
Professionals	3 (13%)	5 (22%)	15 (65%)	ns
Public	80 (19%)	59 (14%)	281 (67%)	
3. Encourage her 13-year-old daughter to be tested now (with parental permission)				
Professionals	1 (43%)	12 (52%)	10 (44%)	<0.001
Public	265 (63%)	10 (2%)	145 (35%)	
4. Suggest a prophylactic double mastectomy for the 42-year-old woman				
Professionals	3 (13%)	0 (0%)	20 (87%)	<0.001
Public	338 (81%)	2 (0.5%)	80 (19%)	

The percentage of professionals who agreed to tell the patient's 42-year-old healthy sister that she is at risk for cancer was more than 10 times the percentage of the public. There was no significant difference in opinions on point 2. The majority of the public agreed that she should encourage her 13-year-old daughter to be tested now. Over half of the professionals were neutral on this point. The replies reflect the different knowledge and motivations of professionals and public. The public do not want to disturb the life of young girl and they want her to avoid the psychological stress of a mastectomy, but they do not understand the long term consequences of this advice. However, the great majority of the public agreed (81%) that the doctor should suggest prophylactic double mastectomy for the 40-year-old patient (point 4), while only 13% of professionals said they would suggest it.

Case 6. Views on Preconception Sex Selection (see also Table 22)

	(*n* = 36 professionals, 459 public)			*p*
	Agree	Neither agree nor disagree	Disagree	
1. It should be available to everyone, with no restrictions				
Professionals	3 (9%)	2 (6%)	31 (86%)	0.27
Public	60 (35%)	9 (2%)	290 (63%)	
2. It should be available only to married couples				
Professionals	0 (0%)	5 (14%)	31 (86%)	—

Case 6. (continued)

	(*n* = 36 professionals, 459 public)			*p*
	Agree	Neither agree nor disagree	Disagree	
3. It should be available only to families with 3 or more children of the same sex				
Professionals	19 (54%)	9 (26%)	7 (20%)	<0.001
Public	199 (43%)	0 (0%)	260 (57%)	
4. National public health insurance should pay for it				
Professionals	6 (17%)	5 (14%)	25 (69%)	0.004
Public	81 (18%)	10 (2%)	368 (80%)	
5. Its use would be morally justified				
Professionals	7 (10%)	6 (17%)	23 (64%)	
6. It would benefit society				
Professionals	3 (8%)	9 (25%)	24 (67%)	
7. I would use it myself				
Professionals	3 (8%)	6 (17%)	27 (75%)	<0.001
Public	160 (35%)	1 (0.2%)	298 (65%)	

Professionals were more conservative on this topic than the public (see points 1 and 7). This is reflected especially in point 3.

Case 7. Views on Nontreatment of Handicapped Newborn (see also Table 35)

	(*n* = 34 professionals, 1042 public)	
	Support their decision	Would you suggest giving up baby for adoption? yes
If the disorder were		
Trisomy 13 (professionals)	30 (88%)	6 (20%)
45,X (professionals)	5 (15%)	18 (56%)
Trisomy 21 (professionals)	14 (41%)	13 (42%)
Severe, open spina bifida		
Professionals	29 (85%)	7 (24%)
Public	1019 (98%)	82 (8%)

Among the public, 96% said "the parents have the right to decide the life of their severely affected child," and 85% said "the ethics committee has a right to overrule the decision of the parents to protect the life of their severely affected child."

Here the trend was similar in the replies of professionals and public on the question of supporting the parents' decision. Our two extra questions, however, show a conflict in the public's thinking. In their opinion, parents have a right to decide about the life of their severely affected child. However, the majority agreed with the right of the ethics committee to overrule a parental decision if it is possible to save the life of a severely affected child. It is difficult to understand their final point. Is it possible that people agree that parents have the right to decide only with supervision?

Case 8. Views on Parenthood for a Woman with Fragile X (see also Table 33)

	Experts (n = 34) Yes	Public (n = 652) Yes	p
1. Long-acting contraceptives, such as Norplant, Depo-provera, or an intra-uterine device (IUD)	32 (94%)	188 (29%)	<0.001
2. Surgical sterilization	11 (32%)	500 (77%)	<0.001
3. An intense sex education program focussing on contraceptive methods such as pills or barrier methods	20 (59%)		
4. Forbidding unaccompanied male guests in her apartment	1 (3%)		
5. Providing social support so that she can carry out her wish to have children	11 (32%)	111 (17%)	0.009

Most professionals favored long-acting contraceptives. Their answers make sense; there is less support for permanent sterilization than reversible contraception. Trying to forbid male guests would be an infringement on her rights and/or too difficult to enforce. The public favored surgical sterilization. In Hungary this method is rarely used. The public's attitude may reflect the belief that the birth of potentially retarded babies is too burdensome to society.

VIEWS ON AUTONOMY

Responses to the case of the bus driver possibly posing a public danger explicitly acknowledged that the public is willing to diminish the autonomy of one person to protect the health and/or life of many persons, and truth-telling is appreciated more than the right to confidentiality or the patient's secrets. Thus, people seemed to believe, at least in this case, that social interest has a priority in comparison with individual autonomy. This was not the case for the majority of the responding Hungarian geneticists.

The public seemed to accept the concept of personal autonomy, e.g., the unconditional right to self-determination, including the right to make decisions among reproductive choices. Thus, the public was willing to provide more rights for patients and/or prospective parents, to allow them to use available technical options, and to limit the physician's rights to refuse to perform a procedure. The choice of action for the woman with fragile X (case 8) showed that the public (and the majority of professionals) preferred to diminish the autonomy of an incapacitated person in order to protect the right of her child to be born healthy or to prevent what they considered an undesirable, wrongful birth. In such a case, people indicated a stronger concern for the rights of the potential offspring than the rights of the parents. This opinion seems to support the physician's responsibility to provide protection for potential offspring against maternal requests, at least if the prospective mother is mentally retarded. The public does not seem to recognize that to set a precedent may harm the moral order, or may be the beginning of a slippery slope. They need a more balanced sense of rights and possible benefits and harms to prospective parents and their potential children.

There were significant differences in the choices of action among professionals due to their different experiences, activities (e.g., obstetricians interested in prenatal diagnosis and pregnancy termination), professional and personal experience with disability, religion, and educational backgrounds. Due to their knowledge of available regulations, ethical conviction, and higher education, choices of action by professionals are nearer to rule-based principles than choices by the public, who generally did not have such knowledge or background.

The public is more extreme than the Hungarian geneticists in their views on the prevention of genetic disorders, and more willing to suggest the introduction of new technologies such as preconception sex selection.

The public seemed unaware of the economic consequences of their of health care choices. The public thought all available services should be provided on request and that there should be entitlement to full insurance coverage for all services, even though this would diminish the economic interest of the taxpayers. Thus, the concept of patients having a right to any service that they can pay for out of pocket is not popular among the Hungarian public, who believe that the government should pay.

CHAPTER 16

INDIA

I.C. Verma and Kusum Verma

POVERTY AND ETHICAL PRINCIPLES

The fundamental ethical principles that govern the practice of genetic medicine are patient autonomy, beneficence, non-maleficence, and justice. In India and other developing countries the application of these principles is influenced by poverty and numerous social factors. Table 1 compares the demographic indicators in India with those in Thailand, Japan, US, and UK (UNICEF, 1998). It emphasizes the huge population in India, high birth rate, and an infant mortality rate 9–18 times more than that of the US or Japan. Most striking is the per capita income, which is 79 times less than that in the US, and 116 times less than that in Japan. Adult literacy is also almost 1/2 of that observed in US/Japan. It is thus easy to see that the majority of the people are struggling for survival, and have limited resources for their basic needs like food, housing, education, and health. For them "survival is more important than ethics" (Siva Subramanian, 1986). This position is not well appreciated in the West. For example, child labor in India has been the cause of much hue and cry in the West (Flor Cruz *et al.*, 1996). The fact remains that these children have to engage in labor in order to survive. For them "no labor" means no or little food, malnutrition, misery, and slow death. Therefore attempts in western countries to boycott the products of those industries which employ children are ill-founded (Table 1).

Table 1. Demographic Indicators in Different Countries

	India	Thailand	Japan	US	UK
Population (millions)	944.5	58.7	125.4	269.4	58.1
Annual births (millions)	24.3	0.99	1.28	3.83	0.70
Infant mortality rate/1000	73	31	4	8	6
Under 5 mortality rate	111	38	6	8	7
Life expectancy	62	69	80	76	77
Adult literacy	52	94	>95	99	>95
GNP (US$/capita)	340	2,740	39,640	26,980	18,700

Modified from "The State of the World's Children 1998," UNICEF.

CONSUMER PROTECTION ACT AND MEDICAL GENETIC PRACTICE

In India, traditionally there is great respect for the doctor, who is considered almost a demi-god (Advani, 1980). The patient thus relies heavily on the doctor's advice to decide what should be done. The doctor's word is rarely questioned and is scrupulously followed. Only in recent times have the doctors been brought under the purview of the Consumer Protection Act (1986). The Act aims to protect the rights of consumers through settlement of disputes by consumer tribunals at the district, state, and national level. The Act defines a consumer "as any person who buys goods or hires services for a consideration." This definition is broad enough to encompass a patient as a consumer who obtains professional services from a physician for a fee (Singh, 1994). The medical profession, expectedly, opposed the application of the Act to services provided by doctors. However, the Supreme Court upheld the contention that the Act should apply to medical professionals. The Government hospitals, because they do not charge fees for their services, or charge only for the materials used for patient care, are not under the purview of this Act.

It is generally believed that it was the corrosion of medical ethics in medical practice that made Parliament place the medical profession under this Act. Doctors were seen as behaving as traders. The doctors argued that this would slash the thread of faith between the doctor and the patient. It would lead to the practice of defensive medicine, as in western countries, and raise the cost of medical care even further.

This Act has now been applied to genetic practice, especially prenatal diagnosis. In a couple of cases a child with Down syndrome was born, while a normal child had been predicted on analysis of chorionic villus samples. These cases were settled out of court. Recently in Madras a woman gave birth to a child with limb defects. These had not been reported on three ultrasound studies carried out during pregnancy. The parents sued the ultrasonologist and displayed the child with the limb defects in court. Rs. 5,000,000 were claimed in damages, but the judge allotted Rs. 500,000 as damages. This has induced another couple who bore a child with limb defects, which were also missed on ultrasound studies, to file a case. The result of the case is awaited.

The ethics of ultrasound clinics and genetic laboratories running without any check on their competence or their performance is clear. The Department of Science and Technology has developed criteria for basic requirements for laboratories engaged in different types of tests. There will be accredition by request, and the committee hopes that patients will patronize those labs which are accredited by a central board to get a reliable and accurate service. In the opinion of the present author, registration and approval of the laboratories or the ultrasound clinics should be a mandatory requirement.

GENETIC COUNSELING

Let us consider the case of a young couple from a village. The wife had a pregnancy of 15 weeks. Maternal serum alpha-fetoprotein was raised to 4 MOM (multiples of the median). Ultrasound showed that the baby had a small spina bifida. In keeping with usual practice, non-directive counseling was given, the burden of the disease explained, and the couple were asked to make their choice. They were unable to decide and asked to be told in clear terms whether they should abort or continue the pregnancy. It was explained to them that most persons in India under similar circumstances would choose abortion. One can see, therefore, that if the patient is illiterate, and has poor understanding of genetics, non-directive counseling does not work, and the genetic counselor has to give more directive counseling. In view of the Consumer Protection Act, geneticists would tend to give counseling in a non-directive fashion. However, as admitted by most genetic counselors, it is practically impossible to be non-directive in genetic counseling. Once the counselor has gathered information about a client's problems, needs, and aspirations he or she is obliged to provide information to suit the client's particular set of circumstances. Therefore it is in the interest of the client to get directive counseling, but he or she is free to accept it or reject it. It should be mentioned that in the West the major objection to directive counseling comes from advising abortion in the presence of a fetus with a defect. This is not an issue in India, where the presence of a fetus affected with some disease is generally considered an indication for abortion by the majority of the people.

PRENATAL DIAGNOSIS

The burden of genetic disease is high in developing countries, due to the limited facilities for rehabilitation, and the socio-economic cost of looking after these children. The quality of life is more important than sanctity of life. Therefore prevention, prenatal screening, and prenatal diagnosis of genetic disease are important for the people (Verma *et al.*, 1992). Table 2 lists the reasons for genetic counseling over a 6-month-period in Sir Ganga Ram Hospital. This is a private trust hospital, and about 30% of patients are treated free of charge through the general out-patient department. It should be noted that the commonest indication for seeking genetic counseling is "reproductive genetics," of which prenatal diagnosis is the most important. In fact, many parents will not fully investigate their child with genetic disease until they decide to have another child, and it becomes necessary to define the exact cause in the affected child. It is so because in their perception "why investigate the cause of a disease if there is no treatment for it."

Table 2. Genetic Counseling at Sir Ganga Ram Hospital ($n = 1370$)

Indication	No.	%
Reproductive genetics	525	38.9
Prenatal diagnosis	288	
Recurrent spontaneous abortions	119	
Infertility	63	
TORCH infections*	30	
Teratogenic drugs	25	
Mental retardation	132	27.8
Thalassemia, hemophilia, leukemia	52	10.9
Muscular dystrophy, spinal muscular atrophy	41	8.6
Spina bifida, malformations	32	6.7
Short stature, hypogonadism, pubertal problems	19	4.0
Skeletal dysplasias, craniosynostosis	14	2.9

*TORCH = toxoplasmosis, rubella, cytomegalovirus, and herpes simplex.

It would be useless to do prenatal diagnosis, unless one can carry out abortion of an affected fetus. In 1971, India passed the Medical Termination of Pregnancy Act. It was enacted not for encouraging family planning, but for preventing and stopping the large scale illegal and unhygienic abortions which were being done, resulting in many deaths from complications of the procedure. The Act liberalized the abortion laws, and abortion is currently readily available in government and non-government clinics, avoiding many maternal deaths (Verma and Singh, 1991).

Soon the technology for identification of the Y chromosome by fluorescence became available, along with ultrasonography to visualize the fetus. Access to fetal tissues by chorionic villus sampling and amniocentesis became quite easy. This immediately led to this technology being used for diagnosis of fetal sex in couples who had many daughters and no son. The reasons for having at least one son in India have been well documented by Jairup Singh (1998). The practice of prenatal sex determination for social indications became wide-spread in Punjab and Maharashtra (Verma, 1992; Patel, 1995). It was common to see advertisements in newspapers by persuasive argument they used was "spend Rs. 500 now and save Rs. 500,000 to be spent as dowry." This resulted in a large number of abortions of female fetuses. There were fears that the already decreasing sex ratio of females as compared to males would worsen further. Jairup Singh showed this in a sample in Amritsar district where the sex ratio of females to 1000 males declined from 883 in 1990 to 770 in 1996 (Singh, 1998). The Government had therefore to act and show its commitment. There were people, including scientists and doctors, who argued that we should allow prenatal diagnosis of sex for couples with 2–3 daughters. However, due to large-scale female foeticide in the cities, the Government of India passed the Prenatal Diagnostic Techniques (Regulation and Prevention of Misuse) Act, 1995. The Act requires that all prenatal diagnostic centers should be registered and that records be maintained of all prenatal diagnostic procedures performed. The indication for carrying out the test is to be specified and recorded in each case. It stipulates heavy penalties for misuse. But it has been only partially successful. Recently a report appeared in a prominent national daily, "The Statesman," entitled "Calcutta doctor conducting fetal sex tests on the sly." Legislation to end deep-rooted cultural practices and preferences are not very effective in practice, especially when the machinery to enforce and monitor compliance is weak (Kabra, 1997).

The current trends in prenatal diagnosis can be gauged from the authors' experience with thalassemia. Almost all parents who have a child affected with thalassemia realize the severe social and economic burden of looking after such a child. Repeated blood transfusions (each transfusion costing about Rs. 1,000) with a risk of transfusion-transmitted infections, and cost of iron chelating agents means that the total expense of care per year amounts to about Rs. 100,000. Therefore, literally all those who have the knowledge about the facilities for prenatal diagnosis for this disorder opt for it. Those who have no normal child and only an affected child are satisfied by the diagnosis of thalassemia and do not desire diagnosis of sex. Many of those who have one daughter and an affected child often desire that sexing be done at the same time. Since law does not permit us to do this, they would come to us late in pregnancy after confirming the sex of the baby on ultrasound elsewhere. Many who know about bone marrow transplantation, which when successful is a definitive cure for the disease, desire that HLA typing of the fetus be done, as they desire a child who would have an

HLA match to the affected child. Would they abort the fetus if it did not match? Only the future will tell. So far we did this in only one case, and although the fetus was only haplo-identical and did not match completely, the parents decided to continue the pregnancy.

ARTIFICIAL INSEMINATION

Assisted reproductive technologies have been a great boon to many childless couples. Artificial insemination is one such method. So far there has been absence of clear guidelines and explicit legal sanction. Realizing the surreptitious use of this technique and the risk of exploitation, the Delhi Legislative Assembly enacted the Delhi Artificial Insemination (Human) Act in 1995. It legalizes donation of semen and ovum, and provides for compulsory registration of the banks for storage, sale, donation, and supply of semen. It lays down safeguards for preventing transmission of infections and prohibits segregation of X and Y chromosome bearing sperm. The Act states that complete records of the donor should be kept, while assuring complete secrecy. The enactment of the law is not the end of all problems, but can also be the beginning of many. Only its working over a period of time will prove its utility (Kusum, 1995).

ETHICS OF GENOME RESEARCH

India offers an ideal setting for genome studies – a vast population, a large number of castes and communities (4,635), eight religious categories, and 325 languages (Singh, 1992). Consanguineous marriages occur in large parts of India. In South India, about 40% of couples among all religious groups marry consanguineously. In North India consanguineous marriages are mostly among Muslims. These families provide an ideal setting to search for new genes by the technique of "homozygosity mapping." Families in China, which also has a large population, have only 1–2 children and are thus no longer attractive for such studies. Africa has families with large numbers of children, but paternity problems invalidate linkage studies. Thus India has become the hunting ground for such studies (Verma, 1997). Since laboratories with facilities for identifying new genes are very few in India, samples from these interesting families are being sent to scientists in western counties. Many argue that this represents a loss of "genetic resources." Already the Indian Government has shown great concern at the biological resources in the country being plundered by foreign companies and scientists (Jayaraman, 1997). Draft legislation has been drawn up to protect its biological resources from being exploited by foreigners without sharing the benefits with local people. Indian scientists who transfer commercially valuable research material to foreign nationals without official approval will be punished. With this background, the Government of India issued a memorandum that permission of the Indian Council of Medical Research (ICMR) would be required before samples are sent abroad, either for research or for diagnostic purposes (ICMR, 1992). More recently, these rules have been liberalized (Memorandum, 1997). The exchange of material for diagnostic or therapeutic purposes for individual cases has been permitted, if this exchange is considered necessary by the doctor concerned.

No permission is required from any authority for this purpose. Where exchange of material is envisaged as part of a collaborative research project, the project proposal as a whole must be routed through the appropriate authority for evaluation and clearance. The exchange of human material should be an integral part of a collaborative project, which should be approved by the institutional review and ethical committee. In order to protect the rights of the Indian study subjects as well Indian scientists and organizations, memoranda of understanding and/or agreements on material transfer should be entered into between the collaborating partners, covering all aspects such as purpose of transfer, research to be carried out, filing of patents, arrangements for future commercial exploitation etc. Premier and national medical institutions in the country have been given the authority to approve the sending of samples abroad. However, ICMR has to be informed and this organization will maintain a register of all such joint projects and permissions. In case of involvement of smaller or private institutions the application for approval has to be routed through the ICMR. Hopefully, these rules would allow foreign collaboration as well as ensure that the country will get its due when these samples are sent abroad.

ETHICS OF HUMAN DIVERSITY STUDIES IN INDIA

The benefits of human diversity studies are obvious – one learns about the origin and inter-relationship of the ethnic groups, and can identify resistance/susceptibility genes for infections like tuberculosis or malaria. However, for the local community there is inconvenience and loss of time. Does the community get any real benefit? In most countries like US, Canada, and Australia the aboriginal tribes have refused to participate in these studies saying "our forests and lands have been taken away and now you want to take our blood." These communities mistrust the motives of the investigators as there have been numerous attempts to patent "natural resources and plants" native to these areas.

In India, tribal communities, who are the likely subjects for diversity studies, live in remote areas, are often illiterate and suffer from nutritional and parasitic disorders. Extensive studies are planned by the Department of Biotechnology, Government of India. In the numerous studies performed by the author, these were linked with health care benefits for the community (Verma, 1997). For example, members of the Kondh tribe in Orissa were studied for sickle cell disease, and a local centre for diagnosis and management of this disorder was established. The negrito tribes of the Andaman islands were evaluated to find the cause for their demographic decline, as the Onges number only 100. This was linked with providing them the necessary medical care and reproductive advice during the study. Studies among such isolated and illiterate tribes have raised ethical concerns. The Department of Biotechnology, as a safeguard for such studies, requires "informed consent" of the subjects. Objectives of the study are to be explained in the local language, there has to be voluntary participation, there should be compensation for loss of time and loss of wages, and consent must be obtained from the individual and the community, either written or tape-recorded.

IMPROVING ETHICAL PRACTICES IN MEDICAL GENETICS

Medical genetics is essentially a part of medicine. Therefore, steps afoot to improve ethical practices in medicine would automatically lead to inprovment in ethical practices in medical genetics (Verma, 1998). A consortium of four nationally important medical institutions formulated inquiry-driven strategies for innovations in medical education with financial support from WHO (Verma *et al.*, 1992, 1995). A number of instruments were used for identifying curricular deficiencies during the inquiry. One of the important findings of the above inquiry was that the ethical, moral and humanistic aspects of disease are neither taught nor assessed. The consortium proposed a formal course in medical ethics in all medical colleges. It stressed that this course should be taught preferably by case examples, case discussions and role models rather than as lectures. Hopefully it will be soon introduced by the medical colleges in the country.

ETHICAL GUIDELINES FOR GENETIC RESEARCH

In 1994, the ICMR issued two documents – a policy statement on ethical considerations involved in research on human subjects and guidelines on responsibility for clinical trials of contraceptives conducted in India. These guidelines have recently been re-drafted (ICMR, 1997). A consultative document has been prepared and circulated to 500 prominent persons in diverse fields – doctors and scientists – for their comments. Regional discussions will take place before the document is accepted for national implementation. These guidelines cover general and specific principles, with special sections on human genetics, organ transplantation including fetal tissue transplantation, clinical evaluation of drugs, diagnostics, vaccines, herbal remedies, epidemiological research, and assisted reproduction technologies. These guidelines draw heavily on those prepared by WHO and Nuffield Council of Bioethics, but do take note of the local scene. A recommendation has been made for having an ethical committee in every medical institution. In the words of the chairman, who is a retired Chief Justice of India, and was the Chairman of the Human Rights Commission in India "the draft guidelines are presented for a purposeful and informed public debate to serve as a feedback before they can be given final form."

CONCLUSION

Dr. Halfden Mahler, a former Director General of WHO once said "Is it realistic to expect rational ethical decisions in this world? We are asking for rational decisions in an irrational world, honesty in a dishonest world, generosity in a greedy world, altruism in an egoistic world, intelligence and humility in an arrogant world." The area of human genetics is fast expanding in range and depth, and ethical guidelines for it would need to be constantly updated to keep up with the new technologies and changing needs of society.

REFERENCES

Advani, M. *Doctor–patient Relationship in Indian Hospitals*. Jaipur: Sanghi Prakashan, 1980.

Consultative Document on Ethical Guidelines on Biomedical Research involving Human Subjects. New Delhi: Indian Council of Medical Research, 1997.

Francis, C.M. Ethics in medical education. *Health For the Millions* 1996; 4: 3–6.

Flor-Cruz, J.A., Ganguly, M., and Thompson, D. The Young and the Damned. *Times Weekly* April 15, 1996: 34–37.

ICMR Guidelines on Ethical Considerations involved in Research on Human Subjects. New Delhi: Indian Council of Medical Research, 1994.

ICMR Guidelines on Responsibility For Clinical Trials of Contraceptives Conducted in India. New Delhi: Indian Council of Medical Research, 1994.

Indian Council of Medical Research, Office Memorandum. *Transportation of Biological Samples Out of the Country for the Purpose of Analysis, Diagnosis and Research Work*. New Delhi: Indian Council of Medical Research, 1992.

Jayaraman, K.S. India drafts law to protect bio-sources. *Nature* 1997; 390: 108.

Kabra, S.G. Reproductive health care for women: a saga of excess, exploitation and violation. *Issues in Medical Ethics* 1997; 5: 47–48.

Kusum. Artificial Insemination Act – cutting through natal legalese. *The Hindustan Times* 18 October 1995.

Memorandum No. L.19015/53/97-IH(Pt.). *Guidelines for Exchange of Human Biological Material for Biomedical Research Purposes*. Ministry of Health & Family Welfare. New Delhi: Department of Health, Nirman Bhawan, 1997.

Pandya, S.K. Hospital Ethical Committees. *Issues in Medical Ethics* 1996; 4: 51.

Patel, V. The ethics of gender justice. *Medical Ethics* 1995; 3: 65–67.

Radcliffe-Richards, J., Daar, A.S., Guttman R.D. *et al.* The case for allowing kidney sales. *Lancet* 1998; 351: 1950–1952.

Singh, K.S. *People of India – an Introduction*. Calcutta: Anthropological Survey of India, 1992.

Singh, J. *Medical profession and Consumer Protection Act*. Jaipur: Bharat Law Publications, 1994.

Singh, J.R. Bioethics of sex preference. In: Fujiki, N. and Macer, D.R.J. (eds), *Bioethics in Asia*. Christchurch, New Zealand: Eubios Ethics Institute, 1998, pp. 398–402.

Siva Subramanian. In India, Nepal and Srilanka quality of life weighs heavily. Hastings Center Report. Institute of Society, Ethics and the Life Sciences. No. 10706. Aug 1986: 20–22.

The Prenatal Diagnostic Techniques (Regulation and Prevention of Misuse) Act 1994; and Rules, 1996. New Delhi: Akalank Publications, 1996.

The Medical Termination of Pregnancy Act, 19971 (Act no. 34 of 1974). Gazette of India, Extraordinary, Part II, Section IA, no. 30, Vol. IX, dated 6th July 1973. pp. 515–519.

UNICEF. *The State of the World's Children 1998*. Oxford and Delhi: Oxford University Press, 1998.

Verma, I.C. Bioethical developments in India: 1989–91. In: Lusting, B.A., Brody, B.A., Engelhardt, Jr, H.T., and McCullough, L.B. (eds), *Bioethics Yearbook. Vol. 2. Regional Developments in Bioethics: 1989–91*. Boston: Kluwer Academic Publishers, 1992, pp. 309–341.

Verma, I.C. Ethical concerns in genome diversity studies in developing countries. In: Knoppers, B.M., Laberge, C.M., and Hirtle, M. (eds), *Human DNA: Law and Policy. International and Comparative Perspectives*. The Hague: Kluwer Law International, 1997, 257–263.

Verma, I.C. Ethics and human genome studies and genetic services in India. In: Fujiki, N. and Macer, D.R.J. (eds), *Bioethics in Asia*. Christchurch, New Zealand: Eubios Ethics Institute, 1998, pp. 215–220.

Verma, I.C. and Singh, B. Ethics and medical genetics in India. In: Wertz, D.C. and Fletcher, J.C. (eds), *Ethics & Human Genetics, A Cross Cultural Perspective*. Berlin: Springer-Verlag, 1989.

Verma, I.C., Fujiki, N., Marwaha, R.K., Ahuja, Y.R., Malhotra, K.C., Parikh, A.P., and Sharma, S. The common people's viewpoint on handicaps and heredity. In *Proceedings of 2nd International Bioethics Seminar*, Fukui, Japan, 1992.

Verma, K., D'Monte, B., Adkoli, B.V., and Nayar, V. (eds). *Inquiry Driven Strategies for Innovation in Medical Education in India*. New Delhi: All India Institute of Medical Sciences, 1992.

Verma, K., Nayar, U., and Adkoli, B.V. (eds). *Inquiry Driven Strategies for Innovation in Medical Education in India – Curricular Reforms*. New Delhi: All India Institute of Medical Sciences, 1995.

Verma, K., Mehta, M., Adkoli, B.V., and Paul, V.K. *Issue in Health Care Ethics – Perspectives from India*. KL Wig Centre for Medical Education & Technology, New Delhi: All India Institute of Medical Sciences, 1997.

CHAPTER 17

JAPAN

Ichiro Matsuda

Japanese populations are genetically different from other populations. Mass screening of newborns in Japan revealed that the frequency of hyperphenylalanemia (PKU) was one-tenth and that of galactosemia one-twelfth of the frequency in other countries.[1,2] Genetic diseases specific to the Japanese are acatalasia, Fukuyama muscular dystrophy and a few others. The incidence in Japan of other common single gene disorders, including Duchenne muscular dystrophy, hemophilia A, urea cycle disorders, 21-hydroxylase deficiency, and chromosomal abnormalities is similar to that in other countries. Therefore, it is likely that the absence of some major genetic diseases, such as sickle cell anemia, thalessemia, or cystic fibrosis contributes to the lack of debate on genetic testing and prevention of genetic diseases, along with factors related to Japanese culture and religion. In general, the Japanese people are reluctant to discuss this kind of subject. Nevertheless, the recent expansion of knowledge and technology in genetics and reproductive medicine has brought to light new aspects of biomedical ethics, which are now discussed widely and officially, even in Japan. The Eugenic Protection Act of 1948 was abolished in Japan in 1995, and in 1997, the Health Science Council of the Ministry of Health and Welfare established a Committee of Assessment for Advanced Technology to discuss ethical, legal and social implications of genetic technology as applied to clinical practice.

The Japanese government has proposed a "Seven Year Plan" (from 1997 to 2003) to promote the care of disabled persons with inherited and non-inherited disorders. The total budget will be approximately US$ 8.2 billion.

I. GENETIC SERVICES AND GUIDELINES OF MEDICAL SOCIETIES IN JAPAN

The general goal of genetic services is to aid persons with genetic disadvantages to live and reproduce as normally and as responsibly as possible. This can be done in the presence of measures to provide correct diagnosis, appropriate treatment, and possible prevention of inherited disorders. According to the guidelines proposed by WHO,[3] prevention includes prevention of damage to genes or chromosomes arising from environmental or occupational exposures to mutagens, avoidance of conception, use of donor gametes, and prenatal diagnosis and selective abortion as alternatives. In such circumstances, it is important that genetic services be accepted by users on a voluntary basis and not be imposed by medical professionals, by society, or by the law. The Japan Society of Obstetrics and Gynecology and the Japan Society of Human Genetics have published guidelines to provide genetic services rationally, although guidelines, unlike a governmental act, do not carry the authority to admonish Society members who do not comply.

1. The System of Genetic Counseling in Japan

Even though most couples (88% of more than 1,500 individuals in one study, regardless of profession) wish to have genetic counseling after conception (pre-conception counseling is not popular in Japan, because of limited incidence of inherited diseases), the system of genetic counseling is still at a premature stage in Japan. Board examinations in clinical genetics were established in 1994 by the Japan Society of Human Genetics, and 63 doctors were certified. In addition, there are 317 doctors certified as clinical geneticists on the basis of their history of genetic training. In Japan, only medical doctors can conduct genetic counseling. The greatest disadvantage in promoting a genetic counseling system in Japan is cost, which is not covered by government health insurance.

2. Alternative Reproduction

The Japan Society of Obstetrics and Gynecology proposed several guidelines concerning "*in vitro* fertilization (IVF)". In 1983, it was decided that IVF techniques should be used only for married couples, only for the treatment of infertility, and should not include any genetic manipulation.[5] As described below, the latter two limitations were reviewed in 1997 to allow research on preimplantation diagnosis. In 1984, the duration of time for preserving a frozen embryo was limited to the natural reproductive age of the woman. In 1986, the Society required registration of the names of hospitals performing IVF and also required them to report results annually. For the year 1998, 36,865 women in Japan received IVF or allied technologies, including *in vitro* fertilization–embryo transfer (IVF–ET), gamete intra-fallopian transfer (GIFT), zygote intra-fallopian transfer (ZIFT), and intracytoplasmic sperm injection (ICSI) at 377 registered hospitals.[6] Pregnancy rates and live birth rates per pre-embryo or gamete transfer were 23.1% for IVF, 22.6% for GIFT, 18.7% for ZIFT, and 25.4% for ICSI respectively, and the multiple pregnancy rate per number of pregnancies was 16.9–25.0%. In addition to resulting from IVF technology, multiple pregnancy can occur when hormones are prescribed. The problem with multiple pregnancies is possible births with fetal abnormalities, the rate increasing with the increasing number of fetuses; i.e., 3.6% for triplets, 10.2% for quadruplets, 30.8% for quintuplets.[7] Therefore, fetal reduction seems essential to avoid untoward outcomes. For this purpose, the Society suggested limiting the number of oocytes transferred to the uterus to three, and to reduce dosage of ovulation-inducing agents.[7] A direct technique for reducing fetal numbers, for example using potassium chloride, with or without prenatal genetic testing, has not been officially allowed in Japan, and the guidelines did not address this issue. A survey supported by the Ministry of Health and Welfare found that several hospitals did use this technique to reduce fetal numbers in multiple pregnancies of more than four fetuses, the objective being to avoid fetal abnormalities.

Artificial insemination with donor semen is another choice for couples with infertility problems. In 1997, the Society proposed that the technique be available only to married couples, and that donors should be screened for family history of genetic disease and should also be tested for sexually transmitted diseases, including HIV.[8] For protection of the donor's privacy, semen must be donated anonymously, but the donor's name should be kept on file in an official list in case of future problems. Egg donation, embryo donation, and surrogate motherhood are currently under the discussion by Government Committee.

3. Preimplantation Diagnosis (PID)

Use of oocytes for research purposes was approved only within two weeks after IVF, according to the guidelines proposed by the Japan Society of Obstetrics and Gynecology in 1985.[9] PID might be useful for couples with severe genetic risks, who oppose abortion. PID is morally commendable as an alternative to selective abortion, which carries documented psychological trauma for the couple and especially for the woman.[10] In 1997, the Society[11] decided that PID was still in the experimental stage, therefore (1) researchers should have permission from the Board of Directors of the Society, (2) indications should be limited to serious inherited diseases, and severity of each disease should be evaluated by the Board, (3) information for users should include an outline of the procedure, its safety, the likely outcome and limits of the diagnosis, and the difference between prenatal diagnosis and PID. The high cost and low success rate of the technique (low cost-effectiveness in terms of social resources) are other problems that need attention, in addition to ethical implications of the genetic manipulation itself.

4. Prenatal Diagnosis

The Japan Society of Human Genetics published guidelines for genetic counseling in 1994,[12] and guidelines for genetic testing using DNA analysis in 1995.[13] Both have been endorsed by five other medical societies, including the Japan Society of Obstetrics and Gynecology. Principles of the guidelines for prenatal diagnosis are as follows: (1) prenatal diagnosis should be voluntary in nature, based on autonomous decision-making, (2) appropriate genetic counseling should be conducted before and after the prenatal diagnosis, (3) written informed consent is indispensable for conducting the diagnosis, (4) the diagnosis must be performed using only established and practiced techniques, (5) even though a counselee may desire prenatal diagnosis, the doctor can refuse to provide it if the request is against social and ethical norms (as described below), or against the doctor's personal principles (based on his/her religious or pro-life beliefs). The following statements are included as opinions on prenatal diagnosis: (1) invasive prenatal diagnosis using amniocentesis, chorionic villus sampling (CVS) or cordocentesis is considered for pregnancies in cases where: (a) either of the couple is a carrier of a chromosomal abnormality, (b) there is a history or experience of giving birth to a baby with abnormal chromosomes, (c) the woman is of advanced maternal age, (d) the pregnant woman is heterozygous for a serious X-linked disease, (e) both partners are heterozygous for a serious autosomal recessive disease, (f) either partner is heterozygous for a serious autosomal dominant disease, or (g) there is a risk of serious fetal abnormality (e.g., suspected by ultrasonography in an ongoing pregnancy). (2) except for a diagnosis of an X-linked disease, the sex of the fetus should not be revealed. However, nothing related to abortion for fetal anomalies or chromosome abnormalities was stated, as this procedure is illegal in Japan. Abortion following prenatal diagnosis is performed under the article "permitting abortion if continuation of pregnancy or delivery may cause considerable harm to maternal health for physical or economic reasons" (Maternal Protection Act of 1996).

Most available data on prenatal diagnosis in Japan are from a survey conducted by the Ministry of Health and Welfare in 1998,[14] where 5,748 cases (5,557 amniocenteses, 109 CVS, 77 cordocenteses, 5 fetal skin or liver biopsies) were examined in 136 university hospitals and other general hospitals for the year 1997. In 5,557 of amniocentesis, 3,611 underwent amniocentesis because of at advanced age. This figure suggests only 4% of pregnant women above age 35 underwent amniocentesis, when assumed the survey covered more than 80% of all women in Japan.[14] In all of amniocentesis, 83.6% were done to screen for chromosomal abnormalities, 5.8% because of abnormal ultrasonography, 9.8% because of maternal anxiety, and 0.7% because of Mendelian inherited diseases. The rate of chromosomal abnormalities after amniocentesis was 4.2%. A majority of cordocenteses (55.8%) were done for chromosome abnormalities after fetal abnormalities were suspected as a result of ultrasonography. In all, 14% of cordocenteses done for this purpose revealed chromosome abnormalities. In the group having CVS, 22 cases (20.1%) were screened for chromosomal abnormalities; 2 cases (9.0%) were found to have an abnormality. Other reasons for prenatal diagnosis were fetal rubella infection (18 cases) and single gene disorders (69 cases), including Duchenne muscular dystrophy, hemophilia, Fukuyama muscular dystrophy, 21-hydroxylase deficiency, and ornithine transcarbamylase deficiency.

Maternal and child health statistics for Japan revealed that the total fertility rate decreased gradually from 3.65 in 1950 to 1.38 in 1997, and in parallel, the neonatal mortality rate dropped remarkably from 27.1 per 1000 live births in 1950 to 1.9 in 1997.[15] The rate of induced abortion has also decreased from 42.0 (per 1000 females) in 1960 to 110 in 1997.[16] The rate of use of birth control in Japan from 1990 to 1998 is 56.8% (the highest is in China at 83%).[17]

Since 1985, the primary cause of death in the first year of life has been congenital abnormality, which accounts for more than 35%[18] of deaths. The largest numbers of pregnancies with anencephaly and Down syndrome are in 20 to 35-year-old mothers and over 35-year-old mothers, respectively, because 20 to 35-year-olds have the largest number of births, even though the risk is higher among women over 35. The rate of anencephaly in live births has decreased from 10 per 10,000 births in 1979 to 5.4 in 1990 and 1.6 in 1996.[19] Generally, anencephaly is detectable by ultrasonography within 24 weeks of gestation, this being a routine examination in Japan. Thus, the reduction in birth rates for anencephaly is likely to be the result of selective abortion, but exact data are not available. The rate of live births with Down syndrome has increased slightly from 3.4 to 5.6 per 10,000 in 1981–1983 to 6.0–7.0 in 1989–1994, as a result of an increase in the percent of pregnancies occurring in women over 35 years of age from 4.5% in 1980 to 9.4% in 1994.[19] Although the percent of pregnancies in the high-risk group has doubled, the percent of total births with Down syndrome has not. A survey in Kanagawa prefecture disclosed that the birth rate for Down syndrome has decreased from 6.3 in 1981–1983 to 3.7 in 1990–1993. Selective abortion may have been performed in this district, but no data are available.[20]

Multiple marker tests measuring alpha-fetoprotein, estriol, and chorionic gonadotrophin in maternal serum in early gestation have been used in the USA and in European countries to detect Down syndrome, trisomy 18, and neural tube defects.[21] In Japan, four commercial companies have been involved in this test since 1992, and over 22,000 women have been tested to date. The Japan Society of Human Genetics expressed serious concerns about the marketing of such a diagnostic tool in 1998,[22] as follows: (1) information prior to testing is insufficient, and lacks full and unbiased description of the disorders, (2) sensitivity, specificity, positive, and negative predictive values of the test for Japanese populations have not been sufficiently examined, (3) practicing obstetricians at that time did not fully understand the limitations of the test, and follow-up surveys are needed.

The sex ratio of live births (male/female) has not changed in Japan since 1950, holding steady at 1.02–1.06 in 1951–1990, before prenatal diagnosis became available, and 1.05–1.06 in 1992–1996, after prenatal diagnosis became available.[23]

5. *Newborn Mass Screening*

The newborn mass screening program (NMSP) was set up to identify babies with treatable disease at an early (newborn) stage, before irreversible clinical features are manifested, and to provide appropriate care. Thus, NMSP is meant as "preventive medicine", as well as "genetic testing". NMSP is done on a mandatory basis in many states in the US. The reason for mandating screening is that a refusal of a newborn screening test by proxy (the parents) could result in inescapable disadvantage to some infants, since they may lose the opportunity to be treated for a disease detectable by NMSP.[24] In Japan, NMSP has been carried out on a voluntary basis, but plans are underway to make the tests mandatory. At present, almost 100% of newborns are screened for six diseases, including hyperphenylalanemia, maple syrup urine disease (MSUD), homocystinuria, galactosemia, hypothyrodism, and 21-hydroxylase deficiency. More than 70% of the patients detected have been followed, and favorable results were observed, except for classic MSUD. Nevertheless, NMSP requires a new view based on micro- and macro-biomedical ethics. In 1997, the Japan Mass Screening Society published guidelines,[25] which state that (1) neonatal or infant mass screening requires voluntary consent from the guardian. A thorough explanation of the tests purpose and other information about the test must be given before consent. (2) Test results and other information about the individual are protected under confidentiality rights unless there is a substantial reason for disclosure. If clinical information about the individual is provided anonymously for retrospective epidemiological and follow-up studies, a signed document is unnecessary, but it is advisable to inform the guardians in advance. (3) Samples taken for screening must not be used for other purposes. However, if both of the following apply, the sample may be used for research after receiving consent from the ethical committee: (a) the purpose of the project/research is the advancement of medical science, (b) the sample is anonymous and all identifying characteristics have been removed. (4) Initiating a new, mass screening test using public resources requires sufficient pilot studies reviewing its necessity, the risk of testing, effectiveness of testing and treatment, availability of other methods, and cost–benefits ratio.

6. *Research Using Fetal or Neonatal Tissue*

In 1987, the Japan Society of Gynecology and Obstetrics published guidelines permitting medical research on aborted fetuses or neonatal tissues after the infant's death, providing that (1) the research could not otherwise be performed, (2) the research contributes to significant progress in medicine, (3) consent from the couple or mother has been obtained.[26]

II. CULTURAL CLIMATE AFFECTING GENETIC SERVICES: A SPECIAL SURVEY

For the ideal development of genetic testing, including prenatal diagnosis, one needs to understand the attitude of the Japanese population regarding inherited disorders. Accordingly, we surveyed individuals[4] from three groups: (1) medical professionals, including primary care physicians, nurses, laboratory personnel, and medical students in their final year, but excluding geneticists; (2) non-medical persons (members of the public), including housewives, school teachers, kindergarten level teachers, government employees, business people, office workers, and college students; (3) patients, most of whom (88%) had muscular dystrophy. If the patient was a minor, we asked the parents. (Hereafter, the term "patient" refers to both patients and parents.) To approach the public, we asked the local society of school teachers and the local society of kindergarten teachers and their parent–teacher associations in several districts to send questionnaires to their members. In all, 712 medical professionals (70% of those asked to participate), 635 members of the public (response rate unknown), and 126 patients (53% of those asked to participate) responded, for a total of 1,518 persons. The survey used a multiple choice format. This survey was separate from the 36-nation survey and was unique to Japan. Two-thirds of patient and public respondents were women. In all, 59.0% agreed that it will be troublesome if "I (we) gave birth to a child with a hereditary disease"; 39.8% said, "should I or a member of my family have a hereditary disease, I would like to accept the fact and not make a fuss about it"; 36.1% said, "it is a concern as to whether there are any affected persons in the family of the man or woman I am going to marry"; and only 9.6% said, "I would like to hide the fact if a person with hereditary disease was a member of my family". There were no statistical differences among these three groups: medical professionals, the public, and patients. These views differ from an earlier report that "genetic disorders in Japan are often cause for shame that falls on the entire extended family".[27]

The following discussion is based on the survey described above.

1. *Prenatal Diagnosis*

Among all individuals, regardless of profession, including patients, 89.2% said that they would accept prenatal diagnosis, if it is available and if there is a possibility that the fetus might have a (severe) hereditary disease or chromosomal abnormality. If the disease is less serious and is treatable, 81% would carry the child to term. When the fetal disease is severe and untreatable, though the term "severe" is open to discussion,[28] 22.0%, 13.1%, and 30.0% of medical professionals, the public, and patients respectively wanted the child to be born. The significantly high percent in the last group is understandable, for adult patients with muscular dystrophy were included. There seems to be agreement across a number of studies that around 30% of women say that they would not consider termination on the basis of fetal abnormality.[29] In our survey, 21.8% of women supported this idea, regardless of their occupation.

Among all individuals, approximately half of those who would consider abortion (45.7%) favored selective abortion, and the other half (52.5%) withheld a decision, even in the group of patients. An allowance for genetic abortion is lacking in the newly changed Maternal Protection Act. The Japanese people are not so willing to express their opinions frankly, and an ambiguous attitude is more acceptable, based on cultural grounds. There may be a difference, of course, between prospective discussions and actual decisions about genetic abortion.

Nevertheless, there was an interesting response to another question. According to the Maternal Protection Act (1996), "abortion following a prenatal diagnosis may be performed, if continuation of pregnancy or delivery may cause considerable harm to maternal health for physical or economic reasons". The question asked whether an additional phrase, "if the fetus has a severe disease", is necessary. Among all individuals, only 6.5% said the present situation is satisfactory, 24.4% said an additional phrase should be added, because the law must be precise, and 69.1% said further discussion on genetic abortion is necessary. There is a contradiction here, because most people said that the issue of genetic abortion has not been sufficiently discussed, yet they were reluctant to be involved in such discussion. Since some respondents were those wishing prenatal testing and genetic abortion, as defined in the survey, efforts to educate the general population are necessary before a rational law on genetic services can be developed. Another survey revealed that more than 70% of obstetricians thought it desirable to have an additional phrase in the Maternal Protection Act permitting genetic abortion.

Another issue is who should be responsible for making decisions about genetic abortion. A number of factors may contribute to the decision, including: cultural and individual attitudes toward abortion, the nature of the abnormality, and the social climate for disabled children. In all, 51.2% of medical professionals, 50.4% of the public, and 56.0% of patients said that the pregnant woman should have the basic right to decide, as it is usually the mother who takes care of the child after birth. In all, 58.9% of men and 50.0% of women supported this idea, regardless of their profession. The rest did not agree that the mother should make the decision about genetic abortion. Even though it is argued that a decision about abortion, whether for genetic or non-genetic reasons, should be a woman's right, to ask only the woman regarding this decision is difficult under the present Japanese system. In the survey, both men and women said that the couple should be responsible for the decision.[4]

2. *Management of Genetic Information*

Genetic information is both uniquely individual and the shared property of families. Thus, family members have a moral obligation to share genetic information. However, the laws affecting confidentiality, privacy, and rights to information have in general not yet taken into account this unusual medical situation. The survey was done to learn the attitudes of Japanese individuals about responsibilities of sharing genetic information with family members. The survey asked "whether members of the family should be informed of your genetic information". Answers were obtained from two different points of view, personal standing (what they would actually do) and general standing (what they think ought to be done). To ascertain both of these attitudes, we prepared two different questions, which were slightly modified and appeared on different pages. There was a significant difference between the two types of answers in all three groups: medical professionals, the public, and patients (Table 1).

The following questions were intended to seek responses concerning genetic information related to future families. Respondents were asked "whether a person you are going to marry should be informed of your genetic information", and "whether you should be informed of genetic information about a person who will become a member of your family (e.g., spouse, adopted children)". The patients' opinions about disclosure of their own genetic information to other family members were the same as their opinions about disclosure of information about other family members to them, with about 67.5% favoring disclosure in both cases. Among medical professionals and public, however, there were significantly different answers to these two questions (Table 1). This suggests two possibilities: one is that most people would like to know genetic information about their family members, but they are reluctant to tell their own genetic information to their (future) family members. The second possibility is that patients and their parents are trying to be fair in their responses to questions about extending genetic information.

Guidelines proposed by WHO suggest depending on the degree of the magnitude of harm that may occur from non-disclosure, the counselor should discuss the limits of

Table 1. Disclosure of Genetic Information

Do you think members of the family should be informed of your genetic status?

	Yes	
	Personal standing (what you would actually do)	General standing (what ought to be done)
Medical professionals ($N = 712$)	42.0	75.5%
Non-medical professionals ($N = 637$)	56.7	79.9%
Patients and parents ($N = 140$)	69.3	85.5%

Do you think the person you are going to marry should be informed of your genetic status?

	Yes
Medical professionals ($N = 712$)	53.2%
Non-medical professionals ($N = 637$)	57.4%
Patients and parents ($N = 140$)	67.6%

Do you think you should be informed of genetic information about a person who will become a member of your family (e.g., spouse, adopted child)?

	Yes
Medical professionals ($N = 712$)	64.0%
Non-medical professionals ($N = 637$)	69.9%
Patients and parents ($N = 140$)	67.4%

confidentiality.[30] In the guidelines of the Japan Society of Human Genetics, it is stated that however, if sharing the information with another specific person (present or future family member) will avoid serious injury to that person, it is necessary to seek the consent of the subject to reveal that information, and even if agreement cannot be obtained, if it is judged necessary the obligation of confidentiality can be broken. Such an exception must follow the judgement of the responsible ethics committee, not that of the counselor.[31]

3. Re-use of Stored Samples

Among all respondents, 93.8% agreed to store the remainder of the samples obtained for genetic testing for the future benefit of patients and their family members. In principle, the samples should be used only for the original purpose, but several international bodies have said that stored samples could be used for anonymous research or surveillance.[32,33] In all, 82.6% accepted this view, if an explanation about possible future uses was given beforehand. Of these, 91.1% wanted to be informed of the test results, regardless of whether a condition was treatable or untreatable. This presents a dilemma, when anonymous samples are used.[34]

III. CONCLUSION

There was little active public discussion about the bioethical implications of techniques such as alternative reproduction and genetic testing in Japan. Medical societies have proposed several guidelines with the intention of rational control of such techniques. The Eugenic Protection Act, under which those with mental disabilities might possibly be sterilized by force, was repealed in 1995. Instead, a new Maternal Protection Act was passed, an Act wherein abortion is legal if continuation of pregnancy or delivery possibly causes considerable harm to maternal health for physical or economic reasons. In other words, at present genetic (selective) abortion is still illegal. Societies associated with disabled people and their parents are strongly against adding the phrase of the fetus has "severe disease" for legal abortion in the Act, because they are afraid the phrase will produce a potential (social) force for aborting the affected fetus, even when parents wish to deliver. Most people believe the issue of genetic abortion has not yet been sufficiently discussed and that the government has only recently shown concern about this issue. However, there is a dilemma in this issue, for an obstetrician lost a law suit in a case of fetal rubella infection, where parents were not warned of the potential severe risk to the fetus, thus losing the chance to choose selective abortion.

A system of genetic counseling must be fully developed in Japan in the near future. In addition, education of the public to learn basic knowledge of medical genetics should be conducted for appropriate dissemination of genetic services.

REFERENCES

1. Scriver, C.R., Kaufman, S., Eisensmith, R.C., and Woo, S.L.C. The hyperphenylalaninemias and disorders of galactose metabolism. In: Scriver, C.R., Beaudet, A.C., Sly, W.S., and Valle, D. (eds), *The Metabolic and Molecular Basis of Inherited Disease,* seventh edition. NewYork: McGraw-Hill, 1995, p. 998, 1016.

2. Maternal Child Division: Children and families Bureau, Ministry of Health and Welfare: Maternal and Child Health Statistics of Japan 1998, p. 101.
3. Wertz, D.C., Fletcher, J.C., Berg, K., and Boulyjenkov, V. Guidelines on ethical issues in medical genetics and the provision of genetic services. WHO Hereditary Diseases Program, 1995, p. 1.
4. Matsuda, I. Report of The Japan Ministry of Health and Welfare, Genetic services-ethical, legal and social issues in Japan (in Japanese) (in press).
5. Japan Society of Obstetrics and Gynecology. Guidelines for *in vitro* fertilization. *Japan Journal of Obstetrics and Gynecology* 1997; 49: 697–720 (in Japanese).
6. Araki, T. Annual report of *in vitro* fertilization. *Japan Journal of Obstetrics and Gynecology* 2000; 52: 962–958 (in Japanese).
7. Japan Society of Obstetrics and Gynecology. Guidelines for multiple pregnancy. *Japan Journal of Obstetrics and Gynecology* 1997; 49: 31–32 (in Japanese).
8. Japan Society of Obstetrics and Gynecology. Guidelines for artificial insemination with donor semen. *Japan Journal of Obstetrics and Gynecology* 1997; 49: 33–35 (in Japanese).
9. Japan Society of Obstetrics and Gynecology. Guidelines for research using human semen, oocytes and fertilized oocytes. *Japan Journal of Obstetrics and Gynecology* 1985; 37: 465–466 (in Japanese).
10. Fletcher, J.C. and Richter, G. Human fetal gene therapy: moral and ethical questions. *Human Gene Therapy* 1996; 7: 1605–1614.
11. Japan Society of Obstetrics and Gynecology. Report of ethical committee for preimplantation diagnosis. *Japan Journal of Obstetrics and Gynecology* 1997; 49: 269–271 (in Japanese).
12. Japan Society of Human Genetics. Guidelines for genetic counseling and prenatal diagnosis. *Japan Journal of Human Genetics* 1997; 42: 375–377.
13. Japan Society of Human Genetics. Guidelines for genetic testing using DNA analysis. *Japan Journal of Human Genetics* 1997; 42: 377–379.
14. Matsuda, I. and Suzumori, K. Prenatal diagnosis in Japan. *Community Genetics* 2000; 3: 12–16.
15. Supra note 2, P21.
16. Supra note 2, P84.
17. Supra note 2, P110–117.
18. Supra note 2, P60–61.
19. Sumiyoshi, Y. Report of congenital abnormality monitoring – 1996 (in Japanese).
20. Kuroki, Y. Report of congenital abnormality monitoring in Kanagawa district, in a Study of Environmental Effects on Children's Life, The Ministry of Health and Welfare, 1995, pp. 194–198 (in Japanese).
21. Wald, N.J., Kennard, A., Hackshaw, A., and McGuire, A. Antenatal screening for Down's syndrome. *Journal of Medical Screening* 1997; 4: 181–246.
22. Japan Society of Human Genetics. Guidelines for multiple marker test for pregnancy 1998 (in Japanese).
23. Supra note 15, P42–43.
24. Bobinski, M.A. Genetics and reproductive decision making. In: Murray, T.H., Rothstein, M.A., and Murray, R.F. (eds), *The Human Gene Project and the Future of Human Care*. Bloomington, IN: Indiana University Press, 1996, pp. 79–112.
25. Japan Society of Mass Screening. Guidelines for mass screening. *Japan Journal of Mass Screening* 1998; 8: 1–5.
26. Japan Society of Obstetrics and Gynecology. Guidelines for research using fetal and neonatal tissues after death of these subjects. *Japan Journal of Obsterics and Gynecology* 1997; 42: 22 (in Japanese).

27. OhKura, K. and Kimura, R. Ethics and medical genetics in Japan. In: Wertz, D.C. and Fletcher, J.C. (eds), *Ethics and Human Genetics: A Cross-Cultural Perspective*. Heidelberg: Springer-Verlag, 1989, pp. 294–316, at 314.
28. Supra note 3, P40.
29. Green, J. and Statham, H. Psychosocial aspects of prenatal screening and diagnosis. In: Marteau, T. and Richards, M. (eds), *The Troubled Helix: Social and Psychological Implications of the New Human Genetics,* Cambridge: Cambridge University Press, 1996, pp. 140–176.
30. Supra note 3, P40.
31. Supra note 12, P379.
32. Council for International Organizations of Medical Sciences (CIOMS). *International Ethical Guidelines for Biomedical Research Involving Human Subjects*. Geneva: WHO, 1993.
33. World Health Organization. *Proposed International Guidelines on Ethical Issues in Medical Genetics and Genetics Services*. Geneva: WHO, 1998.
34. Supra note 3, P75.

CHAPTER 18

PERU

Teresa Perez

DEMOGRAPHICS

Peru is the fourth most populous country in South America, after Brazil, Colombia, and Argentina, with 22,639,444 inhabitants, according to the IX Population Census in 1993[1]. Most of the population (about 70%) lives in large cities, such as Lima, Trujillo, Cuzco, Piura, and Chiclayo, with the remaining living in rural villages. The total population is evenly divided between men and women. Currently, age distribution indicates that 37% of the population is under 15 years of age. The annual average growth from 1981 to 1993 was 2.2%, which continues the declining growth rate in the last 30 years. Fertility has declined in Lima (the capital) and in other major cities, but it is still high in rural villages among uneducated women. According to the most recent demographic surveys[1], by the end of their reproductive lives, women without education will have four times more children than those with higher education, and women in rural areas will have three times more children than the residents of the capital. Infant mortality has declined from 73 per 1000 to 55 per 1000 in urban areas. However, in rural areas, infant mortality is still over 70 per 1000. These statistics show that Peru has not yet resolved some basic health problems. The primary factors are very little education, poverty, an unhealthy environment, and a lack of political resources to solve these problems.

EPIDEMIOLOGY OF GENETIC DISORDERS IN PERU

Maternal age is the principal cause of non-disjunction, and "*de novo*" translocation is the other. The frequency of Down syndrome is 1 in 800 newborns. Down syndrome is the most common cause of mental retardation seen in Peru. It is frequently found among children of older mothers. Other common causes of mental retardation are fragile X and several types of metabolic disorders. These disorders occur frequently according to worldwide statistics, but are difficult to diagnose in Peru, due to an absence of molecular, biochemical, and metabolic techniques.

Major neural tube malformations are the second most frequent type of disorder. Defects from the closure of the neural tube, such as anencephaly, meningomyelocele, and spina bifida, are found with an incidence of 1.2 per 1000. Interruption of pregnancy is becoming earlier, due primarily to prenatal diagnosis through sonography rather than by a biochemical analysis.

GENETIC SERVICES

There are no large-scale biochemical maternal screening programs in the major hospitals. The triple test is used mainly in private practice. Chromosomal abnormalities, especially trisomies, such as Down (21), Patau (13) and Edwards (18) syndromes are found through cytogenetic analysis in amniotic fluid of patients at risk. Prenatal, cytogenetic, and sonographic studies are conducted at Social Security hospitals in Lima, but none are performed in the other cities.

Problems occur when fetal abnormalities are found and the parents ask for interruption of the pregnancy in the second trimester (16 weeks and after). Restrictive abortion laws in Peru make all genetic abortions illegal, and sometimes hospitals cannot accept families' demands for abortion. Therefore, families have to search for an illegal way to resolve their problems.

Neonatal screening programs are very rare, and only a few university hospitals and private clinics can detect an inborn error of metabolism, such as PKU or hypothyroidism.

Specialists in human and medical genetics practice only in Lima. When genetic cases need medical attention, they must be referred to a public or private practice in the capital for diagnostic treatment and genetic counseling. Slowly, the National Social Security Institute is becoming interested in starting cytogenetics laboratories in the second largest city of Peru, Arequipa.

After nearly 30 years, genetic services in Peru have increased modestly. They now include clinical genetics, cytogenetics (which includes prenatal diagnosis through amniotic fluid but not chorionic villus sampling, due to technical rather than ethical reasons), biochemical genetics, and genetic counseling. Private genetic services attend to middle and upper-class patients, while public services attend to the working and middle-income classes. The lower class and unemployed are almost completely without genetic attention. Very little is being done by the government to attend to these needs, because other problems have priority, such as decreasing infant mortality, which is still very high.

ABORTION

Maternal mortality, according to the latest reports[2], is due primarily (in 85% of cases) to abortion, infection, and hemorrhage, which are mainly the results of abortions performed under illegal conditions. These illegal abortions are due to a law in Peru that mandates a penalty for all abortions, except when a mother's life is at risk.

Abortions done for genetic reasons in Peru are still very low in number, in the context of the total number of abortions provided, 271,000 per year, or 4 to every 10 births. These are primarily due to unplanned pregnancies, principally among teenagers or young women who do not attend fertility centers. Twenty percent of maternal deaths due to an abortion occur in pregnant teenagers.

There is a controversy over whether a patient who has a fetus with a diagnosis of a lethal neural tube defect, such as anencephaly, should be able to have a legal abortion. At present, she does not have the right to do so, since Peruvian law says her life is not at risk. Does she wait nine months to have this baby? Of course not. She finds a way to interrupt the pregnancy under illegal conditions. This also happens when prenatal diagnosis is positive for autosomal trisomies, for example, 21, 13, or 18. The parents choose to have an abortion in the same illegal manner.

In both cases, an abortion must be performed in the second trimester. There are no genetic abortions in the first trimester, due to the absence of diagnosis through chorionic villus studies. The earliest studies after amniocentesis are performed at 15 weeks, and the latest, at 25 weeks of pregnancy.

Almost 97% of parents choose to have an abortion after they know the fetus has a serious chromosomal abnormality or a major malformation. Sexual trisomies are less frequently found; they are not considered serious problems by medical professionals. However, if the parents want a small family, they usually insist upon asking for an abortion, since they are afraid of the future for this kind of child.

Before a prenatal test, couples should receive information regarding the risk of this test, and they should have the freedom to choose a continuation or an (illegal) interruption of the pregnancy after they receive the results.

ETHICAL PROBLEMS

Medical geneticists are included under ethical standards codified by the Peruvian Medical Association or *Colegio Medico del Peru*[3]. Interactions with patients and their families should follow the principles of respect, privacy, and freedom to make an autonomous decision.

Abortion

The main ethical problem in medical genetics is how the professionals confront the issue of abortion after prenatal diagnosis, due to the conflict between the need and the reality. Laws forbid abortion for genetic reasons, and this puts the families and their physicians in a difficult situation, especially in public hospitals supported by the government. Private practice is a bit more flexible.

Reproductive Options

Another ethical problem regards reproductive options. Adoption is not a popular or usual method of resolving this type of problem, due to very obsolete and impractical laws. Insemination of donor sperm is the most popular option, due to its low cost. The second most popular is *in-vitro* fertilization, which has the lowest cost in Latin America. To use either of these options, the patients must have some very expensive medical exams, which the majority cannot afford.

Genetic Services

An extremely low proportion of the Peruvian population receives genetic attention. It is more a service for individuals and families than a public health program, for several reasons. First, there are not sufficient genetic services throughout Peru, since genetic specialists are concentrated in the capital. Second, genetics is a new medical discipline and is not well known by doctors in general practice. They are not well aware of genetic diseases and are not very helpful when attending to these kinds of patients. Third, poverty and a lack of education for women are great handicaps in preventing malformations related to nutritional factors, or in avoiding pregnancy at early or late reproductive ages.

Finally, the government does very little to solve these problems. However, it has created a law on family planning, which is part of a National Population Program. This program ran from 1991 to 1995, and encouraged women to reduce the size of their families from 3.3 children to less than two. This program has eight primary areas: research, statistics, communication to the general public, education, women's health promotion, youth health promotion, population limitation, and improvement of the environment. The national program expects to see changes in the reproductive lives of the population. Through programs of intregal attention to mother and child, and giving priority to rural areas and peripheral urban areas, the government has created the P.A.I.S. (*Programa de Atencion Integral de Salud*). This program is free of charge for people with very low incomes, who are the great majority of the women in rural areas.

However, there is nothing in this program concerning primary care in genetics, and no basic recommendation for ending pregnancy for women late in their reproductive lives, or of testing to screen for the most common malformations. For all of these reasons, it is not possible to forsee a substantial decrease in birth defects, as is already being seen in other countries. The majority do not receive prenatal testing; they go through all nine months of pregnancy before they are aware of the condition of their babies.

Almost all women (98%) who have amniocentesis and receive a positive result for a chromosomal anomaly ask for an interruption of their pregnancy during the second trimester. These types of abortions are still very few. However, they are increasing in private practice, but are still very far from becoming an important cause of abortion in Peru.

In the future, more genetic services will be provided, since patients want this attention. They know that genetics is important not only for them, but also for their families and future generations. Teaching human and medical genetics not only in medical school, but also to other health professionals, such as psychologists and nurses, will provide more people who are well trained to be part of a multidisciplinary team to provide genetic services. It is very important to know how to handle a genetic case, especially in those situations when a patient feels guilty. In this case, the physician or counselor has to talk to them long enough to clear their doubts. To do this, professionals need to be very well prepared. So far, medical geneticists perform the diagnosis, treatment, and counseling. There are no special programs for counselors, which means that this job is assumed by the medical geneticists.

In Peru, to achieve all of these goals, it is necessary to increase interest in genetics, but there are still more urgent unsolved public health problems, such as maternal and infant mortality.

Women's status in Peru, as in other developing countries, is unequal and unfair to women. Their human rights are not respected, not only in education and work, but also at home and in social recognition. There is still a long way to go to solve this problem and to put women in the equal place where they belong. This situation is a handicap in promoting public education to encourage couples, and especially women, to have freedom of choice about parenthood. This freedom of choice is essential for the development of genetics.

REFERENCES

1. Inei. *Peru: Resultados Definitivos*. Perfil Socio Demografico Coleccion Analisis Censal No. 7 Lima, 1994.
2. Ferrando, D. *Aborto Inducido Una Realidad Latinoamericana*. New York: The Alan Guttmacher Institute, 1994.
3. Colegio Medico Del Peru. *Salud Poblacion Y Desarrollo*. Peruvian Society of Obstetrics and Gynecology, 1994.

PART III

ETHICS AND GENETICS IN GLOBAL PERSPECTIVE

CHAPTER 19

FROM GENETICS TO GENOMICS AND BEYOND

This chapter discusses a cluster of changes and other issues in the transition from genetics to "genomics" and "proteonomics." When the first survey was completed in 1985, direct DNA testing was in an experimental stage. Older interactions between genetics and society are in the process of rapid change, especially in affluent nations. At its start and finish (1990–2001), the Human Genome Project (HGP) evoked competing metaphors to convey its impact on society. Alongside praise for the HGP, misplaced hopes, misunderstanding, and mistrust obstruct integration of genetic discoveries into mainstream medicine. If physicians are to use these discoveries in practice, traditional medical ethics needs to refit its "ship of morality" (Benjamin, 2003, 112) for more turbulent seas. Would a code of ethics help the world's geneticists earn more public trust? Or is there a more practical way?

FIVE MAJOR CHANGES IN THE FIELD

An account of changes in the field must begin with the HGP. What does the HGP mean for science and society? Metaphors abound to convey the significance of the project. Four types compete for attention: literary, mechanistic, exploration, and religious. It is difficult to ascertain who first used the term "Rosetta Stone" (Petsko, 2001), a landmark in hieroglyphics. Like many other scientists, Jasney and Kennedy (2001) describe "a library of books that help define us and our place in the tapestry of life." Others use language about a powerful tool that unlocks secrets of genetic heritage, a computer program, or a collection of order forms in the body's factory. Martin Bobrow (2001) said that the HGP was like cresting a mountain pass to view a fertile plain below rich with possibilities of further discovery. Skeptics like James Watson (2003) use religious metaphors like "the Holy Grail" or the "Bible of biology" for the HGP. Walter Gilbert, molecular biologist and Nobel laureate, was the first to use the "grail" as a metaphor for sequencing the human genome (Cook-Deegan, 1994, 88). Religious believers see the handiwork of an "intelligent Designer." Metaphors aid understanding but do not rise above their cultural sources (Keller, 1995). Metaphors can also be self-serving. More than one metaphor is needed to convey modern biology as one part of a whole spectrum of disciplines and methods that study life in the universe.

We propose a mixed metaphor of exploration, prospecting, and mining. The human genome can be compared to a mountain in a chain of mountains, i.e., other genomes. The mountain looms over a frontier town of descendants of explorers who found routes to its foothills. One must imagine this mountain as a storehouse of vast resources of knowledge for human health and disease, rather than deposits of precious metals or ore. To locate, extract and use these resources is very expensive. The first step, after earlier and painstaking expeditions, was to make a physical map of all locations of extractable resources. With precise and efficient tools, the map makers finished two years ahead of schedule. Older settlements supported the exploration, but many have mixed feelings about the venture. To some it provokes fear, and others project near-religious ideas of a "quest." The new knowledge challenges some worldviews, inherited beliefs about health, and traditions of healing. True to their history, wealthier sectors of the older communities are using the mountain's resources before majorities can agree on how their beliefs and ways of life ought to change to insure the benefits of the new knowledge.

This extended metaphor seems consistent with the vision of Collins and Guttmacher (2003) for genomic research and social change. At the end of the HGP, Collins (2001) quoted Churchill on "the end of the beginning." The story is not of war but of stages of discovery and increasing social responsibility to share the benefits fairly. The HGP's first impact is on genetics, the study of single genes. "Identification of the genes responsible for human mendelian diseases, once a herculean task requiring large research teams, many years of hard work, and an uncertain outcome, can now be routinely accomplished in a few weeks by a single graduate student with access to DNA samples and associated phenotypes, an Internet connection to the public genome databases, a thermal cycler and a DNA-sequencing machine" (Collins, 2001, 835). Tools created in the HGP are now useful in exploring the "multifactorial" diseases that involve several genes and their interactions with environments. Using new microarray technology, expressions of thousands of genes can be studied "in a single afternoon" (Collins, 2001, 835).

"Genomics" is the study of all genes in the genome and applies to common diseases, like cancers of the breast and colon, HIV infection, and neurodegenerative disorders like Parkinson's and Alzheimer's disease (Guttmacher and Collins, 2002). "Proteonomics" is the large-scale analysis of the proteins in a cell. By mining other mountains in the chain, the differences between human and other genomes will appear at the level of proteins. This step will make the individualization of drug treatment and other therapies possible, as well as safe and efficient vectors for human gene therapy. Without advances in just distribution of health care resources, the clinical results could be limited to the most affluent in wealthy nations. Signs already point to this result in terms of change in the economics of genomic research.

The second change is a major shift in the economic infrastructure for research and development in genetics and genomics from governments and academic institutions to commercial enterprises (Cook-Deegan, 2001). Pharmaceutical and biotechnology firms, which are also beholden to stockholders, now fund more research than national governments. Questions arise about the power and accountability of these global corporations in setting directions for research and offering services through commercial laboratories. Patenting of genetic discoveries is needed to raise capital, but this practice begs a question about whether DNA is property to be traded in the marketplace. Or, as some see it, is the human genome a heritage of humanity to be shared by all peoples and not to be owned in part or in whole by any one entity? Commercialism, a relatively small force in the mid-1980s, now raises a host of issues about professional integrity and conflicts of interests, especially in the context of genetic research. Trust in the integrity of the health professions is a critical issue in developed nations (O'Neill, 2002a).

The third change is increased professional emphasis on ethical issues, spurred by the HGP and by recognition that the trust of the public and of patients is at stake. Geneticists routinely convene national and international meetings devoted to ethical, legal, and social issues. The International Federation of Human Genetics Societies,[1] founded in 1996, now serves as the main body to consider ethical issues and ratify statements generated by member organizations. The genetics community is much stronger and more attentive to ethical issues than in the past. Attention to ethics and self-regulation are marks of a maturing profession.

A fourth change is the rapid entry of new people into the field of genetics. More than half of those who answered the 1995 questionnaire were not in the field in 1985. In 1985, the median age was 45, with 14 years experience in genetics. In 1995, the median age was 40 (42 outside the US), with nine years of experience. A new generation of geneticists is entering the arena of practical ethics. Their training and continuing education needs to include attention to ethical issues in genetic services.

Finally, users of genetic services have raised their voices, partly out of fear that they will be denied fair access to important goals, such as employment and health care, partly to demand protection of privacy and rights to employment, insurance, and schooling. The rise of genetic support groups, including some in developing nations, may do as much to influence the course of genetic services as efforts of genetic professionals. Support groups help to empower their members by education and strengthening their capacity for autonomous choices. These groups also make the wider public aware of the work that society needs to do through policy and legislation to insure the benefits of genetic discoveries and minimize the harms to persons and their opportunities.

GENETIC ESSENTIALISM AND "GENOHYPE"

A transition from genetics to genomics depends in part on an informed and educated public. The public learns about scientific issues through the media. The media have tended to focus heavily on the HGP's potential effects on medicine and thereby raised hopes. Costa (2003) reports a content analysis of articles about the HGP in leading newspapers in Italy, France, Great Britain, and the US from 1990 to 2001. The study asked two questions: How was the article "framed" as to theme? Was the "valence" positive, negative, or objective? Of four frames (medical, scientific, business, ethical/legal), the study found medical themes far more numerous. The medical articles tended to have a "positive valence." The *New York Times* far outstripped others[2] in total articles on the HGP.

Public support of the HGP was premised on expectations of clinical applications. Aided by the media, the public's hopes are vulnerable to disillusionment. This state of

---

[1] <http://www.ifhgs.org>

[2] *Corriere della Sera, La Republia, Le Monde, The Independent*

affairs contributes to broader erosion of public trust in institutions and professions (O'Neill, 2002b). Some academics have also indulged in sensationalism. For example, an article in a medical law journal described knowledge erupting from the HGP that would revolutionize medicine and "profoundly affect the lives of *present* [italics ours] and future members of society, bringing into question the very identity of the human individual" (McLean and Giesen, 1994).

Many in the public attach excessive hopes and fears to genetic discoveries. Genes becomes responsible, not only for heart disease and cancer, but for smoking, "novelty-seeking behaviors," violence, and divorce. The reality is very different. According to British geneticist Steve Jones, "In some ways, the new genetics is no more than another form of high-tech medicine, of crucial importance to a few but irrelevant to the many. At present it suffers from too much publicity and too few results." Jones claims that "geneticists are in part to blame for overstatements and false hopes. But others are also at fault: those, for example, with ethical or political objections to the science, who also stand to gain from exaggerating what genetics might do." (Jones, 2000, 17).

Medical care appears to take on a new dimension if the word "genetic" becomes attached to a test, diagnosis, or treatment. For many people, genetic information has a special or even sacred status. What was once routine becomes fraught with uneasiness, a fear of "knowing too much" about dark secrets or forbidden things. A family history is usually provided to a physician or an insurance company without much concern; a "genetic test" arouses anxiety or calls for protective legislation, even though it may provide less information, overall, than the family history. People nonchalantly give out Social Security numbers on request, to shopkeepers, utility companies, and supermarkets. Ask them for a sample of their DNA, and some may panic. At this point in time, far more can be done against an individual with a Social Security number (including identity fraud) than is likely with a sample of DNA, but somehow the genes are sacred. Many think that prying into genes is capturing their essence, their soul. "I am my genes" is a deep-rooted belief that underlies genetic essentialism. This belief is that genes govern what we are, predict what we will be, and construct our identity, consciousness, and individual uniqueness. A corollary belief is that genes are immutable. The individual is therefore exempt from blame, responsibility, and punishment. As the center of one's being, genes deserve a special protection for inner privacy. Until recently, many intellectuals attributed far more power to environments to shape human ill or well-being than to heredity (Pinker, 2002). In contrast, the public has consistently leaned towards genetic essentialism. According to sociologist Dorothy Nelkin, throughout the 20th century, Americans attached more importance to genetics than environment, except for a brief period after World War II when there was hope for a fresh start in society (Dreyfuss and Nelkin, 1992; Nelkin and Lindee, 1995).

We do not know whether genetic essentialism also holds sway outside the West. In some Asian cultures, there may be less differentiation between natural and artificial, genetic and non-genetic. The concept of continual change may compete with conceptions of genetic stability or immutability. The Buddhist Law of Eternal Change says that we are not the same as we were only a minute ago. Nothing on earth – the world of illusion – is fixed. However, "Asian ethics" is far from a unity; there may be more schools of thought than in the West (Sen, 2000). We do not know how various religious, cultural, and philosophical traditions will respond to genetics or whether genetic essentialism will become a worldwide phenomenon.

Genetic essentialism encourages expectations after every small genetic discovery, a state of frantic overestimation best described as "genohype" (Holtzmann, 1999). Genohype may start with over-hopeful researchers who overestimate the value of discrete pieces of information. It is fueled by the media. Genetics sells, even though "genetic" cures for genetic disorders are still in an early and experimental stage. The profession and the media have a responsibility to curb a trend that distorts the public's understanding of the issues and the climate of policy debate.

IS GENETICS BEING INTEGRATED WITH GENERAL MEDICINE?

Will the results of the HGP soon be integrated into general medicine by primary care physicians, who will join geneticists on the front lines of genetic services? Although this process has begun (Collins and Guttmacher, 2003), we see such integration taking considerably longer, because effective treatment at the molecular level will take many more years to develop. Physicians and psychologists (Patenaude, *et al.* 2002) are gradually learning to use genetic information in following patients at higher risk for cancer, heart disease, neuropsychiatric disorders, and stroke. Current measures are mainly preventive and not curative. In the future, geneticists and physicians will cooperate in the management of a patient's disease over a lifetime, but that future is far from realization. Meanwhile, we are awash in a sea of genetic information that resembles the flotsam after a hurricane. Tantalizing bits of data swirl around us, each of which may point to a new opportunity for treatment, or, like many scientific discoveries, may increase our knowledge without improving our health.

Many geneticists expected that the HGP would propel many new genetic tests into medicine very quickly. What is the scorecard so far? In many countries, newborn screening for treatable diseases has been a part of primary care for many years, as has carrier screening in some populations for mutations that cause sickle cell anemia or thalassemia. The carrier test for the mutations that cause cystic fibrosis has penetrated primary obstetric care in the United States, but not in the United Kingdom or most of Europe. Prenatal diagnosis has become widespread in English-speaking nations, Europe, and Japan, but not in the rest of the world. Adult genetic testing is still in the first stage of its history.

Geneticists also assumed that most people would want to know their genetic status, and that there would be a stampede to take the new tests. In fact, most people do not want to take genetic tests for adult-onset disorders, either because the tests have low predictive value or because no acceptable or useful treatment is available (Shiloh, 1996). Epidemiological studies take many years to provide accurate background data to establish risks. The two mutations that make Ashkenazi Jews susceptible to breast cancer were identified in 1994 and 1995. Epidemiology and genetic testing across an entire population of patients and families has now established that the carriers' lifetime risks of breast and ovarian cancer are 82% and 23%, respectively (King *et al.*, 2003). Women with a mutation have a stark choice (Couzin, 2003). Prophylactic double mastectomy can clearly lower risk, but debate persists about cancer in remaining tissue. Oophorectomy appears to reduce ovarian cancer risk by 85–95% and breast cancer risk by 50%, possibly because of lowered estrogen levels. Treatment for breast cancer remains elusive.

The genetic test for Huntington disease is close to 100% predictive, but it cannot tell a person when the disease will begin. Far fewer than expected have sought the test.

Only a minority want to be tested to make life plans or inform their options for reproduction. The prospect of learning one's fate when there is no treatment creates dread. Many with relatives who died from Huntington disease prefer not to know. When the fear of genetic discrimination by insurers or employers is added to the lack of treatment, people have good reasons to abstain from testing.

At present, genetic testing has little to offer the average adult who has no family history of a known genetic disease. Noted geneticist Arno Motulsky suggests that the only test that most of us might find worth taking is for hemochromatosis, a metabolic disorder causing iron overload that can damage the heart or liver (Motulsky, 1999). Even this test is controversial because many, perhaps most, of those with the mutation never develop symptoms. The first population-wide genetic tests will likely be to estimate people's reactions to drugs. Some are "slow metabolizers," who have a high risk of severe reactions; others are "fast metabolizers," who may not respond at all to the usual dosage. Given that prescription drugs kill 100,000 people a year in the United States alone, identifying the 8–10% who may overreact would be a major advance.

Genetic discoveries are also entering medicine quietly and through a backdoor of genetically engineered drugs and vaccines based on genetic research. Already, specialists in hematology, oncology, and cardiology are likely to know more about the genetic aspects of treatment than most geneticists. Doctors may not even use the words "genetic" to describe these modalities. Now, when doctors prescribe genetically engineered growth hormone to treat short stature in children, they do not use the word "genetic," as it may frighten people away. They simply prescribe growth hormone.

Genetics and general medicine will gradually converge, but the process will be slow and halting. In this process, genetic information ought not be regarded as sensitive and dangerous, because of its importance to human health. However, the reality is that genetic information is very sensitive (privacy issues are real) and many feel that it can be dangerous to their insurance, employment, or schooling. In some societies it could reduce chances of marriage or taint entire extended families. Without effective and vigorously implemented laws to prevent discrimination, there can be no real integration of genetics into medicine. To achieve a measure of social and economic reform in access to health care and effective laws, it could well take another 30 years for medicine and society to adapt to genomics. In such a future, a health care system would include:

- Comprehensive genomics-based health care (including diagnosis and treatment at the molecular level by successful gene therapy in somatic cells).
- Preventive medicine tailored to the individual's genome.
- Early detection of most illness by genetic testing.
- Disease predisposition profiles identified at birth.
- The possibility of safe germline gene therapy (as a result of technology that can alter DNA in the gametes) so that transmission of a disease mutation to future children would not occur. The debate about the ethics of such therapy would still be vigorous, despite the benefits of prevention.
- An extended average lifespan of more than 85 years.

In this process of change, the widely shared perception of the "dangerousness" of genetic information will subside. In time, genetic information will be as routine in medicine as one's blood type or other common test results. DNA samples will be banked routinely as part of medical records. Genetics will be assimilated into all subjects taught in medical school. Primary care physicians will be expected to counsel patients about their genetic risks for disease and also about their reproductive risks and options.

DO PHYSICIANS NEED A "NEW ETHICS" FOR GENETIC ISSUES?

Even if integration will be a lengthy process, will primary care physicians need a "new" ethics to deal with genetic issues? If genetic information is not inherently dangerous or different in kind from other information relevant to health care, we see no intrinsic reason to wall off genetics issues from other ethical issues in medicine and treat them differently. Primary care physicians work with families and should have no trouble understanding the issues that arise when genetic information about an individual also provides information about other family members. Until society prevents genetic discrimination in insurance and employment, physicians will need to be vigilant in guarding patient privacy and confidentiality of genetic information. Physicians will also need to learn how and when to use the exception (when justified) to warn family members at genetic risk, as discussed in Chapter 6, when not doing so would lead to significant harm and the harm could be averted by the warning.

There is strong evidence that a traditional outlook in medical ethics is unprepared for complexities of counseling before and after genetic testing (Geller, 1993). Our research, limited only to the United States, showed major differences between the approaches of primary care physicians and geneticists. Table 90 shows how medical paternalism would probably influence primary care physicians in counseling patients on reproductive choices. This orientation differs sharply from the non-directive approach of US geneticists. We do not know whether such a division exists in countries with a less well-established tradition of genetic non-directiveness. Medical paternalism has a bad reputation in North America, because it is associated with authoritarianism. However, the concept of "good" paternalism is linked to physicians' obligations to increase health benefits for patients and families and to prevent or minimize harm. When the patient is incapacitated or a minor, as in the case of a child of Jehovah's Witness parents who needs a life-saving blood transfusion, society permits judgments and actions, even over the religious objections of parents, to save the child's life. However, if persons are capable of making health care or reproductive decisions, US society is generally opposed to permit others, even physicians motivated by beneficence, to override them arbitrarily. Bioethics came into being, in part, to protest violations of autonomous choice by paternalistic physicians in research and treatment (Jonsen, 1998). Given the strong commitment to reproductive freedom in the US, many would object to an unmodified paternalism on the front lines of genetic services. Since genetic counseling requires time that many physicians do not have, it is possible that the profession of genetic counseling will expand in concert with the movement of genetics into mainstream medicine.

PRIMARY CARE PHYSICIANS' KNOWLEDGE ABOUT GENETIC CONDITIONS

Physicians' greatest needs may be for scientific and technical education in human genetics. Studies of primary care physicians in several nations and the US have confirmed these needs. In one study, family physicians discussed family history in only one in four patient visits (Acheson *et al.*, 2000). Four in five of 1,100 allied health professionals had no formal training in genetics (Lapham *et al.*, 2000). A comprehensive survey of physician education and knowledge about genetics in Europe found their knowledge inadequate in the Czech Republic, Germany, Hungary, Italy, Norway, Poland, Portugal, Spain, Sweden, Turkey, and the United Kingdom (Harris and Reid, 1997). A French survey found that only 54% of physicians were aware that BRCA mutations contributed to breast and ovarian cancer (Julian-Reynier *et al.*, 1999). A US study (Doksum *et al.*, 2003) showed that many physicians talk with patients about genetics and breast cancer, and order tests for BRCA mutations but lack basic knowledge of the genetics of breast cancer.

In our survey, the physician questionnaire included brief knowledge tests about some aspects of genetic conditions that parents or prospective parents might want to know and would expect their physician to know: e.g., life expectancy, presence of mental retardation, ability to complete regular high school, to work, or to become a biological parent, mode of genetic transmission (autosomal dominant, recessive, X-linked), and availability of prenatal diagnosis. None of these questions asked about complex diagnoses, locations of a gene on a chromosome, or details of scientific procedures.

Most physicians got most of the answers right. However, most readers would probably not want to be a parent visiting one of the 13% who thought most children with Down syndrome could finish regular high school (a few have done so, but this is extremely rare), or those who said that boys with cystic fibrosis (64%) or Klinefelter syndrome (45%) could father biological children when in fact they are sterile. Nor would most want to be the patient of the 30% who said life expectancy with cystic fibrosis was under 20 (it is now well over 30), or the 17% who said Turner syndrome was usually associated with mental retardation, or the 23% who said that Huntington disease was recessive. Finally, there were the 30% who said that trisomy 52 was prenatally diagnosable. This was a trick question, since humans have only 22 pairs of autosomes plus 2 sex chromosomes, 46 chromosomes in all. Respondents could have said "don't know," which was an option offered, but many physicians hate to admit not knowing. In sum, there is a great need for physician education in the basics of human genetics, starting in medical schools.

DO GENETICISTS NEED AN INTERNATIONAL CODE OF ETHICS?

There is a history of "one-way" monologue about ethical issues in genetics. Discussion began in the West and was exported to other settings, because professional societies in most of the world looked to the West and used its publications for their own discussions. This history has raised suspicion about Western cultural imperialism. There are two reasons not to fear this outcome. First, there is a healthy growth of interest among geneticists everywhere in ethical issues, and they are drawing on their own cultural resources. Scholars in ethics and religion in many nations now contribute, which tends to balance the weight of Western sources. The result is a much richer dialogue. Second, Western medical ethics has much to offer, especially in encouraging individual decision-making. Physician–patient relationships in many parts of the world are taking on Western characteristics, with patients assuming more autonomous roles. As patients participate more actively, they will gain power and their health will improve.

Recent years have seen a spate of international pronouncements on genetics, ranging from lofty, overarching proclamations on the dignity of the human genome to specific recommendations against particular, usually new, reproductive technologies. Many international organizations now believe that in order to make their mark in the modern world, they must make a statement on genetics, usually in the areas of cloning or patenting. In view of the plethora of existing statements and many more to come, does the world really need a universal code of ethics in genetics? In 1993, we argued for such a code (Wertz and Fletcher, 1993). Now we are not so sure.

Before discussing the issues, we need to distinguish between "codes of ethics" and "codes of practice." Codes of ethics are usually based directly on moral principles assumed to be universal, such as respect for persons. In genetics, UNESCO's Universal Declaration on the Human Genome and Human Rights (1997) is perhaps the best example of a code of ethics. One cannot disagree with the sweeping principles of the Declaration. With the exception of a last-minute addition of a ban on human cloning, which may not stand the test of time, the UNESCO Declaration was designed to last for many years and should survive the rise and fall of a cascade of new genetic technologies.

The problem with the Declaration is that it offers nothing to the average clinician. Knowing that the human genome is imbued with "dignity" and should not be bought and sold does not help guide medical practice. Most international statements are quite removed from practice. The 1990 Declaration of Inuyama (Fujiki *et al.*, 1991, 102–103), which arose from a conference sponsored by the Council for International Organizations of Medical Sciences (CIOMS) may be an exception, but has received little attention.

"Codes of practice," as opposed to "codes of ethics," recommend actions in specific types of situations. Statements by professional societies are the most common examples. Often these are directed at specific topics, for example genetic testing of children or disclosing information to relatives against a patient's wishes. Most genetics professional societies now have statements on at least some topics, and a few, such as the German Society of Human Genetics (see Chapter 14), have made comprehensive statements. Professional statements are not legally binding and carry no penalties for non-adherence. Their effects on practice are unknown. Medical provincialism hinders their implementation. Specialists rarely read statements by societies representing other medical specialties, and general practitioners probably do not read specialty statements at all. In the US, it is likely that few professionals outside genetics have read any of the 25 policy statements of the American Society of Human Genetics.[3]

On the one hand, there are sweeping statements at the highest level that offer little help with the dilemmas of practice. On the other hand, there are country-specific, topic-specific, and profession-specific statements that cannot lay claim to worldwide adherence. The need is for an international code of practice sufficiently universal to elicit widespread general agreement while allowing room for individual nations to work out their own approaches to controversial issues. There may never be agreement

---

[3] <www.ashg.org>

on some issues, especially on the value of the embryo and fetus, but there could be agreement on nondirective counseling, privacy, and informed consent. Such a code is needed because (1) in some nations, especially in the developing world, discussion of ethical issues in genetics is in an early stage, and policymakers and public health officials are seeking guidance, and (2) in the absence of a code, laws could become the default position. Laws have a place in protecting people's rights to employment, affordable health care, and truthful information (especially if a test is sold or advertised commercially). However, laws may be the worst way to approach reproductive technologies or emergent research. Laws are frequently a "knee-jerk" reaction to new developments (if you are against it, ban it) and often follow hasty, ill-considered fact-gathering by jurists who do not fully understand the science. An example is Germany's Embryo Protection Law, which has been interpreted as forbidding preimplantation embryo diagnosis. Now that the procedure, and the needs of people desiring it, are better understood, some German professionals have decided that the law was a mistake (DW-World, 2003). Reversing it is easier said than done. This is true of most laws. Without an accepted code of practice that may help assure the public that geneticists are ethically responsible persons, public fears will mount and demands for laws will increase. The practice of genetics could become ossified by a steady intrusion of regulations that remove flexibility from professional–patient relationships. Surely a code would be preferable.

There are some welcome developments in the middle ground between abstract statements and specific recommendations for practice. One approach works through genetics professional societies. The European Society of Human Genetics[4] is developing policy statements in general areas such as genetic screening, insurance and employment, archived specimens, and genetic services, with the goal of formal acceptance by each of ESHG's member professional societies. Eventually, these statements may go to the International Federation of Human Genetics Societies for approval. IFHGS includes genetics professional societies around the world.

Another approach aims at nations where genetics is relatively new. The World Health Organization's (WHO) "Proposed International Guidelines" (1998) (Appendix) is the most comprehensive code of practice to date. Its 16 pages cover most aspects of genetic counseling, testing and screening, and prenatal diagnosis. This document emerged from a consensus meeting of WHO advisors (including practicing medical geneticists, genetics researchers, ethicists, and lawyers) from around the world. From an ethical perspective of principles that have very wide scope, it lays out general guidelines for practice (e.g., prenatal diagnosis should be available, voluntary, and preceded by unbiased, informative counseling) without being so overly specific as to exceed some countries' resources (e.g., it sets no maternal age limit for prenatal diagnosis). The WHO Proposed International Guidelines are aimed at public health officials, primary care physicians, and policymakers as well as genetics professionals. Findings of our survey, especially with regard to international consensus or absence of consensus, informed many of the deliberations underlying the document. The guidelines are supported by a comprehensive review of ethical issues in medical genetics (Wertz *et al.*, 2001).

---

[4] <www.eshg.org>

Together, these two documents come close to a worldwide code of practice. Previously, we had envisaged a relatively brief statement of general principles, such as "genetics services should be voluntary." Such a statement is no longer necessary, because most of these principles have now been accepted as part of worldwide discussion. What is needed is a comprehensive code of practice. The WHO Guidelines approximate that need at present. In the future, people with genetic conditions and the public – the most important stakeholders of all – must necessarily play important roles in developing new codes of practices.

REFERENCES

Acheson, Louise S., Georgia Wiesner, Stephen Zyzanski, Meredith Goodwin, Kurt Stange Family history-taking in community family practice: implications for genetic screening. *Genetics in Medicine* 2000; 2: 180–185.

Benjamin, Martin. *Philosophy and This Actual World*. Lanham: MD, Rowman & Littlefield, 2003.

Bobrow, Martin. Comments. The Wellcome Trust, Sanger Institute. <http://www.sanger .ac.uk?HGP/draft2000, commentary.shtml>.

Collins, Francis S. Contemplating the end of the beginning. *Genome Research* 2001; 11: 641–643.

Collins, Francis S. and Alan E. Guttmacher. A vision for the future of genomics research. *Nature* 2003; 422: 835–847.

Cook-Deegan, Robert M., Amber Johnson, and Carmie Chan. *World Survey of Funding for Genomics Research*. Final report to the World Health Organization and Global Forum for Health Research, September 2000. *Journal of BioLaw and Business Special Supplement* 2001. <http://www.stanford.edu/class/siw198q/genomics>.

Cook-Deegan, Robert. *The Gene Wars. Science, Politics, and the Human Genome*. New York: Norton, 1994.

Costa, Tullia. The Human Genome Project and the Media. Case Study: The Relation Between Genetics and the Media. *International Journal on Scientific Communication*. 2003; N4. <http://Jekyll.sissa.it/Jekyll_comm/articoli/art04_03_eng.htm>.

Couzin, Jennifer. Choices – and uncertainties – for women with BRCA mutations. *Science* 2003; 302: 592–593.

Doksum, Teresa, Barbara Bernhardt, and Neil A. Holtzman. Does knowledge about the genetics of breast cancer differ between nongeneticist physicians who do or do not discuss or order BRCA testing? *Genetics in Medicine* 2003; 5: 99–105.

Dreyfuss, Rochelle C. and Dorothy Nelkin. The jurisprudence of genetics. *Vanderbilt Law Review* 1992; 45: 313–324.

DW-World. De Deutsche Welle, Germany hints at Loosening of Stemcell Law," October 30, 2003. <http://www.dw-world.de/english/0,3367,1430_A_1016499_/_A,00.html>

Fujiki, Norio, Viktor Bulyzhenkov and Zbignew Bankowski (eds). *Medical Genetics and Society*. Amsterdam: Kugler Publications, 1991.

Geller, Gail, Ellen Tambor, Gary Chase, Karen Hofman, Ruth Faden, Neil Holtzman, Incorporation of genetics in primary care practice. Will physicians do the counseling and will they be directive? *Archives of Family Medicine* 1993; 2: 1119–25.

Guttmacher, Alan E. and Francis S. Collins. Genomic medicine – a primer. *New England Journal of Medicine* 2002; 347: 1512–1520.

Harris, Rodney and M. Reid. Medical genetic services in 31 countries. An overview. *European Journal of Human Genetics* 1997; 5: S2.

Holtzman, Neil A. Are genetic tests adequately regulated? *Science* 1999; 286: 409.

Jasney, Barbara and Donald Kennedy. The human genome. *Science* 2001; 291: 1153.
Jones, Steve. *Genetics in Medicine: Real Promises, Unreal Expectations. One Scientist's Advice to Policymakers in the United Kingdom and the United States*. New York: Milbank Memorial Fund, 2000.
Jonsen, Albert R. *The Birth of Bioethics*. New York: Oxford University Press, 1998, pp. 336–337.
Julian-Reynier, Claire, F. Elsinger, Moatti, J.P., Hagay Sobel. French physicians' knowledge about hereditary breast/ovarian cancer: the need for continuing vocational training in genetics. *Community Genetics* 1999; 2: 165–172.
Keller, Evelyn F. *Refiguring Life: Metaphors of Twentieth Century Biology*. New York: Columbia University Press, 1995.
King, Mary-Claire, Joan Marks, Jessica Mandell, for the NewYork Breast Cancer Study Group. Breast and ovarian cancer risks due to inherited mutations in BRCA1 and BRCA2. *Science* 2003; 302: 643–646.
Lapham, Evelyn V., Chahira Kozma, Joan Weiss, Judith Benkendorf, Mary Ann Wilson. The gap between practice and education of health professionals: HuGEM survey results. *Genetics in Medicine* 2000; 2: 226–231.
McLean, Sheila A.M. and Dieter Giesen. Legal and ethical considerations of the Human Genome Project. *Medical Law International* 1994; 1: 159–175.
Motulsky, Arno. G. If I had a gene test, what would I have and who would I tell? *Lancet* 1999; 354: 35–37.
Nelkin, Dorothy and Susan M. Lindee. *The DNA Mystique: The Gene as a Cultural Icon*. New York: Freeman, 1995, pp. 34–35.
O'Neill, Onora. *Autonomy and Trust in Bioethics*. Cambridge: Cambridge University Press, 2002a.
O'Neill, Onora. *A Question of Trust*. Cambridge: Cambridge University Press, 2002b.
Patenaude, Andrea F., Alan E. Guttmacher, and Francis S. Collins. Genetic testing and psychology. New roles, new responsibilities. *American Psychologist* 2002; 57: 271–282.
Petsko, Gregory A. The Rosetta stone. *Genome Biology* comment 2001; 2: 1007.1–1007.2.
Pinker, Stephen. *The Blank Slate. The Modern Denial of Human Nature*. New York: Viking, 2002.
Sen, Amartya. East and West: the reach of reason. *New York Review of Books* 2000; 33–37.
Shiloh, Shoshana. Decision-making in the context of genetic risk. In Theresa Marteau and Martin Richards (eds), *The Troubled Helix: Social and Psychological Implications of the New Human Genetics*. Cambridge: Cambridge University Press, 1996, pp. 82–103.
Watson, James D. *DNA: The Future of Life*. New York: Knopf, 2003.
Wertz, Dorothy C. and John C. Fletcher. Proposed: an international code of ethics for medical genetics. *Clinical Genetics* 1993; 44: 37–43.
Wertz, Dorothy C., John C. Fletcher, and Kare Berg. *Review of Ethical Issues in Medical Genetics*. Report of Consultants to WHO. Geneva: World Health Organization Human Genetics Programme, 2001.
World Health Organization. *Proposed International Guidelines on Ethical Issues in Medical Genetics and the Provision of Genetic Services*. Geneva: WHO, 1998.

CHAPTER 20

MORAL PRIORITIES, CASES, AND PRACTICES IN MEDICAL GENETICS

The goal of this discussion is to promote dialogue in forums for geneticists and policy makers. Examining ethical issues in a field like medical genetics is comparable to navigation and repairs during a sea journey (Benjamin, 2003, 112). After locating position, the crew must alter or maintain direction. Faults in the structure of the ship or equipment that make it unseaworthy or unprepared for trouble must be corrected. These corrections have to be done on a ship already underway. There are different priorities, but all are important. In ethics, as in all human ways of knowing, the ship cannot return to port for refitting (Benjamin, 2003, 112) once the journey has started.

Ranking moral priorities require criteria derived from the goals of clinical genetics as a branch of medicine within societies. Priorities must take account of new goals in diagnostic, therapeutic, and public health implications of common diseases. However, older goals are still intact. Clinical genetics exists to help people and families at higher risk to reproduce as normally as possible. Its social goals include education in genetics at all levels and to espouse and guard voluntary societal measures to prevent the harms of genetic diseases, with exceptions like newborn screening for some diseases that are treatable.

The content of the international survey reflected the field at the time, during the initial impact of the "new genetics" on DNA testing. The questionnaire was pre-HGP and its relevance to lifelong disease management or public and global health (Merikangas and Risch, 2003). The proposals for moral priorities below need re-examination in the present context. The chapter also discusses hard cases and practices in medical genetics. Below, we give an overview of our perspective in ethics.

We are mindful that majority opinion, even of those with special knowledge, is not a basic source of ethics. Majorities can be misinformed, bigoted, or disdainful of minorities or minority views, as is evident from the history of race and ethnic relations. Survey data inform us about how genetic professionals *probably* think about cases and practices in genetic services. The data may also help in weighing the practicality of changing a practice. However, survey results are not equivalent to important values and principles for moral deliberation on cases or practices.

ETHICS AND MORAL DELIBERATION

There is no prefabricated way to work through an ethical problem. Geneticists are familiar with paths into and through many problems. For example, most try to be empathetic, non-biased, and honest in disclosing ambiguous fetal test results to anxious parents-to-be. However, some cases that test the limits of autonomy or responsibility to society can throw experienced geneticists nearly or completely off-balance. In these situations, Buber's (1965, 164) metaphor of a "narrow rocky ridge between the gulfs" can ring true. Ethics helps human beings to keep enough balance, when taking old or new paths in social and personal life, to maintain moral insight and equilibrium. Like law or medicine (Benjamin, 2003, 119), ethics is a body of practical knowledge about human problems, i.e., conflicts of moral principles, duties, or loyalties in relationships. Individuals, families, and communities experience these struggles in all spheres of their lives and activities, except when emotions are numbed or ideology overwhelms minds. Then the greatest evils are done or tolerated.

Because ethics is a living and evolving body of knowledge, open societies require a public process of debate about continuity and change in the moral aspects of social practices. Cross-cultural issues complicate debate. For example, in counseling people at higher genetic risk who want children, adoption is an important option. However, cultural differences in adoption practices raise moral, social, and legal problems. Can Islamic practices that permit only foster parenting of orphans, so as to protect the paternal bloodline, fit into cultures with traditions of full adoption? Contrary to what a legally adopted child could expect, Islamic teachings hold that foster children cannot have the family name or inherit any of its estate. What is best, all things considered, for orphans in multicultural societies? How do counselors find their way through this complicated option?

Rational deliberation, not majority vote, is the substantive heart of the public process that nourishes ethics. Daniel Callahan (1995, 274) finds that the authority of ethics "lies, in the end, in the plausible insight and persuasive rationality of those who can reflect thoughtfully and carefully on moral problems." What does rationality mean? Philosophers give different answers. Stephen Toulmin points to evolutionary factors. Human beings show rationality "not by ordering their concepts in tidy formal structures, but by preparedness to respond to novel situations with open minds – acknowledging the shortcomings of their former procedures and moving beyond them" (Toulmin, 1972, vi). Views in ethics compete for influence in the practical world. Societies that are pluralistic and democratic grant no privileged place to any moral tradition or theory. However, moral authority is not a fiction; it is cumulative, i.e., earned by the merits of balanced arguments and judgments on particular issues.

We take a broadly pragmatic approach to ethics and to cross-cultural issues in ethical debates. Interpretations of moral norms and ways of life are relative to culture and personality. The international survey confirms relativity in ethics. But cultural and psychological relativity do not drain ethics of content or moral force. Although ethics is a phenomenon within and not independent of human experience, it has objectivity in two senses. First, ethics is objective as a growing and self-correcting body of knowledge about practices in the actual world (Benjamin, 2001, 2003). Second, the basic injunctions of ethics are "no less objective than law or medicine," (Benjamin, 2003, 119) when it comes to destructive acts like murder, torture, genocide, rape, etc. Such acts violate "the rock bottom of our convictions" (Wittgenstein, 1969) about the intrinsic value of human life and persons. "If these things aren't wrong, nothing is" (Benjamin, 2003, 115). The quest for objectivity in ethics can go too far. For some philosophers (Beauchamp, 2003) there must be a set of universal norms to which all human behavior is accountable, i.e., *the* "common morality." One can dispute this claim on empirical and logical grounds but not surrender objectivity in ethics. There is a middle way that combines relativity and objectivity in ethics.

Methods of deliberation about ethical problems (Arras, 2003) also compete for selection. The method of "wide reflective equilibrium" has proven useful in bioethics, when compared with highly theoretical or case-by-case approaches (Benjamin, 2001, 29–30; 2003, 112–123). In this approach, one weighs ethical problems in cases by creating a dialogue (internally or externally) that moves between three interactive elements: the values and principles at stake, the problem at hand, and relevant background beliefs and theories.[1] A dialectical method of analyzing problems in a framework of *prima facie* principles of biomedical ethics (Beauchamp and Childress, 2001) has influenced work in international forums, including WHO publications on clinical genetics (Wertz *et al.*, 2001). If a method moves only between values or principles and the problem at hand, it may lack larger constraints and correctives. Choices of principles and how to put them to work are vulnerable to cultural and personal bias. Our judgments about past cases are fallible and need evaluation. Background beliefs and theories create some critical distance from the other two elements and constrain them. Benjamin writes: "Background beliefs and theories include beliefs about the nature of persons; the nature of the world as revealed by science and metaphysics (including whether there is a God, and if so, God's role in our lives); human psychology, sociology, and political and economic behavior; the nature of non-human animals, and so on" (116). Intelligible background beliefs (e.g., biological evolution by natural selection, moral pluralism, etc.) and high standards of knowledge (e.g., psychology, social sciences, etc.) are sources of criticism on other elements in the dialogue of ethics.

John Dewey (1988, 164) wrote: "life is a moving affair in which old truth ceases to apply." To make a judgment in a novel situation can lead to change of background beliefs and practices. Benjamin recalls (2003, 118–119) how beliefs about who is a living person changed in response to crises involving hopelessly unconscious patients and decisions about organ transplants. Some were already advocating for neurological criteria to diagnose death (Beecher, 1968) before the public process of debate and reform began. A cluster of changes reduced harms of unwanted critical care and misuses of scarce medical resources. However determination of death required an impartial process to protect dying patients from premature organ harvesting. The debate progressed into new territory about states of unconsciousness and personhood (Jennett and Plum, 1972).

One can trace similar events in general medicine and clinical genetics. Historically, there was a strong belief that a good doctor, like a good parent, ought to do nothing to disrupt therapeutic benefits to the patient of implicit trust in the doctor. Good doctors shielded patients from the shock of disclosure of a sensitive diagnosis. Changing beliefs about paternalism and evidence of benefits of disclosure have steadily caused practice to shift towards respect for a patient's right to know. In 1985, geneticists were almost evenly divided on disclosing test results to a woman with

---

[1] Benjamin refers to sources in the work of John Rawls (1971) and Normal Daniels (1996) for his discussion of "wide reflective equilibrium."

androgen insensitivity syndrome and a 46-XY karyotype (Wertz and Fletcher, 1989, 18). By 1995, there was strong consensus worldwide that women over 18 ought to know their diagnosis, if they want to know it (Table 15).

We move now to discussion of moral priorities, cases, and practices.

INEQUITIES IN ACCESS TO GENETIC SERVICES

Sidney Brenner, a leading figure in genetics, recently wrote: "Somehow, in our abstruse discussions about the high technology of the developed world, we need to balance the equation by thinking about the rest of the world as well" (Brenner, 2003, 533). We propose that equitable access is the highest moral priority in genetic services today, due to the magnitude of the gap between human needs and available resources. Authors of reports from nations in the survey identified four kinds of inequities: financial, geographical, educational, and cultural. Financial inequities are by far the most important and are found even in countries with national health care systems (Harris and Reid, 1997). Geographical inequities exist even in wealthy nations, such as Norway, Sweden, and Switzerland, with rural people less likely to receive services. All authors reported that people with less education were less likely to use genetic services. In addition, there are sometimes cultural barriers, with minorities receiving fewer services. For example, Indians (Native Americans) in Peru, Argentina, Brazil and Mexico, Turks in Germany, Arabs in Israel, and African-Americans and Hispanics in the United States are less likely to receive services than the majority groups.

In developing nations, other aspects of health care will understandably have priority over genetics services at present. One cannot expect a nation that spends only a few dollars per capita on health care to give priority to genetic services. However, in the future scientists will probably find a genetic as well as a social component for most causes of death and disability. In the United Sates, 9 of the 10 leading causes of death have known genetic components: heart disease, cancer, stroke, chronic obstructive pulmonary disease, pneumonia/influenza (through the immune system), diabetes, suicide (through tendency to depression), kidney disease, and chronic liver disease. Together, these account for 76% of deaths in the USA. At the same time, genetics offers hope for effective treatment or cure. The most modern drugs for diseases ranging from cancer to influenza are based on discoveries in molecular biology. When we exhaust our supply of effective antibiotics, as is already happening with tuberculosis, we will turn to genetically based treatments. At that point, genetics would become a basic part of public health worldwide. Long before then, we need to arrive at equitable solutions guaranteeing access for all.

Some inequalities can probably never be resolved, especially those rooted in geography. The more educated a population, the better its health will be. Universal education and sensitivity to the beliefs and needs of minorities are prerequisites to equitable access to health care and genetic services. Reform requires a moral perspective in which loyalty to the principle of distributive justice limits and balances loyalty to other principles. In most countries, responses to our survey did not show such a perspective at work. In most nations, the majority of geneticists focused primarily on the needs and rights of individual patients and families, without considering the needs of the wider society. Many would undoubtedly consider references to social concerns "eugenics" in a pejorative sense. The exceptions were China and Cuba, where socialist traditions still govern health care and other social goals, but individual autonomy is limited. Geneticists will need to undertake the difficult task of maintaining fundamental loyalty to, and respect for, patients, while at the same time keeping in mind the requirements of equity and justice.

Genetics professionals are in a better position than many others to argue and work for change, due to their generally high status in society and their knowledge of the cultural factors that contribute to illness and disease. At a minimum, professionals ought to aspire to the ideal of being good citizens of their society and servants of the wider task of promoting health. Shimm and Spece (1996, 52) discuss health professionals' social role as a "good citizen." In the same volume, Buchanan (1996, 129) argues society has a right to expect public-spirited behavior in return for the special economic and self-regulatory privileges afforded to health professionals. As good citizens, geneticists and counselors can carefully document the harms done by inequitable access to genetic services and identify needed reforms. Geneticists are a better position than most to do this work and to make the results known to policymakers and the public. Cultivating the virtue of good citizenship can be done in harmony with practicing clinical virtues.

There are further global inequities. Some nations have few or no genetic services at all and were therefore not included in the survey. These nations include most of Africa and the Middle East, parts of Asia, and Central America. Technologically speaking, sub-Saharan Africa appears to be dropping off the face of the map as social conditions worsen. Extraordinary efforts by international organizations like the WHO or UNESCO, together with contributions from private industry, would be required to lay groundwork for genetic services in these regions. International redistribution of the benefits of genetics should go beyond providing post-trial medications to participants in successful clinical trials.

In view of the likelihood that most benefits from the HGP may accrue to rich people within rich nations, the Ethics Committee of the Human Genome Organization (HUGO), an international organization of genetics researchers, urges that for-profit organizations engaged in genetic research share at least 1–3% of their net profits in the form of benefits to communities or to the health care infrastructure of entire nations. The statement argues that "benefit is a broader concept than profit. Benefits may include information, therapies, improved environments (schools, libraries, sports facilities, clean water), improved healthcare, increased human respect. Sometimes researchers, communities, and the public differ on what constitutes a benefit. Therefore, community participation (if a community can be delineated by a disease-related voluntary association, by ethnicity, or by geography) is essential from the outset." (HUGO, 2000, 49)

HUGO used three reasons to support the argument. The first two are specific to genetics; the third applies to health research generally.

Common Heritage. We share 99.9% of our genetic make-up with all other humans. In the interests of human solidarity, we owe each other a share of common goods, such as health. However, genetic research has focused public interest on the 0.1% of our genetic make-up that is unique. Genetics may be divisive as well as exclusive.

It remains to be seen if the HGP will lead to sharing of resources or to increased emphasis on ethnic and individual differences. Awareness of human genetic commonality as a species may not lead to human social solidarity.

The Genome as a Common Resource. Starting with Hugo Grotius' law of the sea in the 17th century and proceeding to the international law governing air and space in the 20th century, such global resources are viewed as common and equitably and peacefully available to all humankind. International law may therefore set a precedent for regarding the human genome as a common resource. HGP researchers, by placing their sequencing data on publicly available websites, have in effect recognized it as such. It may therefore follow that profits and other benefits should be publicly shared. International conventions provide precedents in biodiversity (UNEP, 1992), and food and agriculture (CGFRA, 1991).

Justice. In general, justice has three aspects. The first is compensatory and mandates that recompense should be in proportion to contributions. The second is procedural, maintaining that the process of deciding who receives what must be impartial and must include all stakeholders, and the third is distributive, in that resources should be allocated fairly. Often, participants in genetic research provide only cell samples. Arguments from compensatory justice would suggest that most should receive little because they have contributed little. The Committee chose to take a broader view, concentrating on distributive justice. When there is a vast difference in power between an organization carrying out research and the people providing material for that research, and when the organization stands to make a substantial profit (albeit taking the risk of investment), concerns about exploitation arise. Considerations of equity support the desirability of acting to meet needs, including needs of the healthcare infrastructure. Much genetic research differs from research involved in clinical trials, in that it may benefit only future generations, thus complicating the possibility of just distribution unless directed to the wider society.

The HUGO (2000) statement is one small step in attempting to redress global inequities. Much more will be needed in order to prevent the benefits of genetics from accruing mainly to well-off people in rich nations.

RESPECT FOR AUTONOMY

The 1995 survey results show that geneticists are more likely to respect the autonomous choices of patients than they did in the previous decade. This trend is reassuring insofar as it supports respect for persons and a norm of nondirectiveness in counseling. New developments include the spread of parent groups in many nations, including India and Japan. In time, these parent groups will exercise more influence over provision of genetic services and will increase respect for individuals and families with disabilities.

A moral outlook with a strong role for autonomy encourages people to become morally mature. People who stand up for themselves can also participate more effectively in the struggle for economic and social equality. The moral concept of autonomy is perceived as an export of English-speaking nations, especially the United States, but this perception is wrong. Sen (2000) and Glendon (2001) show that autonomy has deep roots in Asian political and moral thought. At a basic level, respect for autonomy can be in unity with ideals of harmony found in Asian ethics, because of shared fundamental respect for humanness.

Critics of American bioethics see a "triumph of autonomy" (Wolpe, 1998; Fox and Swazey, 1984) as dominance of one philosophical view or a surrender to cultural bias and individualism. There is much substance to this criticism, but it may focus too narrowly on one aspect of autonomy. Autonomy is a complex moral concept. O'Neill's (2002) rigorous criticism of autonomy as only independence is correct. Autonomy also means personal authenticity, effective deliberation, and moral reflection (Miller, 1981; Arras, 2003, 27). The strong meaning of the term involves all of these aspects. Simply to choose freely is its weakest meaning. It is not unusual to see patients in genetic settings who are autonomous in the sense of choosing freely but non-autonomous in a sense of deliberating on the choice in a framework of well-considered values that also structure one's part in society. Genetic counseling that is concerned with the whole patient and family will seek to strengthen autonomy in the full sense of the word.

We propose that a second moral priority for genetic services is to protect the freedom of individuals and families to be fully autonomous, augmented by expert counseling that invites open discussion of moral problems, including the social and economic aspects of choices. Among the information that counselees need are facts about the society's ability to assist them in support of a child with a genetic condition. Parents do not leave their economic needs or their membership in society behind when they seek genetic counseling. To affirm this priority means to oppose laws that prohibit or unfairly restrict parental choices, especially with regard to access to genetic services or the option of termination of pregnancy. The rationale of legal prohibition of abortion is to protect innocent human life and encourage respect for the sanctity of life. However, with prenatal care and genetic services, more planned and wanted children are born and fewer maternal deaths occur (from illegal abortions) than in settings where these services are scarce, anti-abortion laws are enforced, and illegal abortions are common. The sections on Brazil, Peru, and Poland in Chapter 9, "Geneticists Speak Out," support this argument.

DANGERS OF UNBALANCED AUTONOMY

Although respect for autonomy helps people to mature, to gain increased self-respect, to obtain long-overdue rights, and to start social revolutions for the betterment of themselves and their fellows, nevertheless individual autonomy can be dangerous if not balanced by solidarity and responsibility to communities and society. Overemphasis on autonomy can lead to fragmented societies in which the various segments do not care about each other, their country, or the political process necessary to ensure fair health care. Worse still, exaggeration of individual autonomy may lead to total loss of any sense of community. In the United States, where the thin sense of autonomy prevails, individual isolates think mainly of themselves. In medicine, this means not really caring whether other people have access to the best health or preventive care, as long as "I get mine." It explains in part why the US has no groundswell of support for national health insurance.

In genetics, autonomy may mean asking for, expecting to receive, and expecting one's private insurance to reimburse non-medical services such as prenatal diagnosis solely to relieve maternal anxiety. It may also mean making what many consider outrageous choices after prenatal diagnosis and expecting to receive unconditional support from genetics professionals (Hsia, 1979). The myth of selection for "the perfect baby" – which remains a myth because most Americans do not wish to use abortion for this purpose – is founded on future projections of autonomy gone awry. Nevertheless, many Americans do see themselves as "consumers" of genetic services who should have the power to pick and choose whatever they want. The irony is that in a commercial system "consumers" may become the unwitting victims of companies that not only sell tests and treatments, but also provide much of the information available to the public. In medicine, money and information mean power, and the "autonomous" individual may be at a disadvantage.

The worldwide trend towards respect for autonomy, even thinly understood, is an improvement on past constraints on patient decision making. It is better to let some people make what most may consider "immoral" decisions (as long as living people are not directly harmed) than to stifle freedom by law, regulations, or political or medical paternalism. Rather than try to limit individual autonomy in countries where it may have gone too far, it may be better to try to foster feelings of solidarity. Two concepts that currently receive comparatively little attention in bioethics may help: (1) patient responsibility, and (2) the alliance between respect for autonomy and increasing public health.

People who want self-determination need to remember that they have a responsibility to be honest in relationships, to disclose relevant personal and family information to the professional, to learn their own genetic status (if this is possible), and to disclose this status to their partner and family (as long as cultural conditions permit such disclosures without harm). O'Neill's (2002, 73–95) discussion of "principled autonomy" encourages bioethics to move in this direction.

In societies that respect individual and family autonomy, these practices can be allied with other social values and traditions that enhance the well-being of populations and individuals. These include universal access to basic health care, universal basic and higher education, literacy, economic and scientific development, freedom of speech and scientific inquiry, and democratic political institutions. Respect for autonomy is only one value in a commonwealth of values required for the flourishing of genetic services (CORN, 1996). Given the right balance among these other important values, respect for autonomy has a place but is not overriding.

THE LIMITS OF AUTONOMY: FOUR HARD CASES

Most geneticists were more than uneasy about four cases that test the limits of autonomy: (1) "which disorders to abort" (a covering phrase for the 24 fetal conditions listed in Tables 38–44); (2) sex selection through prenatal diagnosis (Tables 16–20); (3) the deaf couple who want a deaf child (Table 23); and (4) parenthood for people with transmissible mental disabilities (the woman with fragile X in Table 33). Except for the deaf couple (where most geneticists outside the US found it relatively easy to make up their minds), these cases topped the list of questions that respondents found most difficult to answer. The four cases must be considered together, because they share the same underlying issues: parental autonomy, defining "disease"; defining "serious" disorders; harm to the child, if born; the goals of medicine, the claims of society; and the uses and disuses of law. Should female sex, which is a severe social and economic "handicap" in much of Asia, be considered a "disease" along with our list of 24 fetal conditions in Tables 38–44? If one believes that parents ought to have free choice about which disorders may warrant abortion, why should not the deaf couple be able to choose a deaf child who may better fit their family and community than would a hearing child? If one believes that purposely selecting a deaf child causes harm to that child, should one support the desire of a woman institutionalized with fragile X, at 50% risk of having a child with more severe retardation than she has, to have children? Saying "no" to parents in any of the last three cases denies their autonomy and calls for a justification of saying "yes" in the first case. If one supports the wish of the woman with fragile X, whose child may be institutionalized for life, how can one justify denying the wishes of the deaf couple, who know how to care for a deaf child that will be welcomed by a caring and functioning community? We first discuss each type of case separately, although it is impossible to separate them into neat, logical compartments.

1. "Which Disorders to Abort"

This issue topped the list of questions that survey respondents found most difficult to answer, with 41% reporting it as number one (Table 83). The issue appeared in two different places on the survey: directiveness or nondirectiveness in counseling after prenatal diagnosis for 24 fetal conditions, and respondents' own willingness to abort for the same conditions.

Another way of asking the question would have been, "For which genetic conditions should prenatal diagnosis be offered?" No doubt, some wish for general guidelines, even though there appears to be a consensus against professional societies making specific lists of disorders that are sufficiently "serious" to warrant genetic services.

In fact, it may be impossible to draw lines between "serious" and "not serious" disorders, except perhaps at the most serious end of the spectrum (anencephaly). Even within the genetics community, there is no consensus. A separate survey of 1,481 geneticists in the US, Canada, Europe, and Latin America showed no agreement about what constitutes a "serious" disorder (Wertz and Knoppers, 2002). Geneticists were asked to list, in their own words, three disorders that they thought were "lethal," three that were "serious" and three that they thought were "not serious." No disorders were named in the questionnaire itself, which contained nine blank lines for respondents to fill in. They listed over 537 disorders in all. Most disorders that appeared on one person's list of "serious" appeared on someone else's list as "not serious" and *vice-versa.* Many disorders appeared on all three lists, including Down syndrome, cystic fibrosis, achondroplasia, and Huntington disease. Geneticists rejected the idea of having professional societies make lists of serious genetic disorders and almost all rejected the idea of governments making such lists. Most thought that the individual patient, together with the individual doctor, should make decisions about the "seriousness" of a disorder. Few supported the idea of a national ethics committee or a hospital ethics committee making such decisions. Most thought that definitions of "serious" should not be used in the provision of genetic services, such as prenatal diagnosis, preimplantation diagnosis, carrier screening, screening of gamete donors, recipients of donated gametes, or gene therapy after the experimental stage. They gave as their reasons for not wanting lists, first, that many genetic conditions have a variable range of expression; second, that making definitions for lists would be unfair to cultural or religious minorities who may view genetic conditions differently; third, that individuals and families hold their own idiosyncratic views of the seriousness of a condition; and fourth, that lists tend to be inflexible and difficult to change, and meanwhile new treatments could become available that would change the categorization of the disorder.

If geneticists cannot agree about what is a serious disorder, it becomes impossible to draw disorder-based lines in offering prenatal diagnosis in societies like the US, where the majority of consumers think they have a right to any (legal) service that they can pay for (Wertz, 2000). If a woman with webbed toes thinks this condition has made her life miserable and wants her fetus diagnosed for the same condition (assuming a diagnosis were possible), it becomes difficult to refuse the procedure, because this would be a denial of her autonomy. Even "normal variations," such as left-handedness and homosexuality (if such could be diagnosed prenatally) would be difficult to separate from the recognized canon of genetic disorders. These variations, which apply to a minority of people, have both health effects (left-handed people have more accidents in a right-handed world and a shorter life expectancy; homosexuals have greater risk of hepatitis B and other sexually-transmitted diseases) and social effects.

And finally, holistic medicine is oriented toward family social and mental health, which users of genetic services may claim would be affected by a child with one of these variations. In the United States, without undesirable legal prohibition, it would be difficult to deny requests from families with means. Gender selection is perhaps the only place where a firm line could be drawn. In this case the human variation is not only normal but biologically and socially necessary, and the population is divided about equally. Many requests for sex selection are based largely on sexism. Even for sex selection, however, there may be rare cases where concerns of family mental health may outweigh arguments against sex selection.

Countries with national health insurance systems find it considerably easier to prioritize and draw lines than those with no effective systems. Even though geneticists could not agree about what constitutes a "serious" disorder, the woman with webbed toes is unlikely to obtain prenatal diagnosis under national health insurance, nor would couples obtain it for left-handedness or homosexuality. Many geneticists in our survey found it convenient to draw the line at public payment for particular procedures, such as sex selection or deaf selection. In many countries, this would effectively make it impossible to obtain such services without traveling abroad. However, for people of means in any country, lack of insurance payment is not an ultimate bar. In the long run, education of the public about various disabilities, and integration of people with these disabilities into the wider society could be more effective in increasing cultural acceptance than are attempts to refuse payment for prenatal diagnosis or abortion services.

2. *Sex Selection by Prenatal Diagnosis (Table 17)*

In most nations, this case is not a frequent or high priority case facing medical geneticists. However, it is perhaps the best example of a case that causes outcries for limits to autonomy. Almost half of respondents had had at least one outright request (Table 19). Sex selection could be the gateway to unbridled "consumer choice" of "desirable" characteristics, though we doubt that in most nations many people would resort to abortion to achieve it. Nevertheless, because sex selection represents a limiting case, we discuss it at some length. It ranked second (29%) on respondents' lists of most difficult questions (Table 83).

In most of the 19 nations previously surveyed in 1985, more would perform prenatal diagnosis for sex selection or would refer in 1995 than in 1985. The exceptions were India (where there is now a law against it), France, and Sweden. Willingness to honor the request, in our view, arises more from geneticists' growing respect for patient autonomy than from any significant moral support for a practice of sex selection (Table 18). In most of the world, geneticists appear unwilling to act as gatekeepers. This does not mean that they approve or support sex selection. Most feel a conflict between maintaining their own integrity and serving what they believe to be people's needs. The typical response presents a compromise: "I don't approve of sex selection, but people have the right to make their own decisions and I owe the patient a referral. I don't think a national health plan should pay for this, but they should be able to have it if they're willing to pay for it themselves."

If this statement and the findings represent a growing trend in medical genetics, we believe that it bears careful scrutiny. Our long-held position (Wertz and Fletcher, 1989) is that cooperation with requests for sex selection serves no goal of medicine or medical genetics. Geneticists' willingness to cooperate with this request makes the profession vulnerable to social criticism.

Most sex selection is based on gender stereotyping, the belief that only members of one sex are capable of certain actions, such as inheriting property, carrying on the family name, holding well-paying jobs, or caring for elderly parents. Sex selection helps to perpetuate gender stereotyping by providing parents with a child whom they consider capable of performing these actions. A "selected" child who does not fulfill the gender stereotype could suffer. Ultimately persons of both genders may suffer because society, with the aid of sex selection technologies, forces people into more rigid gender roles. In most societies that prefer a particular sex, sexual stereotyping has placed greater restrictions on women than on men, and has contributed to gender inequality.

However, in societies that prefer gender-balanced families, parents' desire to have children of both sexes evokes considerable sympathy. Feminist ethicist Mary Mahowald argues that sex selection is not necessarily sexist. Parents, especially those with several children of the same sex, might wish to have the experience of raising a child of the other gender. According to Mahowald (2000, 47), "Selection of either males or females is justifiable on medical grounds and morally defensible in other situations as long as the intention and the consequences of the practice are not sexist." Sexist intentions are those based on the notion that one sex is inferior to the other; sexist consequences are those that disadvantage or advantage one sex over the other. Earlier, Christine Overall (1987, 27) wrote: "sexual similarity or sexual complementarily are morally acceptable reasons for wanting a child of a certain sex." Both Overall and Mahowald argue that parents may be morally justified in desiring the experience of raising a child of a particular sex. Their arguments examine the moral possibility of an exception to a premise against sex selection. Others defend sex selection as a practice from a standpoint of procreative freedom (Savulescu, 1999; Robertson, 2001).

We agree that not all sex selection need be sexist. Most people, however, regard prenatal diagnosis and abortion as going to extraordinary lengths simply in order to have a particular type of child-rearing experience. We suspect that most families seeking prenatal diagnosis for sex selection either (1) did not really want the pregnancy in the first place, or (2) are basing their requests on sexual stereotyping that is a basis of sexism.

In societies with equal preferences for boys and girls, there may be no immediate discernible social harm from sex selection, which in the future is likely to be preconception rather than post-conception (Robertson, 2001). A pluralist society such as the United States would probably be saved from the worst consequences of sex selection by the sheer variety of people's choices: first-born girl, first-born boy, no birth-order preference, all boys, all girls, equal numbers of both, more boys, more girls, or "leave it to nature." Our patient survey suggests that most people do not want to interfere with nature, though a sizable minority thought that others should have the right to do so.

Data on sex ratios in different nations, which depend on census data, are difficult to derive and interpret. Amartya Sen (Moss, 2001), the economist and demographer, distinguishes between "mortality inequality" and "natality inequality." The former refers to ratios of female to male death, which is higher in regions of the MidEast, Africa, and Asia due to neglect in medical care and nutrition for females. The worldwide ratio at birth is 105 boys to 100 girls. In his work, Sen concentrates on India and China, the world's most populous nations. Several nations show a trend of gender inequality at birth. Ratios were presented to the President's Council on Bioethics (2003), reportedly derived by the US Census Bureau's International Programs Center from 1998 censuses. The ratios are of males to females under five years of age: Venezuela 107.5, Yugoslavia 108.6, Egypt 108.7, Hong Kong 109.7, South Korea 110, Pakistan 110.9, Delhi, India 117, China 117, and Cuba 118. In the Caucasus nations of Azerbaijan, Armenia, and Georgia, the sex ratio was 120. The sex ratio in the US has remained at 104.8, but rises were noted for Chinese Americans to 107.7 and Japanese Americans to 106.4.

In nations with a preference for males, sex selection, whether pre- or post-conception, poses a major social threat. For this reason, codes of genetic practice such as the WHO *Proposed Guidelines* and many other national and international bodies find sex selection morally unacceptable.

Declining sex ratios of women to men are data showing pervasive gender inequality and obstacles to women's advancement. Sen (1990) and Coale (1991) estimated more than a decade ago that 60–100 million women are "missing" from the world's population. This figure included 29 million in China and 23 million in India. For a later assessment by geneticists in China and India see Chapters 13 and 16. India appears to be the worst case example. New data released by the United Nations Population Fund show that between 1991 and 2001, the number of girls per 1000 boys fell from 945 to 927, for the 0–6 years age group (Sharma, 2003). A moral perspective that weighs the consequences of actions cannot ignore these facts.

There may be other socially negative consequences of sex selection. It may encourage selection on cosmetic grounds, such as height, weight, or eye, hair, or skin color, if these selections are ever technically possible. Some parents would select for such purposes, especially for weight (Wertz *et al.*, 1991). New discoveries in behavioral genetics may lead to parental selections for personality traits such as risk-taking or risk-avoidance. If one would oppose this direction, it is inconsistent to cooperate with sex selection today.

Respondents frequently cited protecting professional integrity as a reason to refuse requests for sex selection (Table 18). Physicians have no duty to honor some patient requests that are useless (e.g., a request for penicillin for a cold) or harmful (e.g., a parent's request to test a child for a late-onset, incurable genetic disorder). Sex selection does not serve any legitimate goal of medicine and is not a medical service. Professionals do not have to honor requests or offer referrals.

None of the arguments above is based on a background belief that the fetus is a person or has "independent moral status." The societal and interpersonal effects of sex selection, not the status of the fetus, are the strongest global arguments against the practice. Although status of the fetus is an important issue, we disagree with McCullough and Chevenak (1994, 210) who claim that it is the only determining factor.

Many geneticists look to a public policy solution against sex selection. As shown in Table 30, the majority of geneticists worldwide, with the exception of the US, Sweden, Russia, and China, thought that abortion for sex selection should be illegal. Laws would presumably help them act as gatekeepers without appearing paternalistic in the professional–patient relationship. The benefits and harms of law depend on the cultural situation. In nations where selection against female fetuses is a social problem, such as China and India, law has at least the symbolic virtue of reaffirming the equality of women, even if the law is not always enforced. In India, law has effectively prevented genetics professionals (but not ultrasound entrepreneurs) from doing the procedure, thereby reducing the overall number of sex selections. Sex selection laws are not the answer to the problem of the millions of missing women in such nations, however, or to the equality of women in general.

Where sex selection has become a social problem, long-range solutions are education for women and equality in the workforce. Societies generally place a higher value on women if their work is recognized as productive (this usually means work outside the home), if they have some economic rights, including rights to inherit property, and if there is an awareness of the social changes necessary to overcome inequalities (Sen, 1990). In developing nations, women's longevity and the ratio of women to men are in direction parallel with women's participation in the workforce.

In nations where sex selection is not regarded as a social problem, the harms and benefits of laws will depend upon the legal and cultural approaches to abortion in general. In nations where abortion is already forbidden except to save the mother's life or for rape, as in most of Latin America, laws against sex selection could be redundant, and also ineffective. The widespread practice of (illegal) abortion in many such countries has led to disregard of overly restrictive laws. On the other hand, in nations where abortion on request is permitted and anti-abortion groups are not a political force (as in some parts of Europe) laws against sex selection, although limiting freedom of choice, would not necessarily lead to other restrictions on abortion. In countries where abortion rights are threatened, such as the United States, laws against sex selection could be an opening wedge for anti-abortion forces and could lead to laws restricting abortions for some genetic disorders or to laws against abortion in general. Most US opposition to sex selection is fueled by the abortion debate, not the debate about human diversity or equality for women. Overall, law could be a dangerous solution for the sex selection problem in most nations.

3. The Deaf Couple who Desire a Deaf Child

Although this is a rare case and did not make the list of "most difficult" questions (probably because, in most of the world, so many respondents were firmly against the couple's request for prenatal diagnosis), the deaf couple is an extremely unsettling, limiting case (Green, 1997). If hearing parents can abort a deaf fetus (and receive tacit praise from geneticists and society), why should not deaf parents have the freedom to abort a hearing fetus in order to try to have a deaf child? If parents are going to be responsible for deciding which genetic disorders are "serious," it seems only fair that a deaf couple should be able to decide that deafness is not only "not serious," but actually preferable in their family situation. However, it remains to be seen whether "hearing" in a deaf family is a serious disorder justifying abortion. Most people find the deaf couple's request deeply troubling. The immediate emotional reaction is to say, "this is wrong": wrong because (a) the parents are selfish and have no concern for the welfare of the future child; (b) it is wrong to abort a normal fetus; (c) the deaf child that they would eventually have would be harmed by being born deaf; (d) they could carry the normal fetus to term and give the child up for adoption; (e) using prenatal diagnosis for the purpose of having a child with a genetic condition, instead of using it to avoid having a child with a genetic condition, is a perversion of the purpose of prenatal diagnosis, which many people believe is to prevent birth defects; and (f) the couple's request is contrary to the goals of medicine, which are to prevent, cure, or alleviate disease. These are strong objections, but when examined carefully they may not support an argument that the deaf couple is any more wrong than the hearing couple who would abort a deaf fetus. Let us examine each reason:

(a) *Selfishness*. Most people want to have children who are like themselves, especially in regard to social interaction. Is this necessarily selfish or uncaring? The deaf couple argue that they could be better parents to a deaf child than to a hearing child.

(b) *Aborting a normal fetus*. Around the world, millions of normal fetuses are aborted every year. What makes this case special is that the fetus would be aborted specifically *because* it is normal. But what is "normal," and who should decide what is normal? In some deaf families, being deaf is normal and hearing is not. If parents are to be allowed to decide what is normal (after counseling), why should not this right extend to the deaf couple?

(c) *Harm to the deaf child*. The idea of intentional harm to a child arouses indignation. If the parents caused the child to be deaf, this would be a wrong of great magnitude. However, conception of a deaf fetus is not of the same moral order as heavy drinking during pregnancy, where the mother's own actions harm the fetus. Nature (acting through the parents' genes) has caused the child to be deaf from the time of conception. The parents have done nothing, intentional or unintentional, to harm the fetus. They have simply chosen to carry a naturally deaf fetus to term, rather than aborting it. It would be impossible to argue that being born deaf is worse than not being born at all. However, if effective treatment were available for the child after birth, the parents would face a dilemma. Parental refusal of treatment would cause harm to the child, but parental acceptance of treatment would negate their original purpose in having prenatal diagnosis, which was to have a deaf child.

(d) *Adoption for the hearing fetus*. Both a deaf child and a hearing child could be adopted in most developed nations. The same argument applies to hearing parents who abort a deaf fetus and deaf parents who abort a hearing fetus. In those developing nations where most children with disabilities are unadoptable for economic reasons, the deaf parents' request is unlikely to occur, for the same economic reasons.
(e) *Turning the purpose of prenatal diagnosis upside down*. It is true that prenatal diagnosis was developed to identify (and presumably abort) fetuses with genetic or chromosomal disorders, thereby preventing birth defects. However, in some nations the stated policy goal now is to provide information about the health of the fetus so that couples can make their own decisions. If the purpose of prenatal diagnosis is simply to provide information, the deaf couple's request falls within acceptable parameters. However, the WHO *Proposed Guidelines* are more limiting. "The purpose of prenatal diagnosis is to rule out the presence in the fetus of a particular medical condition for which the pregnancy is at increased risk" (Appendix). Still, the *Guidelines* say that parents should make their own decisions.
(f) *The Goals of Medicine*. Purposely having a deaf child may at first glance appear not to fall within the goals of medicine, yet if the goals are broadly interpreted to include family mental health and improved parent–child interaction, it could be argued that for these parents having a deaf child rather than a hearing child is an overall improvement in health.

There are good reasons to support the deaf couple's request. We are not urging geneticists to accede to it; many, in good conscience, will find it unacceptable. The deaf couple's request differs from the second hard case, sex selection, in a major respect. There is unlikely to be a societal effect. Such requests will be rare. Most deaf couples want hearing children. Most people with achondroplasia want children of normal stature. If people have inward hopes that at least one child will be like themselves – a natural desire – few are ready to act on these wishes through the drastic steps of prenatal diagnosis and abortion. Occasional requests as described in our case are unlikely to increase the number of deaf children. Therefore we regard this case as consequentially different from sex selection.

4. Parenthood for People with Transmissible Mental Disabilities

This issue ranked third (28%) on respondents' lists of most difficult questions, just one point below sex selection. Although there was widespread support for a blind couple (presumably with adequate financial means) having a blind child, there was no consensus about our limiting case: the woman with fragile X who has sufficient retardation to require an institutional setting. Each child has a 50% chance of inheriting her "good" X chromosome and presumably being normal. However, the child also has a 50% chance of inheriting the "bad" X chromosome with the "fragile" site. Fragile X is one of several genetic conditions where the genetic "mistake" becomes larger in succeeding generations. This means that a child who inherits the fragile X chromosome will have greater retardation than the mother, who is already in an institution. A boy will be worse off than a girl, but both will be affected. (Fragile X does not follow the standard rules for transmission for X-linked disorders.) If this woman has a child, there is a 50% chance that the child will spend its life in an institution, at a considerable cost to society. There is also a 50% chance that the child will have a normal life; in societies with will and means, or through the help of an extended family, the mother, even if living in an institution, can be supported in caring for the child. This case poses a dilemma between respect for autonomy and the needs of society.

The rise of respect for autonomy is clear in the degree of support (51% worldwide) for this woman (Table 33). These responses are in clear ethical tension with the older goal of genetics to prevent birth defects wherever possible. Responses show that many contemporary geneticists, and not only those in English-speaking nations, are willing to step aside and respect a woman's desire for parenthood, even if her child may have a genetic syndrome. This is an unusual and relatively new degree of moral tolerance associated with the ascendancy of respect for autonomy, especially since the capacity of the woman in the case is impaired to some degree. Our concern is not only with the particular case, but with a moral direction that would lean so far toward respecting autonomy that other ethical considerations would be muted or left out of counseling altogether.

Respecting the wishes of persons with disabilities is commendable; however, the counseling in such cases ought to include information that is responsive to the claims of other principles. Non-directive counseling is premised on freedom of persons and families to make their own choices about reproduction. It does not follow that geneticists must morally neuter themselves in counseling. Professionals can respect the autonomy of a person with a disability – by reassurance of the freedom to make the choice – but the claims of other principles, such as preventing harm, can legitimately be brought into counseling. The consequences of this woman's actions may bring her fulfillment as a parent but may also cause serious harm to her child. There are also considerations of justice. She has a responsibility to the community in terms of a duty to use community resources fairly. She should be fully informed in a way that she can understand. If she cannot understand the consequences of her actions (a situation posited in our case vignette), counseling should follow general rules for counseling persons with diminished mental capacity. Appointment of a legal guardian may be necessary, with the geneticist counseling the guardian about whether to support her becoming a parent. However, no coercion should be used with regard to a decision to become a mother. As in the case of the deaf couple, parenthood for people with transmissible mental disabilities is unlikely to change society. Most people with severe mental retardation are unlikely or unable to reproduce.

With the exception of sex selection, we have supported parental choice constrained by economic and social norms. We do not support insurance coverage for all requests, nor do we favor public policies that would actively support prenatal diagnosis for cosmetic selections or "normal human variations," such as homosexuality, or for selection of deaf children. In some instances individual acts could be justifiable. Sex selection stands apart because of its potential to change societies that favor one sex over the other. Sex selection may require recourse to law, but only in countries where sex selection is likely to exacerbate already existing social problems, such as an unbalanced sex ratio. We are aware of the discriminatory aspects of calling for an international legal "double standard" based, for example, on Western versus some Asian cultures, but in many nations law will do more harm than good.

DIRECTIVENESS IN COUNSELING

We propose that a third moral priority in international medical genetics is to encourage non-directive counseling, especially where genetics professionals affirm it in principle but counsel in a different way. Most geneticists would give purposely-biased information after prenatal diagnosis, except in English-speaking nations and parts of Europe (Tables 38–44). It is not surprising that practices in genetic counseling reflect the cultural conditions within which counseling occurs. In most nations, the cultural construction of the professional–patient relationship is paternalistic. Both parties expect that the professional will give a recommendation about what ought to be done, not only medically but morally, if moral values are at stake. In this construction, paternalism is understood positively as an expression of medical beneficence, or what is in the patient's or families own good. The non-directive counseling that is espoused by genetics professionals in North America and English-speaking nations is an aberration from genetics practice in most other parts of the world. Non-directiveness grows out of the 18th-century Enlightenment belief in individual autonomy and the rights of humans to govern their own destinies. It is also based on a presupposition of universal literacy, basic education, and public participation in the political process. Genetic counseling in most nations is done by physicians. (A separate counseling profession for non-medical professionals exists mainly in North America, but is developing in some other nations.)

Geneticists in the United States, Canada, Australia, South Africa, and the United Kingdom attempt to provide information that is "as unbiased as possible." To their credit, these counselors also try to avoid letting their personal attitudes toward aborting or carrying to term influence how they provide this information. Outside these nations, most counseling is directive; the counselor urges people toward a course of action (abortion or carrying to term) or provides intentionally slanted information in order to influence people's decisions. The information is sometimes presented as "fact." Giving purposely slanted information under the guise of medical/scientific "fact" or "truth" is morally worse than openly urging people toward a particular course of action. People given false "facts" have no opportunity to resist. This type of biased counseling takes advantage of the inequalities of the professional–patient relationship in order to reach outcomes favored by the professional. Even if intended to benefit people, this behavior can be seen as a form of disrespect for persons. Falsification is the foundation of successful propaganda. In some countries with few services for persons with disabilities, providing pessimistically slanted information may serve unspoken and unspeakable social ends. There was, however, also some pessimistic counseling in some nations with more advanced services for people with disabilities, such as the Netherlands and Israel. Many geneticists outside North America said that it was their duty to present information in such a way that people would "do the right thing" or "it would prevent harm." They may have considered themselves nondirective as long as they refrained from actually telling people what to do. Many may not have believed that giving slanted information was "directive" counseling.

Around the world, the morally optimal approach in genetic counseling would be, first, to listen carefully, and second, to empathize with and understand people's needs and situations. The most important clinical virtue in medical genetics is empathy, the ability to listen and to internalize people's needs. "Nondirective" genetics professionals would then proceed on the basis of the individual's and family's best interests, keeping in mind that most people are capable of understanding what they need to know about the genetic condition, regardless of their level of education and income. Counseling also needs to include the psychosocial aspects as well as the medical aspects of a condition. These aspects include effects on a marriage and family life, financial costs, including the level of public support (if any) for the child, and what the child would be like as an adult living with the family. People need to know whether the society has supports for children with genetic conditions and what their fair share of such supports would be. Such information should be conveyed objectively and without threat.

It requires significant economic resources to support persons with disabilities. Clearly, economic information could be used to pressure a decision in a negative direction to conserve costs to society and relieve the family of economic burdens. The major focus in genetic counseling should be on the family, but genetics professionals are also stewards of society's financial resources and need to be familiar with the facts about economic costs.

In many cultures, including North America, people ask the professional what she or he would do in the same situation. If the family seems unable to come to a decision without this information, and if they ask, it may acceptable for the professional to reveal this information, provided that it is done in the context of support for other alternatives that the family may choose and that the professional makes clear that she or he will never actually be in their identical situation.

Suppose the family asks a further question, "What do you think we should do?" Unless treatment is available (and we do not regard abortion as treatment), the best course of action for genetics professionals is to refrain from answering this question with directive advice, because to do so could people's moral responsibility for their own choices. People need to be encouraged to make their own decisions and stand on their own two feet, especially with regard to issues of reproduction.

DISABILITY

Geneticists in most nations take a pessimistic view of disability, as evidenced by:

(1) Directively pessimistic counseling. In those countries where counseling was primarily directive, counseling for most conditions was pessimistic (Tables 42–44). The largest percents that would be pessimistic were in Eastern Europe, Asia (with the exception of Japan), the Near East, and Cuba. Belgium, France, and Greece also held views that were notably more pessimistic than other nations in their regions. In some of these nations, pessimistic counseling may be realistic and fits a current situation of extreme scarcity of resources. For example, in India, there is no treatment for children with thalassemia, except for those with the wealthiest parents. In China, cleft lip and palate are often not treated because most people do not have publicly-financed health insurance or the funds to afford an operation. Pessimistic counseling may be realistic, but it needs to be modified by commitment to a vision of social justice that includes sharing resources with the weak and not only the strongest and fittest individuals.

(2) The majority of geneticists would themselves abort for most genetic conditions (Tables 46–49). There was a statistical relationship between personal views on abortion and the direction of counseling.
(3) Perceptions that social support is not enough and persons with disabilities do not contribute to society: over two-thirds agreed with the statement, "some disabilities will never be overcome, even with maximum social support," and majorities agreed with the statement that "society will probably never provide adequate support for people with disabilities" (Table 51). Few geneticists agreed with the statement that "the existence of people with severe disabilities makes society more rich and varied."
(4) Substantial numbers agreed that knowingly having a child with a disability would be unfair to the child (Table 34), as well as "socially irresponsible in an era of prenatal diagnosis."
(5) About half would support a parental decision to refuse a life-saving operation on a newborn with Down syndrome (Table 35).
(6) A substantial percentage is disappointed when parents decide to refuse prenatal diagnosis when indicated or carry to term a child with a disability.

Geneticists' widely shared pessimism about disability may have an important influence on counseling. The counselors' personal predisposition to abortion is bound to be influential. A counselor's personal views and values may get communicated, sometimes through tone of voice or gesture, even though the counselor exerts strong efforts to conceal these ideas. Our findings support the truth of reports of members of disability groups that they perceive bias against people with disabilities in contemporary genetics (Asch, 1999). We propose that the fourth moral priority of medical genetics is to examine a bias about disability and address the reasons for it.

Economic resources are directly relevant to this issue. In nations with no or few resources to support people with disabilities, pessimistic counseling may be realistic counseling. In nations with sufficient resources, but which choose to expend these resources for other purposes (such as the military or tax cuts), geneticists need to be aware that pessimism about disability may in fact be serving a hidden policy goal. Such societies are in fact saving money by avoiding births of persons with disabilities.

In nations where societal support for persons with disabilities is the norm, geneticists need to keep in mind how their counseling reflects on people with disabilities as a whole. The counseling session should occur against a background standard of respect and care for people with disabilities. Some people in the disability community have long said that genetics supports the idea that persons with disabilities are of lesser worth, that their human potential has been crippled, and that the world would be better off without them. The training of geneticists and counselors should involve spending quality time with families with disabilities in their daily lives, rather than seeing them only in a clinical setting.

EUGENICS

We propose that a fifth moral priority for geneticists, especially in some developing nations, is to reject eugenics as a goal of genetics. Our study found eugenic views, especially in parts of Asia and Eastern Europe; however, most geneticists rejected involuntary sterilization and government involvement (Tables 54–56). One could argue that eugenics persists, but on an individual and voluntary basis. A decision to abort an affected fetus is eugenic in the sense that parents, considering their family situation, decide against the birth of a child with a particular genotype (Kitcher, 1996, 196–197).

It is reassuring that few geneticists worldwide agreed with involuntary sterilization (Reilly, 1991), which was abused in the "old eugenics" (Table 55). However, there is still support for the concept of negative eugenics in some nations. Policy-makers may need to protect persons who are vulnerable to pressure or coercion into sterilization under the guise of its being "voluntary." Eugenics – whether individual or social – has a strong economic component. The countries in our survey with the greatest pessimism about disability, the most directive counseling, and the greater support for government coercion were all developing or East European nations. Eugenic thinking thrives in hard times. The best defense against a resurgence of eugenics is an educated public with an awareness of the meaning of lives with disability.

FAMILY RELATIONSHIPS

We propose that the sixth moral priority in medical genetics is to review practices in genetic services that involve the well-being of family members and children whose parents desire testing for later-onset disorders.

1. Warning Relatives at Genetic Risk

As in 1984, our study found no consensus about telling relatives at genetic risk against a patient's wishes. However, except in English-speaking nations and Western Europe, majorities would tell relatives, especially if they asked. In 1998, the American Society of Human Genetics (1998) published a statement on "Professional Disclosure of Familial Genetic Information." The policy states that "disclosure should be permissible where attempts to encourage disclosure on the part of the patient have failed; where harm is highly likely to occur and is serious and foreseeable; where the at-risk relative(s) are identifiable; and where either the disease is preventable/treatable or medically accepted standards indicate that early monitoring will reduce the risk." A further standard for disclosure is that "the harm that may result from failure to disclose should outweigh the harm that may result from disclosure." The elements of this policy are very similar to one in the WHO *Proposed Guidelines* (Appendix). There is strong agreement at the level of professional guidelines and bioethics commissions that geneticists may warn relatives in cases of significant genetic risks and when the information would be used to prevent or lessen harm.

In genetics, the unit of confidentiality may be the family, as long as the disease is treatable or preventable. As part of their duty to educate patients, counselors should inform them that genetic information may be useful to their relatives and may invite individuals to ask the relatives to seek genetic counseling. One reason for asking the relatives to seek professional counseling, rather than asking patients themselves to give information is that patients may inadvertently misinform relatives about the risks and nature of the disorder. An innovative practice of the Japanese Society of Human

Genetics (1995) guidelines requires that an ethics committee oversee any genetics cases in which patient confidentiality may be overridden. A second level of review provides a check on the views of individual clinicians, who may overestimate the value of genetic information to the relatives.

2. *Telling Spouses*

Does the spouse or partner have a right to know about an individual's genetic risks? If a couple intends to have children, geneticists may need to encourage the individual to share genetic information with the spouse or partner so that both can be involved in reproductive planning. In some societies, however, especially where individuals and entire extended families are blamed for their genetic constitutions, such openness may not be possible. Geneticists need to weigh the risk of family strife or dissolution against the (sometimes remote) possibility that the knowledge might be used to avoid the birth of a child with a genetic condition.

3. *Testing Children*

Geneticists in English-speaking nations would not test children for adult-onset disorders. In Southern and Eastern Europe, Asia, and Latin America, most would test children at parents' request (Table 65). In some nations, such as the US, Canada, and UK, there is a growing respect for the rights of children, reflected in laws and regulations. US regulations on research with human subjects require that children with the capacity to do so give "assent" to participate in research studies. Assent indicates that the child understands that the activity is research and that benefits and risks have been explained in simple terms. In the United States, many states accord privacy rights to minors in regard to HIV testing, treatment of sexually transmitted diseases, contraceptive advice, and abortion. It is likely that the rights of minors, particularly for adolescents with regard to their own health care, will increase in the future, at least in English-speaking nations. Much of the world, however, believes that parents have the right to control their children's health care, including requests for genetic testing. The child is regarded, in effect, as the property of the parents, and the parents may exert control until the child reaches legal majority. Many cultures assume that parents will always know the best interests of the child and will act accordingly.

Everyone agrees that symptomatic children should be tested if a test is necessary to confirm a diagnosis. There is also agreement that asymptomatic children should be tested if a disease is treatable or preventable. The areas of most disagreement are carrier testing and presymptomatic testing for adult-onset disorders. One can distinguish between these two cases in terms of benefits to the child and participation in the decision making. If a young adolescent wants testing for carrier status in order to understand his or her reproductive risks, it could be advisable to proceed. Some prior conditions could include level of education in the family and the degree of risk of stigmatization, which needs discussion in counseling.

There is no evidence that testing children for disorders that occur later in life is of benefit to the child or to the family. The potential for harm is considerable. The harms are such that many professional groups, including the American Medical Association

(1995), American Society of Human Genetics (1995), National Society of Genetic Counselors (1995), the UK Genetic Interest Group (1995), and Institute of Medicine (1994) have urged caution.

Usually, testing is best postponed until the child reaches the age of legal majority, unless there is a medical benefit to testing in the immediate or near future. The same caution applies to adopted children and children awaiting adoption. If adopting parents ask for genetic testing, some professional organizations have recommended that the agency not proceed, because adopted children need to be treated with the same respect that all children deserve (ASGH and ACMG, 2000). They should not be treated as second-class persons or as "commodities." However, the agency needs to provide a full and accurate medical history of both birth parents. In other words, the information available to adoptive parents should be at the same level as that available to biological parents.

4. *Misattributed Paternity*

The choice of words is also a moral choice. "Misattributed" paternity (Ross, 1996) is preferable to the term nonpaternity, a term that carries a moral judgment that the child has no father. Most geneticists would not disclose this *accidental* finding to a husband or partner, even if he asks. There is no uniformly good approach to situations where a test taken for another purpose may also reveal a different biological father than expected. Including information about the possibility of adventitious discovery in the informed consent document, though honest, does not protect the woman or child. In order to prevent familial harm, it would necessary to counsel her first, alone, before the informed consent, so that she can choose to withdraw from testing if she desires. In effect, this shifts the "nondisclosure to husband" approach, favored by almost all geneticists, to pre-testing instead of post-testing. Genetics professionals are usually not in a position to assist couples with the psychosocial results of disclosing this secret. If the facts are disclosed to anyone, it should be the woman. Geneticists owe her a referral to adequate counseling for help in the situation.

A policy of advance discussion will probably not work in cultures where the topic itself would be inflammatory, or women rarely see medical professionals alone, and the risks of harm to mother and child are great. In any culture, test results not directly relevant to health, such as suspicion of adultery, or the sex of the fetus in the absence of an X-linked disorder, may be withheld if this appears necessary to protect a vulnerable party.

PRIVACY AND INSTITUTIONAL THIRD PARTIES

There was worldwide consensus that employers, insurers, and schools should not have access to genetic information without an individual's consent. We propose a moral priority in medical genetics to review and implement practices to protect genetic information from third parties and to encourage laws to prevent genetic discrimination.

There were some reports of refusals of employment or insurance in Europe, but most came from the United States. In the United States, reports were few when compared with patient volume. Others, studying groups of experienced, college-educated consumers in genetic support groups, have found far greater fear of "genetic discrimination" (Lapham *et al.*, 1996; Hall and Rich, 2000) and few cases in the industry. Perceptions of "discrimination" may depend on social locus as well as phraseology of questions; the phenomenon may be rarer than media reports suggest. However, the reports of our geneticist respondents document the existence of at least some discrimination. How these reports are interpreted will depend in part on the personal and political views of the interpreter. At present, genetic information is seen as "dangerous," especially in the United States, where private health insurance is the means of access to basic health care for the majority. The problem is very serious, because persons who would benefit from genetic testing are refusing it for reasons of fear. Peterson *et al.* (2002) found that at least 25% of women served by a large breast cancer service refused testing for this reason. Progress on legal measures is slowly succeeding. Thirty-six states in the US have laws against forms of genetic discrimination by health insurers, although these laws differ in important respects. Unfortunately, some state laws concentrate only on molecular or chromosomal tests, leaving insurers free to obtain and use their single most valuable piece of genetic information – the family history. In some insurance structures, genetic privacy laws may protect only about 5% of people (Reilly, 1997). The US Senate unanimously passed a bill that forbids health insurers and employers to use genetic information to raise premiums or in some contexts of hiring (Zitner, 2003). This social progress required 10 years of effort, and great credit must go to leaders in the HGP and the National Human Genome Research Institute (Collins, 2003).

In the United Kingdom, rejection for life insurance, which is unusually necessary for a home mortgage, spurs fear of discrimination. In the UK, insurers have shifted from a practice of assessment of risk with actuarially-based premiums to a practice of averting higher risks, wherever possible (Harris, 1998, 262). A House of Commons report (2001) recommended a moratorium on any requirements to disclose DNA test results to an insurance company. O'Neill's discussion (2002, 110–117) places the onus on the insurance industry to seek scientifically grounded evidence about all risk factors, including environmental and life style, as well as genetic, before constructing "actuarially-fair" loading of premiums. In order to protect basic human rights, legislatures in most countries will need to make certain that everyone has equal access to basic health insurance and also to a basic amount of life insurance. This goal may be achieved either by mandating provision of insurance or by refusing access for insurers to genetic information. Employment is also in need of protection; any employer request for genetic testing should be limited strictly to ability to perform the job, safely. Tests could be made available, on a voluntary basis, for occupationally-related susceptibilities, but results should be available only to the worker.

SUMMARY

This final chapter has discussed and proposed moral priorities and review of practices in medical genetics. We also discussed our judgment about some of the hardest cases in the survey. Taking into account the changing goals of medical genetics, we recommend that forums for geneticists and policy makers consider these proposed priorities, i.e., to:

- ameliorate patterns of inequity of access to health care and genetic services that exist in every nation and among nations;

- protect and expand autonomy (in its strong meaning) in reproductive decisions, augmented by expert counseling that invites open discussion of moral problems, as well as social and economic aspects of different options;
- encourage respect for persons by teaching the principles of nondirective counseling to trainees and implementing them in practice;
- examine bias in medical genetics about disability and take steps to reduce its influence on decision making;
- review practices in genetic services that protect the well-being of family members, including children whose parents request genetic testing for late-onset disorders;
- review and implement practices to protect genetic information from third parties and encourage laws to prevent genetic discrimination in health and life insurance and in employment practices.

REFERENCES

American Medical Association Council on Ethical and Judicial Affairs. Testing Children for Genetic Status. *Code of Medical Ethics*, Report 66. Chicago, IL: AMA, 1995.

American Society of Human Genetics Board of Directors and American College of Medical Genetics Board of Directors. Points to consider: ethical, legal, and psychosocial implications of genetic testing in children and adolescents. *American Journal of Human Genetics* 1995; 57: 1233.

American Society of Human Genetics. Professional disclosure of familial genetic information. *American Journal of Human Genetics* 1998; 62: 474–483.

American Society of Human Genetics Social Issues Committee and the American College of Medical Genetics, Social, Ethical, and Legal Issues Committee (ASGH and ACMG). Genetic testing in adoption. *American Journal of Human Genetics* 2000; 66(3): 761–767.

Arras, John D. The Owl and the Caduceus. Does bioethics need philosophy? In Franklin G. Miller, John C. Fletcher, and James M. Humber (eds), *The Nature and Prospect of Bioethics: Interdisciplinary Perspectives*. Totowa, NJ: Humana Press, 2003, pp. 1–42.

Asch, Adrienne. Prenatal diagnosis and selective abortion: a challenge to practice and policy. *American Journal of Public Health* 1999; 89: 1649–1657.

Beauchamp, Tom L. and James F. Childress. *Principles of Biomedical Ethics*, 5th edn. New York: Oxford University Press, 2001.

Beauchamp, Tom L. A defense of the common morality. *Kennedy Institute Journal of Ethics* 2003; 13.3: 259–274.

Beecher, Henry K. Ethical problems created by the hopelessly unconscious patient. *New England Journal of Medicine* 1968; 278: 1425–1430.

Benjamin, Martin. Between subway and spaceship: practical ethics at the outset of the twenty-first century. *Hastings Center Report* 2001; 31.4: 24–31.

Benjamin, Martin. *Philosophy and This Actual World*. Lanham: MD, Rowman & Littlefield, 2003.

Brenner, Sidney. Humanity as the model system. *Science* 2003; 302: 533.

Buber, Martin. *Between Man and Man*. Translated by Ronald G. Smith. New York: Macmillan, 1965.

Buchanan, Allen E. Is there a medical profession in the house? In Roy G. Spece, Jr., David S. Shimm, and Allen E. Buchanan (eds), *Conflict of Interest in Clinical Practice and Research*. New York: Oxford University Press, 1996, pp. 105–136.

Callahan, Daniel. Bioethics. In Warren T. Reich (ed.), *Encyclopedia of Bioethics*, 2nd edn. New York: Simon & Schuster MacMillan, 1995, pp. 256–274.

Commission on Genetic Resources for Food and Agriculture (CGFRA). *Code of Conduct on Biotechnology as it Relates to Genetic Resources for Food and Agriculture*, 1991.

Coale, Ansley J. Excess female mortality and the balance of the sexes in the population: an estimate of the number of 'missing females'. *Population and Development Review* 1991, 517–523.

Collins, Francis S. and James D. Watson. Genetic discrimination: Time to act. *Science* 2003; 302(5646): 795.

Council of Regional Networks for Genetics Services (CORN) Committee on Ethics. Code of ethical principles for genetic professionals: an explication. *American Journal of Medical Genetics* 1996; 65: 179–183.

Daniels, Norman. *Justice and Justification: Reflective Equilibrium in Theory and Practice.* Cambridge and New York: Cambridge University Press, 1996.

Dewey, John. *Human Nature and Conduct.* Carbondale, IL: Southern Illinois University Press, 1988.

Fox, Renee C. and Judith P. Swazey. Medical morality is not bioethics. Medical ethics in China and the United States. *Perspectives in Biology and Medicine* 1984; 27: 336–360.

Genetic Interest Group. The genetic testing of children. *GIG Response to the Clinical Genetics Society Report.* London: GIG, 1995.

Glendon, Mary Ann. *A World Made New: Eleanor Roosevelt and the Universal Declaration of Human Rights.* New York: Random House, 2001.

Green, Ronald M. Parental autonomy and the obligation not to harm one's child genetically. *Journal of Law, Medicine, and Ethics* 1997; 25(1): 5–15.

Hall, Mark A. and Stephen Rich. Laws restricting health insurers' use of genetic information: impact on genetic discrimination. *American Journal of Human Genetics* 2000; 66: 293–307.

Harris, Rodney and Margaret Reid. Medical genetic services in 31 countries: an overview. *European Journal of Human Genetics* 1997; 5: S2.

Harris, John. *Clones, Genes, and Immortality: Ethics and the Genetic Revolution.* New York: Oxford University Press, 1998.

House of Commons. Select Committee on Science and Technology. *Report on Genetics and Insurance.* HC 174, 2001.

Hsia, Y. Edward. The genetic counselor as information giver. In Alexander Capron, Mark Lappe, and Robert Murray, *et al.* (eds), *Genetic Counseling: Facts, Values, and Norms.* New York: Alan R. Liss, 1979, pp. 169–186.

Human Genome Organization Ethics Committee (HUGO). Genetic Benefit Sharing. *Science* 2000; 290: 49. <http://www.gene.ucl.ac.uk/hugo/benefit.html>.

Institute of Medicine, Committee on Assessing Genetic Risks. *Assessing Genetic Risks.* Washington, DC: National Academy Press, 1994, p. 276.

Japanese Society of Human Genetics. *Guidelines for Genetic Counseling and Prenatal Diagnosis*, 1995.

Jennett, Bryan and Fred Plum. Persistent vegetative state after brain damage. A syndrome in search of a name. *Lancet* 1972; 1(7753): 734–737.

McCullough, Lawrence B. and Frank A. Chevenak. *Ethics in Obstetrics and Gynecology.* New York: Oxford University Press, 1994, p. 210.

Kitcher, Philip. *The Lives to Come. The Genetic Revolution and Human Possibilities.* New York: Simon & Schuster, 1996.

Lapham, E. Virginia, Chahira Kozma, Joan Weiss, Judith Benkendorf, Mary Ann Wilson. Genetic discrimination: perspectives of consumers. *Science* 1996; 274: 621–624.

Mahowald, Mary Briody. *Genes, Women, Equality.* New York: Oxford University Press, 2000, p. 121.

Miller, Bruce. Autonomy and the refusal of lifesaving treatment. *Hastings Center Report* 11(4): 22–28, 1981.

Moss, Simeon. An alarming trend in gender equality. *Cornell Chronicle* 32(32); 2001: p. 1.

Merikangas, Kathleen R. and Neil Risch. Genomic priorities and public health. *Science* 2003(5645): 599–601.

National Society of Genetic Counselors. Prenatal and childhood testing for adult-onset disorders. *Perspectives in Genetic Counseling* 1995; 17: 5.

O'Neill, Onora. *A Question of Trust*. Cambridge: Cambridge University Press, 2002.

Overall, Christine. *Ethics and Human Reproduction*. Boston: Allen and Unwin, 1987, p. 27.

Peterson, Emily A, Kara Milliran, Karen Lewis, Susan Goold, Sofia Merajver. Health insurance and discrimination concerns and BRCA 1/2 testing in a clinic population. *Cancer Epidemiology Biomarkers Prevention* 2002; 11(1): 79–89.

President's Council on Bioethics (2003). Ethical aspects of sex control. (Staff Working Paper). <http://www.bioethics.gov>.

Rawls, John. *A Theory of Justice*. Cambridge, MA: Harvard University Press, 1971.

Reilly, Philip R. *The Surgical Solution: A History of Involuntary Sterilization in the United States*. Baltimore: Johns Hopkins University Press, 1991.

Reilly, Philip R. Laws to regulate the use of genetic information. In Rothstein, Mark A. (ed.), *Genetic Secrets: Protecting Privacy and Confidentiality in the Genetic Era*. New Haven: Yale University Press, 1997, pp. 369–391.

Robertson, John A. Pre-conception sex selection. *American Journal of Bioethics* 2001; 1(1): 2–9.

Ross, Lainey F. Disclosing misattributed paternity. *Bioethics* 1996; 10: 114–130.

Savulescu, Julian. Sex selection: the case for. *Medical Journal of Australia* 1999; 171: 373–375.

Sen, Amartya. East and West: the reach of reason. *New York Review of Books* 2000; 47(12): 20–26.

Sen, Amartya. More than 100 Million Women are Missing. *New York Review of Books* 1990; 37(20): 61–66.

Sharma, Dinesh C. Widespread concern over India's missing girls. *Lancet* 2003; 362: 1553.

Shimm, David S. and Roy G. Spece, Jr. Ethical requirements of physicians' roles. In Roy G. Spece, Jr., David S. Shimm, and Allen E. Buchanan (eds), *Conflict of Interest in Clinical Practice and Research*. New York: Oxford University Press, 1996, p. 52.

Toulmin, Stephen. *Human Understanding. The Collective Use and Evolution of Concepts*. Princeton, NJ: Princeton University Press, 1972.

United Nations. *The World's Women: Trends and Statistics 1970–1990*. New York: United Nations, 1991.

United Nations Environment Programme (UNEP). *Convention on Biological Diversity*, Article 1, 1992.

Wertz, Dorothy C. and Fletcher, John C. Fatal knowledge? Prenatal diagnosis and sex selection. *Hastings Center Report* 1989; 19(3): 21–27.

Wertz, Dorothy C., James M. Rosenfield, Susan R. Janes, and Richard W. Erbe. Attitudes toward abortion among parents of children with cystic fibrosis. *American Journal of Public Health* 1991; 81: 991–996.

Wertz, Dorothy C. Drawing lines: notes for policymakers. In Erik Parens and Adrienne Asch (eds), *Prenatal Diagnosis and Disability Rights*. Washington DC: Georgetown University Press, 2000, pp. 261–287.

Wertz, Dorothy C., John C. Fletcher and Kare Berg. *Review of Ethical Issues in Medical Genetics*. Geneva: WHO, 2001.

Wertz, Dorothy C. and Bartha M. Knoppers. 'Serious' genetic disorders: can or should they be defined? *American Journal of Medical Genetics* 2002; 108: 29–35.

Wittgenstein, Ludwig. *On Certainty* 1969, § 341.

Wolpe, Paul R. The triumph of autonomy in American bioethics: a sociological view. In Raymond DeVries and Janardia Subedi (eds), *Bioethics and Society: Constructing the Ethical Enterprise*, Englewood Cliffs, NJ: Prentice-Hall, 1998, pp. 38–59.

World Health Organization, *Proposed International Guidelines on Ethical Issues in Medical Genetics and Genetic Services*. Geneva: WHO, 1998.

Zitner, Aaron. Senate blocks genetic discrimination. *Los Angeles Times* 2003; A-1. <http://www.latimes.com/news/printedition/asection/la-na-genetests15oct15,1,6733328.story?coll=la-news-a_section>.

ABOUT THE AUTHORS

THE EDITORS

Dorothy C. Wertz is Research Professor of Psychiatry at the University of Massachusetts Medical School, Shriver, Division, Waltham, MA, USA, and Senior Scientist in the Division of Social Science and Ethics and Law. She received her Ph.D. from Harvard in the Study of Religion and taught sociology and anthropology for 18 years. Since 1981, she has researched social and ethical aspects of human genetics. She has recently completed (with John C. Fletcher and Kåre Berg) "Proposed International Guidelines on Ethical Issues in Medical Genetics and the Provision of Genetics Services" for the World Health Organization (1998). Her books include "Lying In: A History of Childbirth in America," 1989, (with Richard W. Wertz), and "Ethics and Human Genetics: A Cross-Cultural Perspective" (with John C. Fletcher, 1989).

John C. Fletcher, Ph.D. is Professor Emeritus of Biomedical Ethics and Internal Medicine at the University of Virginia School of Medicine. He was the co-editor, with Dorothy C. Wertz, of Ethics and Human Genetics: A Cross-Cultural Perspective, (Heidelberg: Springer-Verlag, 1989) which reported an earlier survey of medical genetics in 19 nations. Dr. Fletcher served as Chief of the Bioethics Program, Clinical Center, National Institutes of Health (1977–1987) and as Director of the Center for Biomedical Ethics, University of Virginia (1987–1997). He recently received the Lifetime Achievement Award of the American Society for Bioethics and Humanities.

ARGENTINA

Dr. Elba M.P. de Giorgiutti is a medical doctor specializing in genetics and bioethics. She received her medical degree from the University of Buenos Aires and has had post-graduate courses in the United States and Europe. She is former President of the Argentine Genetics Society and a member of several international genetic societies. She has published almost one hundred research papers on genetics. She currently runs GENMÉDICA Institute in Buenos Aires, specially dedicated to research, counseling and care of families with genetic disorders. She is professor of Bioethics at the Bioethics Institute at the Argentine Catholic University and an active member of the bioethics committee of the Argentine Society of Clinical Research.

AUSTRALIA

Dr. Alison Colley is a clinical geneticist who trained in New South Wales, Australia and in Manchester in the United Kingdom. Dr. Colley is Head of the Department of Clinical Genetics in South Western Sydney and a Senior Lecturer at the University of New South Wales. From 1996 till 1998, she was Chairperson of the Australasian Association of Clinical Geneticists and from 1998 till present is the Secretary of the Human Genetics Society of Australasia.

Keith Joseph is currently teaching and undertaking research in ethics at the University of Newcastle, New South Wales, Australia, where he also completed a Ph.D. on the ethics of human genetic engineering. From 1993 to 1998, he taught moral philosophy at the Australian Catholic University of Sydney. He has been Secretary of the Australian Association for Professional and Applied Ethics since 1995 and is a member of the International Association of Bioethics.

BRAZIL

Francisco M. Salzano works at the Genetics Department, Biosciences Institute, Federal University of Rio Grande do Sul at Porto Alegre, Brazil. His investigations, developed during the past 50 years, include many aspects of the genetics and evolution of humans and other organisms. He authored or co-authored 12 books, 38 chapters in books, three monographs, and at the last count 346 full scientific papers. More than one generation of Brazilian geneticists has taken shape under his guidance. He received the National Prize Almirante Alvaro Alberto, the highest distinction awarded by the Brazilian government to its scientists (1994), is a Member of the National Order of Scientific Merit, Great Cross Class (1995), and received the 1997 Annual Award of the Ibero-American Society of Human Genetics and the 1999 Franz Boas High Achievement Award of the Human Biology Association. In 1999, he was elected a Foreign Member of the US National Academy of Sciences.

Lavinia Schüler-Faccini is Associate Professor, Genetics Department, Biosciences Institute, Federal University of Rio Grande do Sul and Coordinator of the Information System on Teratogenic Agents which functions at the same university's Clinical Hospital. She obtained her Ph.D. in 1992 and worked in 1996/1997 at Toronto's Hospital for Sick Children. Her main present research interests center on teratogenesis, mutagenesis and congenital malformations.

CANADA

Elizabeth Ives is a clinical geneticist who received her training in medicine at the University of Glasgow, Scotland, and in public health at the University of London before moving to Canada in 1960. Thereafter she completed training in pediatrics before obtaining a Masters degree in Human Genetics at the University of Michigan. Dr. Ives has helped to develop medical genetics services and has worked as a clinical geneticist with academic affiliations in the three Canadian provinces of Saskatchewan, Alberta and Newfoundland. She has recently retired from directing the provincial genetics program in the latter.

CHILE

Ricardo Cruz-Coke received his M.D. from the University of Chile in 1950. He is Professor of Medicine, Professor of Clinical Genetics, and founder of the Genetic Unit, Hospital J. J. Aguirre, University of Chile, Santiago in 1965. He was President of the Association Latinoamericana de Genetica (1979–1981), President of the Programa Latinoamericano del Genoma Humano (1990–1992), Associate Editor of the journal Clinical Genetics (1970–1992), Secretary of the Ethics Committee, Hospital J. J. Aguirre, and is a member of the International Bioethics Committee of UNESCO (1994–1999) and of the Human Genome Organization (HUGO).

CHINA

Renzong Qiu is Senior Research Fellow and Professor, and also Director, Program in Bioethics, Institute of Philosophy, Chinese Academy of Social Sciences; Member, Board of Directors, International Association of Bioethics; Assessor, Executive Committee, International Union of History and Philosophy of Science/Division of Logic, Methodology and Philosophy of Science; Vice-President, Asian Association of Bioethics; Vice-President, Chinese Society for Philosophy of Nature, Science and Technology; Member, Ethics Committee, Human Genome Organization (HUGO); Member, Ethics Committee, Chinese Ministry of Public Health. He has published 12 books, including Bioethics, Patients' Rights, AIDS, Sex and Ethics and more than 200 articles or chapters in Chinese, American, British, German, Spanish, Romanian, Japanese, Hong Kong and Taiwanese journals or books. He organized and coordinated the Chinese section of the international survey.

Xin Mao, M.D., is a senior scientist and Associate Professor at The First University Hospital, West China University of Medical Sciences, Chengdu, Sichuan, China. He received his M.D. degree from West China University of Medical Sciences in 1986 and received postdoctoral training at the Chromosome Research Unit, Faculty of Science, Hokkaido University, Sapporo, Japan from 1989 to 1990. In the past 10 years, he researched molecular pathological studies of p53 gene mutations; molecular genetics, immunology and clinical studies of Alzheimer disease; genetic studies of mental disorders of sex deviations; and social-psychology studies of divorce in China.

CZECH REPUBLIC

Jana Zidovská is a clinical geneticist at the Institute of Biology and Genetics of Charles University in Prague and a member of the Central Committee of Ethics of the Czech Ministry of Health. She is specialized in psychiatry and neurogenetics, especially Huntington disease, and is Chair of the Czech Huntington Association. She works on psychological and ethical aspects and effectiveness of genetic counseling.

Lenka Foretova, M.D., is a pediatrician and a medical geneticist. She spent four years at the Division of Cancer Epidemiology and Control at Dana-Farber Cancer Institute, Boston, USA, working with Professor Frederick Li and Judy Garber. Most of her current work is devoted to the heredity of breast, ovarian and colorectal cancers, genetic counseling and molecular genetic testing. She has been head of the Department of Cancer Epidemiology and Genetics at the Masaryk Memorial Cancer Institute, Brno, since February 1999.

DENMARK

Lene Koch, Ph.D., Dr. Phil. is professor at the Department of Health Services Research, which is a section of the Institute of Public Health, University of Copenhagen. She is a historian by training, and since 1985 she has done extensive research in the area of genetic and reproductive technology. She has participated in several European Union projects on genetics and reproductive technology and is the author of two volumes on the history of eugenics and compulsory sterilization in Denmark: "Racehygiejne I Danmark" (1996) and "Tvangssterilisation I Danmark" (2000). Currently she heads an interdisciplinary project on genetics and disease prevention funded by the Danish Research Council. She was a member of the Danish Ethical Council from 1994 to 2000.

Søren Nørby, M.D., Ph.D., was formerly Associate Professor of Human Genetics in the faculty of Health Sciences, University of Copenhagen. From 1979 to 1989 he was Chair at the Institute of Medical Genetics, with a special interest in polycystic kidney disease. From 1989 to 2000 he was Chair at the Institute of Forensic Medicine, specializing in mitochondrially inherited disease, for which he pioneered diagnostic DNA analyses in Denmark. He later studied normal mitochondrial genetic variation and its application to human population genetics. He was a member of the Danish Council of Ethics 1990–1994.

FRANCE

Claire Julian-Reynier, M.D., M.Sc. is Senior Researcher at the French National Institute for Health and Medical Research (INSERM). She is a public health physician and epidemiologist, working at a social sciences unit "Epidemiology and Social Science of Medical Innovation" located in the Marseille Cancer Center (INSERM, U379-Institut Paoli-Calmettes) belonging to the French National Federation of Cancer Centers. Her present research interests include evaluation of medical innovation in adult genetics/cancer genetics, in particular psycho-social issues in patient provider-relationships and psycho-social issues in cancer treatment and prevention.

GERMANY

Imgard Nippert is a medical sociologist. Since 1997, she has been Professor and Chair, Women's Health Research, at the Medical School, Westfaelische Wilhelms-Universitaet Muenster, which is affiliated with the school's department of human genetics. Her research and publications include impact of prenatal diagnosis upon women and evaluating genetic screening programmes in Europe. Her current research is funded by the German Human Genome Project.

Gerhard Wolff is Professor of Human Genetics at the University of Freiburg im Breisgau, Germany, and a psychotherapist. He is head of the Department of Clinical Genetics and Genetic Counseling at the Institute of Human Genetics and Anthropology and runs a private psychotherapeutic praxis with a focus on genetic diseases. He is Chair of the Committee on Public Relations and Ethical Issues of the German Society of Human Genetics.

GREECE

Christos S. Bartsocas, M.D. is Professor of Pediatrics at the University of Athens Faculty of Nursing and Chair of the Department of Pediatrics of the faculty at the "P & A Kyriakou" Children's Hospital, in Athens, Greece.

Ariadne Malamitsi-Puchner, M.D. is Associate Professor of Pediatrics, University of Athens School of Medicine and Neonatologist at the "Aretaeion" University Hospital in Athens, Greece.

HUNGARY

Andrew E. Czeizel graduated from Semmelweis University of Medicine in Budapest, Faculty of General Medicine. He was formerly of the Laboratory of Human Genetics, and Department of Human Genetics and Teratology, National Institute of Hygiene, also of the National Institute for Health Promotion. Since 1986, he has been Director of the WHO Collaborating Centre for the Community Control of Hereditary Disease in Budapest. He participated in the establishment of the Hungarian genetic counselling network and developed the method of "information guided" counselling. He is author of, "*The Right to be Born Healthy: the Ethics of Human Genetics in Hungary*." By WHO assignment, he elaborated the Optimal Family Planning Model, which since 1989 has functioned as the Hungarian Family Planning Service at 30 centres. He has demonstrated the effect of periconceptional use of folic acid-containing multivitamins for prevention of neural tube-defects and other congenital abnormalities.

INDIA

Dr. I. C. Verma is Senior Consultant and Head, Department of Medical Genetics, Sir Ganga Ram Hospital, New Delhi 110060, India. Dr. Verma has been a pioneer in the field of medical genetics in India. The genetic center he established was designated as a WHO collaborating center in genetics. He has been consistently interested in ethical aspects of genetic services in India, and has conducted a number of surveys to gather the views of geneticists as well as the general public on ethical issues. He was a member of the government committee which drafted the legislation for registration of all prenatal diagnostic facilities and banning of fetal sexing for social reasons. He was also a member of the committee of the Indian Council of Medical Research to revise the ethical guidelines for genetic studies and research.

Dr. Kusum Verma is Professor and Chair, Department of Pathology, All India Institute of Medical Sciences, Ansari Nagar, New Delhi 110029, India. She is Secretary of the Ethical Committee and the Institutional Review Board to review and regulate all research projects on humans undertaken by the faculty, both national and international. She is the principal investigator from India in the W.H.O.–S.E.A.R.O. Project on "Teaching and Application of Health Ethics in South East Asia."

ISRAEL

Zully Gelman-Kohan, M.D. is presently Director of the Clinical Genetics Institute at the Kaplan Medical Center, Rehovot, Israel and is President of the Israeli Society of Medical Genetics. She graduated from the Buenos Aires National University, Argentina. She was a fellow in Genetics at the Moore Clinic, Johns Hopkins University, Baltimore, MD with Professor Victor McKusick and specialized in dysmorphology and clinical genetics.

ITALY

A. Gioia Jacopini, a psychologist and psychotherapist, has worked as a researcher at the Istituto di Psicologia of the National Council of Research in Rome since 1973. She was professionally trained by Professor Seymour Kessler at the University of California, Berkeley. Since 1982 she has investigated the psychological aspects of genetics, dealing first with reproductive decision-making of people at risk for specific genetic disorders and then looking into the individual dimensions of living with genetic risk. She has designed and produced a number of videos on genetic counseling and on Huntington disease (HD) in Italy, the United States and Australia, in order to educate the medical profession about psychosocial aspects of genetics.

Marina Frontali, M.D. is qualified in Medical Genetics and works at the Institute of Experimental Medicine, Italian National Research Council, Rome, Italy. Her main research topics have been: genetic counseling methods and molecular genetics of Huntington Disease (HD) and Autosomal Dominant Cerebellar Ataxias (ADCA). She is a genetic counselor at the Prenatal Diagnosis Center and at the Institute of Mental and Neurologic Disorders of La Sapienza University, Rome, Italy. Since 1989 she has headed, with Dr. A.G. Jacopini, an HD and ADCA predictive testing program.

JAPAN

Dr. Ichiro Matsuda, M.D., Ph.D. was formerly Chair and Professor of Pediatrics at Kumamoto University School of Medicine till April, 1998. He is currently Professor Emeritus at Kumamoto University, Visiting Professor at Johns Hopkins University (Genetics and Public Policy, Director; Dr. Neil A. Holtzman), a Committee Member of the National Committee for Medical Genetics, Chair of the Ethical Committee of the Japan Society of Human Genetics, and an Executive Member of the International Society for Neonatal Screening.

Kaede Tomoeda, M.D. is a pediatrician and a student at the Graduate School of Medicine at Kumamoto University.

MEXICO

Rubén Lisker, M.D. has been practicing medical genetics since 1966 and has written over 250 papers and three books. From 1993 to 1997, he was a member of the International Bioethics Committee of UNESCO. From 1970 to 1971, he was President of the Mexican Society of Human Genetics.

Salvador Armendares, M.D. has been practicing medical genetics since 1966 and has written over 100 papers and three books. He was President of the Vth International Congress of Human Genetics in 1976, and in 1972 and 1973 was President of the Mexican Society of Human Genetics.

Alessandra Carnevale, M.D. has been practicing medical genetics since 1973 and has written over 100 papers, several referring to ethical issues in genetics. During 1982 and 1983, she was President of the Mexican Society of Human Genetics. She is currently the Director of the Mexican National Institute of Pediatrics.

Osvaldo Mutchinick, M.D., Ph.D. has been practicing medical genetics since 1969. Has written over 100 papers and was President of the Mexican Society of Human Genetics during 1984–1985. He is a member of the International Clearing House for Birth Defects Monitoring systems and Director of the WHO Collaborative Center for Genetics and Community Services in Mexico.

THE NETHERLANDS

Leo P. ten Kate is Professor of Clinical Genetics at the "Vrije Universiteit" in Amsterdam, The Netherlands, and Editor in Chief of the journal Community Genetics. He obtained his training at the University of Groningen, The Netherlands (with Professor George JPA Anders) and the University of Washington, Seattle, USA (with professor Arno G Motulsky). His research interest is in epidemiology of genetic diseases and community genetics.

POLAND

Jacek Zaremba, M.D., Ph.D., D.Sc. a neurologist and clinical geneticist, is Professor and Head of the Department of Genetics at the Institute of Psychiatry and Neurology, Warsaw. He is Vice-chair of the Polish Genetics Society, and one of the Vice-chairs of the International Bioethics Committee of UNESCO.

Elzbieta Fidzianska, M.D., Ph.D., D.S. a virologist and medical geneticist, is Associate Professor of the Department of Genetics in the Institute of Psychiatry and Neurology, Warsaw.

PERU

Teresa Pérez, M.D. is Professor of Pathology and Director of the Genetics Institute, Universidad Peruana Cayetano Herdia. She is also Chief of the Pathology Department, Nacional Hospital Guillermo Almenara ESSALUD, Lima.

RUSSIA

Vladimir Ilich Ivanov is Doctor of Biological Sciences and Professor of Genetics at the Russian State Medical University in Moscow. He is a Member of the Russian Academy of Medical Sciences, Director of the Research Centre for Medical Genetics of the Academy, and author and/or co-author of about 300 scientific publications in genetics, biology and bioethics. His current research activities include bioethical aspects of genetic research and service, developmental genetics and human malformations.

Vera Leonidovna Izhevskaya, M.D., Ph.D. is Secretary of the Research Centre for Medical Genetics, Moscow. Her field of research is clinical genetics and genetic counseling, especially in haematology, bioethical aspects of genetic research and service. She is author and/or co-author of about 40 scientific publications.

SOUTH AFRICA

Jennifer G. R. Kromberg is a social scientist and Honorary Associate Professor, Department of Human Genetics, South African Institute for Medical Research and University of the Witwatersrand, Johannesburg, South Africa. She is also Principal Genetic Counsellor at the Queensland Clinical Genetics Services and Senior Lecturer, Griffith University, Brisbane, Australia.

Trefor Jenkins is Professor of Human Genetics, Emeritus, University of the Witwatersrand, Johannesburg, South Africa.

SPAIN

Dr. Carmen Ramos, is Associate Chief of the Department of Genetics, Fundacion Jimenez Diaz, Madrid. She has a Ph.D. in Biological Sciences. She has been involved in reproductive issues and cytogenetic prenatal diagnosis for 25 years. She is the Coordinator of the Spanish accreditation committee for cytogenetic prenatal diagnosis laboratories.

Dr. Carmen Ayuso is Associate Chief of the Department of Genetics, Fundacion Jimenez Diaz, Madrid. She is a Doctor of Medicine. She has been involved in cytogenetics, genetic counseling and medical genetics for 21 years.

Dr. Isabel Lorda, M.D. is a postdoctoral fellow. She has been involved in cytogenetics, genetic counseling and medical genetics for 8 years.

SWEDEN

Ulf Kristoffersson, B.A., M.D., Ph.D. is senior consultant in Clinical Genetics at the University Hospital and Associate Professor at the University Medical Faculty, both in Lund, Sweden. He has been involved in genetic research for 30 years. He has made most of his scientific contribution in topics related to cancer genetics, but has also been involved in quality assessment issues. At present he is a member of the Board of the European Society of Human Genetics, and a member of its Performance and Public Policy Committee (PPPC).

Göran Hermerén, Ph.D. is Professor of Medical Ethics and Director of the Department of Clinical Genetics in the Faculty of Medicine of Lund University. In addition to his research in aesthetics, he has published the following books: Värdering och objektivitet [Values and Objectivity] 1972, and Kunskapens pris [In praise of knowledge/The Price of Knowledge] 1986, and a book on research ethics (new revised edition, 1996). He has been a member of many ethics councils over the years, including the National Council on Medical Ethics, the Committee on Ethics in Biotechnology, and the European Science Foundation's Standing Committee for Humanities.

THAILAND

Valairat Dhamcharee, M.D. is a Diplomate of the American Board of Medical Genetics and Pediatrics. She has practiced medical genetics since 1993. She is the first and the only American Board of Medical Genetics certified medical geneticist in Thailand at the present time, and is currently an instructor, clinician, and researcher in genetics at Chulalongkorn University Hospital, Bangkok.

Apiwat Mutirangura, M.D., graduated from Chiang Mai University, Thailand, in 1987 and received his Ph.D. in human and molecular genetics from Baylor College of Medicine, Houston, Texas, U.S.A., in 1993. He is currently working as a principal research investigator at Chulalongkorn University Hospital, Bangkok. He has published more than 25 original articles on molecular genetic research in international journals.

Orasri Romayanan, M.D., graduated from Chulalongkorn University, Faculty of Medicine, Thailand, in 1961. She is a pioneer in counseling in medical genetics and cytogenetic laboratory service in Thailand. She established prenatal chromosome diagnosis in Thailand in 1975. She is currently an honorary lecturer and consultant at Chulalongkorn University Hospital, Bangkok.

TURKEY

Isik Bökesoy, M.D., Ph.D. is Professor of Medical Genetics and Director of the Medical Biology and Medical Genetics Departments of the Medical School of Ankara University. She is Chairperson of the Down Syndrome Association, is a member of the Ethics Committee of the Turkish Medical Association, and a member of the Scientific

Committees of the Ministry of Health for "In Vitro Fertilization" and "Genetics Practice". She is a member of the Turkish Genetics Society.

Berna Arda, M.D., Ph.D., Med. Spec. is Professor of Deontology, Medical School of Ankara University and Chairperson of the Ethics Committee of the Turkish Medical Association and Turkish Society of Bioethics.

UNITED KINGDOM

Angus Clarke is a consultant clinical geneticist in Cardiff, Wales. He studied genetics before clinical medicine, then trained for some years in pediatrics and entered clinical genetics after conducting clinical and molecular genetic research. He has particular interests in the communication processes occurring within genetic counseling, in genetic screening programs and in a range of X chromosome and neuromuscular disorders.

TABLES FOR CHAPTERS 1–8

I. RATIONALE FOR SURVEY

II. METHODS

III. RESULTS

A. Equity

B. Respect for Autonomy

I. RATIONALE FOR SURVEY

Table 1. How Genetics Differs from Other Areas of Medicine

Genetics
1. Provides information about other blood relatives
2. Provides information predicting the future of persons who are now healthy, including children
3. Provides unexpected nonmedical information, such as nonpaternity
4. Has a history of "nondirectiveness" in counseling

II. METHODS

Table 2. Types of Surveys

	Asked to participate	Responded	Response rate (%)
1. International survey of geneticists and genetic counselors (1995)			
37 nations	4,629	2,907	63
(United States)	(1,538)	(1,084)	(70)
2. Primary care physicians (obstetricians, pediatricians, family practitioners, US only)	852	499	59
3. Patients at 12 genetics clinics (US and Canada)	718	476	59
4. US Public	1,000	988	99
5. 1984 survey of geneticists in 19 nations	1,098	682	62

Table 3. Survey Response Rates

Country	Invited to participate	Responded	% Responding
Argentina	57	19	33
Australia	26	15	58
Belgium	40	15	38
Brazil	131	74	56
Canada	212	136	64
Chile	25	16	64
China	392	252	64
Colombia	27	15	56
Cuba	96	18	19
Czech Republic	137	81	59
Denmark	54	28	52
Finland	53	22	42
France	102	75	74
Germany	418	255	61
Greece	12	12	100
Hungary	78	38	46
India	70	23	33
Israel	27	23	85
Italy	23	21	91
Japan	174	113	65

Table 3. (continued)

Country	Invited to participate	Responded	% Responding
Mexico	89	64	72
Netherlands	41	27	66
Norway	18	9	50
Peru	16	14	88
Poland	250	151	60
Portugal	22	11	50
Russia	66	46	69
South Africa	21	16	76
Spain	82	51	62
Sweden	15	12	80
Switzerland	10	6	60
Thailand	38	25	66
Turkey	30	22	73
United Kingdom	217	101	47
United States	1,538	1,084	70
Venezuela	22	16	73
Total	4,629	2,906	63

Table 4. Respondents' Professional Characteristics

Country	M.D. (in parens: % pediatricians)	Ph.D. (%)	Other degree (%)	Training outside own country (%)	Years in genetics (median)	See patients (%)	Patients per week (median)	Professional income (median) US$
English-speaking nations								
Australia	93 (87)	0	7	80	5	93	11	50,000–75,000
Canada	45 (20)	28	26	54	9	84	6	50,000–75,000
South Africa	69 (38)	25	6	6	14	100	6–10	20,000–30,000
United Kingdom	52 (24)	7	41	15	5	98	6–10	40,000
United States	43 (24)	38	19	n/a	9	83	6–10	50,000–75,000
Northern/Western Europe								
Belgium	93 (53)	7	0	60	9	93	11–15	30,000–40,000
Denmark	75 (21)	14	11	15	15	79	1–5	50,000–75,000
Finland	82 (32)	0	18	14	5	77	1–5	40,000–50,000
France	95 (81)	4	1	9	11	99	6–10	50,000–75,000
Germany	68 (23)	28	5	11	10	73	1–5	40,000
Netherlands	74 (48)	19	7	22	9	100	6–10	30,000–40,000
Norway	67 (11)	22	11	38	5	87	1–5	30,000–40,000
Sweden	67 (33)	25	8	25	15	100	1–5	50,000–75,000
Switzerland	100 (60)	0	0	67	23	100	1–5	50,000–75,000
Southern Europe								
Greece	83 (50)	8	9	83	10	92	1–5	20,000–30,000
Italy	80 (65)	15	5	38	16	100	6–10	50,000–75,000
Portugal	60 (60)	20	20	55	7	100	1–5	30,000–40,000
Spain	69 (18)	25	6	33	15	84	10	30,000–40,000

Table 4. (continued)

Country	M.D. (in parens: % pediatricians)	Ph.D. (%)	Other degree (%)	Training outside own country (%)	Years in genetics (median)	See patients (%)	Patients per week (median)	Professional income (median) US$
Eastern Europe								
Czech Republic	69 (35)	10	21	22	11	87	6	<5,000
Hungary	73 (21)	18	9	28	9	84	6–10	<5,000
Poland	52 (30)	17	30	35	9	62	1–5	<5,000
Russia	89 (41)	11	0	5	12	93	6–10	<5,000
Near East								
Israel	61 (39)	22	17	64	12	96	11–15	10,000–20,000
Turkey	73 (32)	18	9	36	14	100	6–10	10,000–20,000
Latin America								
Argentina	93 (36)	7	0	23	14	86	16–20	20,000–30,000
Brazil	85 (30)	4	11	22	7	81	11	<5,000
Chile	94 (63)	0	6	50	12	87	11	10,000
Colombia	46	8	46	57	9	93	6–10	5,000–10,000
Cuba	100 (9)	0	0	7	5	100	30+	<5,000
Mexico	72 (16)	7	21	33	13	81	1–5	10,000–20,000
Peru	93 (14)	7	0	42	15	79	1–5	<5,000
Venezuela	75 (31)	19	6	57	14	100	6–10	10,000
Asia								
China	76 (18)	5	19	12	12	86	1–5	<5,000
India	48 (26)	44	8	30	8	87	6	<5,000
Japan	100 (47)	0	0	5	9	65	1–5	100,000
Thailand	72 (16)	12	16	48	9	71	1–5	20,000–30,000
Total	56 (28)	15	29	25	9	83	6–10	30,000–40,000
Total, excluding US	70 (32)	13	16	25	9	83	1–5	10,000–20,000

Table 5. Respondents' Personal Characteristics

Country	Women (%)	Age (median)	Married (%)	Children			Religion moderately to extremely important (%)	Religious observances/ year (median)	Personal experience with disability (%)	Some experience with support groups ( ) = a lot or extensive (%)
				1	2	3+				
English-speaking nations										
Australia	21	44	93	0	27	60	33	1–2	87	46 (27)
Canada	74	44	79	7	40	33	40	1–2	71	34 (14)
South Africa	38	48	81	0	58	42	75	12	63	89 (22)
United Kingdom	71	41	78	15	36	30	44	1–2	72	35 (14)
United States	72	40	73	17	35	20	54	5	76	30 (12)
Northern/Western Europe										
Belgium	60	43	80	14	43	28	27	0	67	30 (10)
Denmark	39	48	86	11	46	39	22	<1	64	28 (17)
Finland	64	42	77	5	40	50	44	1–2	73	31 (6)
France	45	45	81	6	36	52	55	1–2	63	52 (24)
Germany	51	43	71	18	43	22	43	<1	66	26 (15)
Netherlands	56	38	78	22	22	44	11	<1	67	20 (10)
Norway	50	37	20	0	67	33	60	<1	71	0
Sweden	17	50	83	0	50	33	8	<1	78	50 (13)
Switzerland	17	55	100	17	0	67	50	1–2	80	60 (0)
Southern Europe										
Greece	67	45	83	22	56	12	58	5	42	71 (57)
Italy	29	49	81	44	33	6	62	5	40	50 (33)
Portugal	46	38	64	27	36	18	46	1–2	40	40 (20)
Spain	60	44	67	15	37	41	71	12	61	43 (17)
Eastern Europe										
Czech Republic	67	44	76	25	55	5	16	0	50	22 (6)
Hungary	47	40	73	31	50	13	38	1–2	48	38 (6)
Poland	71	43	69	40	41	12	79	12	62	17 (7)
Russia	78	48	82	61	32	2	52	1–2	40	11 (11)

Table 5. (continued)

Country	Women (%)	Age (median)	Married (%)	Children			Religion moderately to extremely important (%)	Religious observances/year (median)	Personal experience with disability (%)	Some experience with support groups () = a lot or extensive (%)
				1	2	3+				
Near East										
Israel	55	53	83	35	39	13	13	0	50	16 (12)
Turkey	59	40	59	62	15	15	32	1–2	52	62 (31)
Latin America										
Argentina	71	45	71	25	25	50	57	5	69	20 (0)
Brazil	63	41	51	22	29	28	43	<1	63	17 (13)
Chile	53	43	69	20	7	53	81	36	69	31 (8)
Colombia	36	43	79	0	63	25	28	<1	100	25 (17)
Cuba	64	38	50	55	36	9	21	0	54	17 (17)
Mexico	55	42	69	21	34	36	47	5	63	57 (29)
Peru	50	41	64	27	27	46	79	12	80	29 (0)
Venezuela	38	47	69	0	13	73	75	1–2	56	60 (20)
Asia										
China	58	50	96	45	39	10	2	0	40	12 (6)
India	44	42	91	18	32	27	78	1–2	74	33 (28)
Japan	14	45	92	11	51	37	7	1–2	79	13 (2)
Thailand	44	40	64	19	56	25	96	5	71	32 (21)
Total	62	40	76	22	38	25	45	1–2	67	30 (13)
Total, excluding US	55	42	77	23	40	27	39	<1	61	29 (14)

Religious Backgrounds: None, 29%; Catholic, 27%; Protestant, 24%; Jewish, 10%; Buddhist, 4%.

Table 6. Patient Characteristics

Characteristic	North America (n = 476) (Sites: Providence, RI, Buffalo, NY, Detroit, MI, Rochester, MN, Hanover, NH, Charlottesville, VA, Pittsburgh, PA, New York City, Ottawa, Montreal, Quebec, Ontario, Temple, TX, Los Angeles, CA, Edmonton, Alberta)	Germany (n = 593) (Sites: Münster, Freiburg)
Women	91%	90%
Median Age	35 years	33 years
USA	93%	—
Canadian	7%	—
Married	73%	79%
Education (Median)	13 years	
High-School Dropout	13%	
College Graduate	17%	
Post Graduate	11%	
Religion	44%	
Catholic	33% (Includes 11% Fundamentalist)	
Protestant	3%	
Jewish	8%	
Other	11%	
None	—	
Church attendance		
Once a week or more	23%	
1–3 times a month	21%	
Several times/year	16%	
1–2 times a year	16%	
<Once a year or never	25%	
Thought Bible was		
Literal Word of God	26%	
Inspired Word of God, but	54%	
not always to be taken literally	—	
Race		
White	89%	
Black	7%	
Hispanic	3%	
Asian and other	1%	
Occupation		
Professional/managerial	30%	
Clerical, sales	29%	
Service	19%	
Factory worker (production)	8%	
At home (housewife or househusband)	1%	20%
Spouse's occupation		
Professional/managerial	25%	
Clerical, sales	23%	
Production workers	31%	
Service	15%	
Income (US$)		
<5,000	9%	
5,001–25,000	22%	
25,001–45,000	27%	

Table 6. (continued)

Characteristic	North America (n = 476) (Sites: Providence, RI, Buffalo, NY, Detroit, MI, Rochester, MN, Hanover, NH, Charlottesville, VA, Pittsburgh, PA, New York City, Ottawa, Montreal, Quebec, Ontario, Temple, TX, Los Angeles, CA, Edmonton, Alberta)	Germany (n = 593) (Sites: Münster, Freiburg)
45,001–75,000	26%	
>75,000	16%	
Residence		
Small city or town	43%	66%
Suburban metropolitan area	27%	4%
Central city	17%	16%
Rural	13%	16%
Children		
None	8%	47%
1	31%	28%
2	33%	18%
3	13%	7%
4 or more	12%	—
Median age of child patient	5 years	
Most common reasons for visiting clinic:		
"My Doctor Told Me To Come"	47%	
"My Child May Have a Genetic Condition"	46%	
Visiting Clinic for "Advanced Maternal Age Only"	15%	

III RESULTS

Table 7. Summary of Major Findings

1. Most geneticists would give purposely biased information after prenatal diagnosis, except in English-speaking nations and parts of Western Europe
2. Most geneticists around the world take a pessimistic view of disability
3. There is a growing tendency to respect patient autonomy
4. In most of the 19 nations previously surveyed in 1985, more geneticists would perform prenatal diagnosis for sex selection or would refer
5. As in 1985, there was no consensus about telling relatives at genetic risk against a patient's wishes. Except in English-speaking nations and Western Europe, majorities would tell relatives
6. Geneticists in English-speaking nations would not test children for adult-onset disorders. In Southern and Eastern Europe, Asia, and Latin America, most would test children at parents' request
7. There was worldwide consensus that employers, insurers, and schools should not have access to genetic information without an individual's consent
8. Eugenic thinking is alive and well in many developing nations, especially in Asia and Eastern Europe. However, most geneticists reject involuntary sterilization
9. There is a growing tendency toward full disclosure of disease-related information to patients
10. Most geneticists would not disclose an *accidental* finding of nonpaternity to a husband, even if he asks
11. There were few reports of refusals of employment or insurance outside the United States. In the United States, reports were few when compared with patient volume

A. *Equity*

Table 8. Economic and Social Characteristics of Participating Nations[1]

Country	GNP[2] per capita US$	Public health expenditures per capita US$	Population per physician	Infant mortality under 1 year per 1000 live birth	Life expectancy at births	Literacy rate % F/M	Contraceptive[3] use (%)	SES rank[4]
English-speaking nations								
Australia	17,960	947	454	7	77	100/100	76	17
Canada	17,960	1,395	457	6	77	99/99	74	15
South Africa	2,940	105	1,875	48	63	82/82	50	69
United Kingdom	17,680	955	639	6	76	99/99	81	16
United States	25,510	1,504	430	7	76	99/99	73	8
Northern/Western Europe								
Belgium	22,760	1,539	270	7	76	99/99	79	5
Denmark	27,290	1,458	349	7	75	99/99	78	2
Finland	18,340	1,152	375	5	76	100/100	80	13
France	22,800	1,584	360	7	77	99/99	81	8
Germany	25,200	1,430	374	6	76	99/99	78	8
Netherlands	21,500	1,369	391	6	77	99/99	76	13
Norway	25,010	1,837	303	7	77	99/99	84	4
Sweden	21,600	1,331	333	5	78	99/99	78	11
Switzerland	37,500	2,252	332	6	78	99/99	71	1
Southern Europe								
Greece	7,510	305	259	8	78	89/98	—	25
Italy	17,630	1,081	608	7	77	96/98	78	19
Portugal	8,860	355	342	10	75	82/89	66	24
Spain	12,080	694	248	6	78	94/96	59	18

Table 8. (continued)

Country	GNP[2] per capita US$	Public health expenditures per capita US$	Population per physician	Infant mortality under 1 year per 1000 live birth	Life expectancy at births	Literacy rate % F/M	Contraceptive[3] use (%)	SES rank[4]
Eastern Europe								
Czech Republic	3,490	232	278	9	71	99/99	69	38
Hungary	3,910	139	300	12	69	99/99	73	37
Poland	2,450	88	454	13	71	99/99	75	43
Russia	2,510	96	220	—	66	99/99	32	48
Near East								
Israel	14,170	240	439	8	76	93/97	—	22
Turkey	2,120	42	976	47	66	72/92	63	80
Latin America								
Argentina	8,140	174	380	26	72	96/96	74	35
Brazil	3,930	79	749	51	66	83/83	66	56
Chile	3,590	79	920	16	74	95/95	43	51
Colombia	1,730	12	1,115	34	69	91/91	66	83
Cuba	—	—	255	13	75	95/96	78	25
Mexico	3,960	116	1,001	30	71	87/92	53	50
Peru	2,120	17	934	66	66	83/95	59	93
Venezuela	2,690	58	634	31	72	90/92	49	61
Asia								
China	500	10	1,061	23	68	73/90	83	109
India	310	3	2,563	77	60	38/66	43	125
Japan	37,100	1,767	590	5	79	99/99	64	12
Thailand	2,410	29	4,432	23	69	92/96	66	78

[1]Source: Sivard, Ruth Leger, *World Military and Social Expenditures 1996*, World Priorities, Box 25140, Washington, DC 20007, pp. 48–51.
[2]GNP = Gross National Product.
[3]Married women of childbearing age using contraception.
[4]SES = Socio-Economic Standing. Average of ranks for GNP per capita, education, and health.
[5]Estimated for 2000. New York Times, "Freedom's Toll; An Ailing Russia Lives a Tough Life that's Getting Shorter," Dec. 3, 2000, p. 1.

Table 9. Percentage of Population Receiving Genetics Services

Country	% said highest quality services available to all regardless of ability to pay	% Receiving services in 1995	Estimated % will receive services in 2005
English-speaking nations			
Australia	87	25	50
Canada	98	50	65
South Africa	71	15	7
United Kingdom	99	50	67
United States	47	50	70
Northern/Western Europe			
Belgium	100	30	48
Denmark	96	70	75
Finland	100	55	67
France	92	30	55
Germany	98	47	55
Netherlands	100	30	50
Norway	100	40	65
Sweden	100	47	70
Switzerland	100	30	40
Southern Europe			
Greece	80	20	50
Italy	95	27	50
Portugal	82	30	50
Spain	85	27	55
Eastern Europe			
Czech Republic	99	70	78
Hungary	97	46	55
Poland	70	10	25
Russia	51	22	44
Near East			
Israel	78	40	68
Turkey	9	5	24
Latin America			
Argentina	15	10	27
Brazil	23	10	25
Chile	25	20	45
Colombia	7	8	38
Cuba	100	90	95
Mexico	63	10	31
Peru	7	2	10
Venezuela	50	12	30
Asia			
China	26	10	35
India	23	5	25
Japan	50	5	18
Thailand	8	4	15
Total	62	32	50
Total, excluding US	69	25	48

Table 10. Respondents' Ranking of Priorities in Health Care Budgets

	Geneticists	Patients	US Primary Care Physicians
Prevention of low birthweight by providing medical prenatal care	1	1	1
Sex education, including contraception and prevention of HIV	2	2	3
Prevention of low birthweight by treating pregnant women for social risks (drug addiction, alcoholism, homelessness, abuse)	3	4	2
Reducing exposure to known teratogens	4	8	4
Public education about genetics	5	7	7
Birth defects monitoring systems to give early warning about teratogens	6	n/a	5
Testing for genetic predisposition to common diseases such as diabetes, heart disease, cancers, and senile dementia	7	6	6
Carrier testing for recessive disorders such as cystic fibrosis	8	5	8
Gene therapy	10	3	10
Genetic testing for susceptibility to occupational diseases	9	10	9
Research and treatment for infertility	11	9	11

Table 11. Worldwide Distribution of Geneticists

Country	Population, 1994[1]	Genetics service providers asked to participate in survey	Population per genetics service provider
Developed (industrial) Nations[2]			
English-speaking			
Australia	17,841,000	26	686,192
Canada	29,121,000	212	137,363
South Africa	41,591,000	21	1,980,524
United Kingdom	58,088,000	217	267,687
United States	260,529,000	1,538	169,395
Western Europe			
Belgium	10,080,000	40	252,000
Denmark	5,173,000	54	95,796
Finland	5,083,000	53	95,906
France	57,726,000	102	565,941
Germany	82,141,000	418	194,117
Netherlands	15,391,000	41	375,390
Norway	4,318,000	18	239,889
Sweden	8,735,000	15	582,333
Switzerland	7,127,000	10	712,700
Southern Europe			
Greece	10,408,000	12	867,333
Italy	57,154,000	23	2,484,957
Portugal	9,832,000	22	446,909
Spain	39,551,000	82	482,329
Near East			
Israel	5,420,000	27	200,741
Asia			
Japan	124,782,000	174	717,138

Table 11. (continued)

Country	Population, 1994[1]	Genetics service providers asked to participate in survey	Population per genetics service provider
Eastern Europe			
Czech Republic	10,295,000	137	75, 146
Hungary	10,161,000	78	130,269
Poland	38,341,000	250	153,364
Russia	148,366,000	66	2,247,970
Developing Nations			
Near East			
Turkey	60,771,000	30	2,025,700
Latin America			
Argentina	34,180,000	57	599,649
Brazil	159,143,000	131	1,214,832
Chile	14,044,000	25	561,760
Colombia	36,330,000	27	1,345,556
Cuba	10,951,000	70	156,443
Mexico	91,858,000	89	1,032,112
Peru	23,331,000	16	1,458,188
Venezuela	21,378,000	22	971,727
Asia			
China	1,190,918,000	392	3,038,056
India	913,600,000	70	13,051,429
Thailand	58,718,000	38	1,545,211
Totals			
20 Developed Nations in Survey	724,309,000	3,105	233,372
4 Eastern European Nations in Survey	207,163,000	531	390,137
12 Developing Nations in Survey	2,615,222,000	967	2,704,407
Total in Survey	3,546,694,000	4,603	770,518
Nations not in Survey had fewer than 10 Geneticists			
5 Developed Nations	98,846,000		
9 Eastern European Nations & 14 from Former USSR	208,435,000		
94 Developing Nations	1,731,699,000		
Total not in Survey	2,038,980,000		

[1]Source for population data: Sivard, Ruth Leger, *World Military and Social Priorities 1996*. World Priorities, Box 25140, Washington, DC 20007, pp. 45–47.
[2]"Developed" and "Developing" classifications are based on World Bank classifications.

B. *Respect for Autonomy*

Table 12. Views on Patient Autonomy: Rights to Services. % agreeing

Country	Withholding any requested service is paternalistic	Patients are entitled to any service they request and can pay for out-of-pocket	PND should be provided regardless of intended use	Women should have unqualified right to abortion before viability
English-speaking nations				
Australia	50	27	40	67
Canada	38	11	28	71
South Africa	56	38	38	50
United Kingdom	51	4	20	46
United States	56	36	47	85
US Primary Care Physicians	57	26	44	63
US patients (n = 476)	69	59	80	n/a
Northern/Western Europe				
Belgium	53	0	21	39
Denmark	67	19	44	70
Finland	29	0	41	38
France	45	8	86	14
French patients	48	54	68	n/a
Germany	28	13	23	29
German patients (n = 593)	66	48	94	n/a
Netherlands	37	7	37	46
Norway	56	22	22	56
Sweden	58	8	17	75
Switzerland	67	33	50	50
Southern Europe				
Greece	50	33	31	58
Italy	55	35	52	30
Portugal	36	27	18	46
Spain	41	20	18	32

Eastern Europe				
Czech Republic	48	41	42	51
Hungary	15	29	6	74
Poland	67	68	61	35
Russia	55	67	65	88
Near East				
Israel	59	57	50	55
Turkey	63	23	5	91
Latin America				
Argentina	50	43	39	43
Brazil	30	48	30	58
Chile	47	33	27	7
Colombia	54	31	23	79
Cuba	36	46	7	93
Mexico	54	36	8	74
Peru	31	43	14	43
Venezuela	40	38	13	50
Asia				
China	23	28	13	61
India	43	26	27	30
Japan	32	64	6	12
Thailand	32	36	29	18
Total (geneticists)	46	33	35	59
Total, excluding US	40	30	29	46

Table 13. Views on Autonomy: Rights to Referral. % agreeing

Country	It is my responsibility to offer a referral if I am unwilling to perform a procedure for moral reasons	If law forbids a procedure, provider should refer outside country or state	Provider who refuses sex selection should offer referral
English-speaking nations			
Australia	100	47	13
Canada	94	53	28
South Africa	88	63	20
United Kingdom	93	44	14
United States	90	65	55
US Primary Care Physicians	82	55	41
US patients (n = 476)	86	50	50
Northern/Western Europe			
Belgium	100	71	29
Denmark	87	38	15
Finland	91	33	29
France	100	39	4
French patients	88	33	26
Germany	81	17	6
German patients (n = 593)	86	29	29
Netherlands	93	56	26
Norway	78	22	11
Sweden	92	42	8
Switzerland	83	50	0
Southern Europe			
Greece	83	75	42
Italy	95	90	15
Portugal	100	100	18
Spain	88	63	37
Eastern Europe			
Czech Republic	79	51	19
Hungary	80	38	29
Poland	91	62	30
Russia	81	77	60
Near East			
Israel	96	64	36
Turkey	73	82	23
Latin America			
Argentina	54	83	21
Brazil	88	57	39
Chile	56	31	19
Colombia	79	64	50
Cuba	54	64	15
Mexico	95	87	54
Peru	64	79	43
Venezuela	100	81	47
Asia			
China	59	25	2
India	86	50	9
Japan	43	13	4
Thailand	68	60	20
Total	84	51	30
Total, excluding US	80	44	19

Table 14. Views on Autonomy: Rights to Know, Rights to Decide. % agreeing

Country	Parents should be told sex of fetus if they ask, at any time during pregnancy	After taking a test, patients have right *not to know* results	A woman's decision about abortion should be her own, without intervention by anyone	Counselor should support patients' decisions even if disagrees	Parents should have the right to choose children's sex
English-speaking nations					
Australia	47	87	73	87	7
Canada	76	90	81	86	7
South Africa	81	75	75	100	0
United Kingdom	56	88	69	99	3
United States	91	82	85	90	10
US Primary Care Physicians	87	62	65	n/a	11
US patients (n = 476)	92	41	60	79	13
Northern/Western Europe					
Belgium	87	80	40	80	0
Denmark	85	93	67	74	0
Finland	55	87	73	91	0
France	53	78	38	49	1
French patients (n = 381)	86	27	39	75	4
Germany	15	89	63	74	1
German patients (n = 593)	91	35	52	92	4
Netherlands	74	96	48	93	0
Norway	67	78	78	100	11
Sweden	67	75	83	83	0
Switzerland	33	100	83	100	0
Southern Europe					
Greece	67	50	67	83	0
Italy	67	81	52	81	10
Portugal	73	100	36	91	0
Spain	84	50	24	65	0

Table 14. (continued)

Country	Parents should be told sex of fetus if they ask, at any time during pregnancy	After taking a test, patients have right *not to know* results	A woman's decision about abortion should be her own, without intervention by anyone	Counselor should support patients' decisions even if disagrees	Parents should have the right to choose children's sex
Eastern Europe					
Czech Republic	69	63	61	45	1
Hungary	57	86	46	63	3
Poland	68	62	76	54	7
Russia	82	37	73	41	41
Near East					
Israel	86	86	59	86	14
Turkey	18	55	32	55	0
Latin America					
Argentina	93	57	7	79	7
Brazil	89	69	67	72	20
Chile	88	44	13	75	6
Colombia	93	57	36	71	21
Cuba	86	79	14	64	14
Mexico	77	56	42	77	13
Peru	93	57	21	64	7
Venezuela	81	69	13	100	13
Asia					
China	10	29	40	14	6
India	13	43	48	39	4
Japan	10	80	40	65	1
Thailand	44	60	13	74	0
Total geneticists	65	73	66	71	7
Total, excluding US	50	68	55	62	6

Table 15. Full Disclosure of Ambiguous or Sensitive Information to Patients. % would disclose

Country	XY status in a female, age 15, if parents want her to know	XY status in a female age 15, if parents don't want her to know	XY status in a female age 18	XY status in a female age 25		Ambiguous findings after PND[2] 1995	Colleague disagreement about meaning of test results	
				1985[1]	1995		1985	1995
English-speaking nations								
Australia	73	25	93	58	93	86	50	80
Canada	68	23	78	68	84	87	66	80
South Africa	60	7	87	—	93	87	—	79
United Kingdom	66	11	80	37	83	91	52	69
United States	74	14	88	64	92	93	75	83
US Primary Care Physicians	—	—	—	—	—	95	—	97
US Patients	—	—	—	—	—	92	—	95
US Public	—	—	—	—	—	82	—	39
Western Europe	64	8	60	—	67	73	—	50
Denmark	85	27	96	40	95	82	47	70
Finland	90	27	100	—	100	71	—	50
France	38	14	54	13	65	53	47	49
French patients	n/a	n/a	n/a	n/a	n/a	72	n/a	83
Germany[3]	50	18	78	52	82	75	77	77
Netherlands	44	11	83	—	86	93	—	65
Norway	43	14	88	17	83	100	33	56
Sweden	57	17	56	52	71	33	48	30
Switzerland	40	33	80	0	80	50	20	33
Southern Europe								
Greece	38	20	80	14	100	92	43	66
Italy	41	8	67	27	57	78	46	63
Portugal	36	10	64	—	60	82	—	70
Spain	40	18	54	—	70	81	—	58

Table 15. (continued)

Country	XY status in a female, age 15, if parents want her to know	XY status in a female age 15, if parents don't want her to know	XY status in a female age 18	XY status in a female age 25		Ambiguous findings after PND[2]	Colleague disagreement about meaning of test results	
				1985[1]	1995	1995	1985	1995
Eastern Europe								
Czech Republic	61	24	84	—	89	74	—	44
Hungary	39	15	88	20	92	89	60	71
Poland	56	21	75	—	79	72	—	71
Russia	51	18	77	—	87	85	—	53
Near East								
Israel	55	5	64	27	85	96	57	48
Turkey	31	13	44	100	47	76	100	80
Latin America								
Argentina	23	0	79	—	86	67	—	17
Brazil	36	9	45	9	56	77	72	79
Chile	20	8	29	—	50	56	—	81
Colombia	43	23	39	—	62	55	—	60
Cuba	62	20	75	—	92	64	—	46
Mexico	22	16	28	—	39	54	—	42
Peru	64	36	64	—	80	90	—	50
Venezuela	57	15	58	—	58	50	—	82
Asia								
China	60	18	84	—	91	71	—	46
India	81	15	88	52	89	91	56	70
Japan	40	8	50	40	65	63	61	67
Thailand	57	11	90	—	100	83	—	50
Total geneticists	59	16	76	51	82	81	66	69
Total, excluding US	52	16	71	41	78	75	60	63

[1]1985 survey did not specify age, but assumed she was an adult.
[2]1995 survey described a CVS showing an apparently balanced *de novo* translocation in 5% of the cells.
[3]1985 figures are for West Germany.

Table 16. Five Cases of Sex Selection. % Would Perform PND

Country	Single woman wants girl	Couple with 4 girls want boy	Poor couple with 5 boys want girl	Nonwestern couple wants boy	Couple in 40's want girl
English-speaking nations					
Australia	21	21	29	29	36
Canada	17	17	21	21	38
South Africa	14	21	20	21	29
United Kingdom	8	12	14	17	22
United States	35	34	38	38	57
US Primary Care Physicians	29	34	36	33	51
US patients*	55	59	63	61	62
US public	35	38	41**	n/a	n/a
Western Europe					
Belgium	36	36	20	43	53
Denmark	22	22	27	26	46
Finland	14	19	32	33	57
France	4	8	9	10	24
Germany	11	13	14	25	23
Netherlands	0	0	0	12	16
Norway	13	13	11	13	25
Sweden	11	11	25	22	22
Switzerland	0	0	0	0	17
Southern Europe					
Greece	25	33	30	50	50
Italy	29	25	27	30	35
Portugal	46	36	46	36	55
Spain	23	23	15	27	42
Eastern Europe					
Czech Republic	37	49	54	56	68
Hungary	26	63	78	64	61
Poland	29	30	28	48	46
Russia	78	90	72	85	91
Near East					
Israel	67	68	70	82	82
Turkey	0	10	14	5	18
Latin America					
Argentina	33	25	31	25	55
Brazil	35	34	37	32	53
Chile	7	13	13	7	25
Colombia	25	25	23	50	67
Cuba	46	62	57	54	85
Mexico	39	38	28	39	53
Peru	39	39	36	57	50
Venezuela	50	50	13	63	63
Asia					
China	34	24	28	35	29
India	27	32	19	32	46
Japan	10	18	22	19	19
Thailand	8	8	4	8	16
Total geneticists	27	29	31	35	44
Total, excluding US	23	26	27	33	39

*On public and patient questionnaires: "should the doctor perform PND?"
**In public questionnaire, couple has 3 boys instead of 5.

Table 17. Sex Selection: Trends 1985–1995. A couple with 4 healthy daughters desire a son

Country	Perform PND		% Would offer a referral		Total would perform or refer	
	1985	1995	1985	1995	1985	1995
English-speaking nations						
Australia	9	21	8	29	17	50
Canada	30	17	17	34	47	51
South Africa	—	21	—	14	—	35
United Kingdom	9	12	15	27	24	39
United States	34	34	28	38	62	72
Western Europe						
Belgium	—	36	—	7	—	43
Denmark	13	22	0	4	13	26
Finland	—	19	—	5	—	24
France	7	8	6	1	13	9
Germany*	6	13	1	7	7	20
Netherlands	—	0	—	23	—	23
Norway	17	13	0	0	17	13
Sweden	28	11	10	11	38	22
Switzerland	0	0	0	0	0	0
Southern Europe						
Greece	29	33	0	8	29	41
Italy	18	25	0	10	18	35
Portugal	—	36	—	46	—	82
Spain	—	23	—	2	—	25
Eastern Europe						
Czech Republic	—	49	—	4	—	53
Hungary	60	63	0	14	60	77
Poland	—	30	—	6	—	36
Russia	—	90	—	5	—	95
Near East						
Israel	13	68	20	14	33	82
Turkey	0	10	20	0	20	10
Latin America						
Argentina	—	25	—	0	—	25
Brazil	21	34	9	11	30	45
Chile	—	13	—	7	—	20
Colombia	—	25	—	17	—	42
Cuba	—	62	—	0	—	62
Mexico	—	38	—	3	—	41
Peru	—	39	—	0	—	39
Venezuela	—	50	—	0	—	50
Asia						
China	—	24	—	3	—	27
India	37	32	15	0	52	32
Japan	6	18	0	2	6	20
Thailand	—	8	—	0	—	8
Total	25	29	17	20	42	49
Total, excluding US	18	25	8	9	36	34

*1985 figures are for West Germany.

Table 18. Geneticists' Reasons for Their Actions

	Total % giving reason
Against Sex Selection	
1. Misuse of Technology ("Sex Is Not a Disease")	43
2. Oppose Abortion of Normal Fetus	16
3. Prevent Sex Discrimination	7
4. Professionals' Right to Refuse a Service	7
5. Maintain Balanced Sex Ratio	3
6. Avoid "Slippery Slope" Toward "Cosmetic" Selections or Eugenics	1
In Favor of Sex Selection	
1. Respect for Patient Autonomy	29
2. Patients Have a Right to Whatever Services They Can Pay for	8
3. Respect Patients' Culture or Religion	6
4. Preserving Family Unit	3
5. Limiting Population	2

Table 19. Requests for Sex Selection by Prenatal Diagnosis

Country	% Reporting requests	
	Outright request	Suspected request
English-speaking nations		
Australia	67	87
Canada	46	58
South Africa	33	44
United Kingdom	32	37
United States	62	75
US Primary Care Physicians	18	17
Western Europe		
Belgium	50	33
Denmark	41	36
Finland	33	36
France	39	53
Germany	47	46
Netherlands	42	42
Norway	11	22
Sweden	75	75
Switzerland	50	67
Southern Europe		
Greece	42	25
Italy	32	32
Portugal	27	10
Spain	20	6
Eastern Europe		
Czech Republic	27	24
Hungary	67	50
Poland	9	6
Russia	56	31
Near East		
Israel	62	87
Turkey	55	50

Table 19. (continued)

Country	% Reporting requests	
	Outright request	Suspected request
Latin America		
Argentina	0	50
Brazil	10	7
Chile	13	6
Colombia	0	31
Cuba	43	23
Mexico	13	2
Peru	14	0
Venezuela	13	6
Asia		
China	79	73
India	70	61
Japan	32	22
Thailand	0	20
Total geneticists	47	50
Total excluding US	40	38

Table 20. Counseling About Prenatal Diagnosis for Sex Selection*. Counseling approach

Country	Dissuade	Information only	Support patients' decisions
English-speaking nations			
Australia	47	40	13
Canada	30	53	17
South Africa	36	43	21
United Kingdom	51	44	5
United States	24	47	28
US Primary Care Physicians	38	30	29
US patients	31	—	—
Western Europe			
Belgium	80	7	13
Denmark	41	37	22
Finland	43	52	5
France	75	18	7
Germany	71	21	8
Netherlands	65	31	4
Norway	56	44	6
Sweden	40	60	0
Switzerland	83	17	0
Southern Europe			
Greece	58	17	25
Italy	60	35	5
Portugal	36	55	9
Spain	52	44	4
Eastern Europe			
Czech Republic	42	28	30
Hungary	20	26	54
Poland	61	29	10
Russia	9	23	68
Near East			
Israel	14	59	27
Turkey	55	41	4

Table 20. (continued)

Country	Dissuade	Information only	Support patients' decisions
Latin America			
Argentina	58	33	8
Brazil	59	24	17
Chile	80	7	13
Colombia	43	43	14
Cuba	29	36	36
Mexico	42	44	14
Peru	61	29	10
Venezuela	44	44	12
Asia			
China	64	14	22
India	59	18	23
Japan	50	8	42
Thailand	0	100	0
Total geneticists	44	35	21
Total, excluding US	54	29	17

Table 21. Disclosing Fetal Sex. A couple aged 27 requests prenatal diagnosis. They say that they had a child with down syndrome, who died. They show an unusual interest in learning the sex

Country	Would ask if PND is for sex selection	% Would reveal sex within time limit for legal abortion	Had similar case
English-speaking nations			
Australia	80	47	87
Canada	75	69	58
South Africa	80	67	44
United Kingdom	61	59	37
United States	60	85	75
US Primary Care Physicians	74	82	17
Western Europe			
Belgium	79	64	33
Denmark	52	71	36
Finland	57	26	36
France	73	22	53
Germany	64	7	46
Netherlands	76	52	42
Norway	56	56	22
Sweden	56	56	75
Switzerland	67	50	67
Southern Europe			
Greece	92	46	25
Italy	63	53	32
Portugal	70	44	10
Spain	78	59	6
Eastern Europe			
Czech Republic	78	40	24
Hungary	91	53	50
Poland	72	44	6
Russia	78	91	31

Table 21. (continued)

Country	Would ask if PND is for sex selection	% Would reveal sex within time limit for legal abortion	Had similar case
Near East			
Israel	70	62	87
Turkey	91	27	50
Latin America			
Argentina	86	64	50
Brazil	82	60	7
Chile	94	13	6
Colombia	86	83	31
Cuba	70	73	23
Mexico	87	46	2
Peru	92	20	0
Venezuela	80	40	6
Asia			
China	81	16	73
India	100	4	61
Japan	68	22	22
Thailand	92	16	20
Total geneticists	69	53	50
Total, excluding US	74	37	38

Table 22. Preconception Sex Selection. In absence of X-linked disorder, an accurate method of preconception sex selection, such as sperm separation. % Agreeing

Country	Should be available to everyone, no restrictions	Should be available only to families with 3 or more children of same sex	Would be morally justified	Would benefit society	I would use it myself
English-speaking nations					
Australia	27	7	13	0	7
Canada	28	11	17	4	10
South Africa	19	25	13	13	0
United Kingdom	4	8	6	1	7
United States	38	9	12	2	12
US Primary Care Physicians	24	14	15	7	13
US patients	35	13	—	11	26
US public	40	—	—	—	26
Western Europe					
Belgium	0	7	0	7	7
Denmark	7	7	15	0	7
Finland	27	18	18	5	10
France	4	11	3	0	4
French patients (n = 387)	22	26	n/a	13	20
Germany	4	7	7	0	4
German patients (n = 593)	11	11	n/a	4	10
Netherlands	0	0	0	11	0

Table 22. (continued)

Country	Should be available to everyone, no restrictions	Should be available only to families with 3 or more children of same sex	Would be morally justified	Would benefit society	I would use it myself
Norway	11	0	22	1	0
Sweden	17	0	8	0	0
Switzerland	17	17	0	0	17
Southern Europe					
Greece	18	25	22	9	9
Italy	5	10	5	0	0
Portugal	0	27	18	0	9
Spain	16	14	13	0	8
Eastern Europe					
Czech Republic	13	27	30	5	13
Hungary	8	54	20	8	8
Poland	25	28	23	7	20
Russia	42	57	81	24	45
Near East					
Israel	39	15	17	5	22
Turkey	14	29	5	10	0
Latin America					
Argentina	7	36	7	—	7
Brazil	35	13	22	8	20
Chile	19	20	27	13	27
Colombia	43	21	36	7	36
Cuba	36	39	36	14	29
Mexico	27	45	19	11	24
Peru	40	21	21	31	36
Venezuela	33	38	13	7	38
Asia					
China	5	30	15	17	11
India	5	46	24	29	29
Japan	6	15	5	15	3
Thailand	28	56	60	48	24
Total geneticists	23	16	15	6	12
Total, excluding US	15	21	17	8	12

Table 23. PND for a Deaf Couple Wanting Deaf Child. A deaf couple seek prenatal diagnosis in order to have a deaf child. They will abort a "hearing" fetus

Country	% Would dissuade in counseling	% Would perform PND	% Would refer for PND	% Would refuse PND	% Had similar Case
English-speaking nations					
Australia	40	14	21	64	21
Canada	34	18	33	49	7
South Africa	40	20	13	68	6
United Kingdom	51	9	22	69	1
United States	30	35	33	32	7
US Primary Care Physicians	54	21	29	51	0
US patients	38	47	n/a	43	n/a

Table 23. (continued)

	% Would dissuade in counseling	% Would perform PND	% Would refer for PND	% Would refuse PND	%Had similar case
Northern/Western Europe					
Belgium	86	7	13	80	0
Denmark	40	28	0	72	4
Finland	55	29	5	67	5
France	83	1	1	97	4
French patients (n = 304)	36	7	n/a	58	n/a
Germany	66	8	6	86	2
Netherlands	74	4	18	78	9
Norway	57	0	0	100	25
Sweden	40	9	0	91	8
Switzerland	67	17	0	83	0
Southern Europe					
Greece	55	9	0	91	8
Italy	38	38	10	52	10
Portugal	55	18	9	73	20
Spain	66	13	0	87	8
Eastern Europe					
Czech Republic	62	19	3	78	0
Hungary	75	20	9	71	6
Poland	50	12	0	88	0
Russia	49	33	5	62	5
Near East					
Israel	43	38	5	57	0
Turkey	32	9	14	77	9
Latin America					
Argentina	79	8	8	84	0
Brazil	71	21	7	72	2
Chile	75	12	0	88	7
Colombia	39	31	15	54	0
Cuba	46	43	0	57	0
Mexico	56	16	2	82	3
Peru	64	29	7	64	0
Venezuela	81	6	6	88	7
Asia					
China	72	11	1	88	4
India	85	5	0	95	0
Japan	59	15	1	84	5
Thailand	76	4	0	96	4
Total geneticists	51	22	17	61	5
Total, excluding US	61	14	7	79	4

Table 24. Prenatal Paternity Testing at Mother's Request. A woman is involved with two men and wishes to find out which one is the father so that she can decide whether to have the child. % Would

Country	Perform PND without comment	Perform PND only with comment on possible negative consequences of her actions or requirement of psychological counseling	Refer	Refuse and not refer	Had similar request
English-speaking nations					
Australia	13	20	33	13	40
Canada	20	18	40	18	48
South Africa	0	64	29	0	33
United Kingdom	4	15	44	28	30
United States	32	24	35	6	66
US Primary Care Physicians	21	41	29	9	22
US patients	31	56	n/a	13	n/a
Northern/Western Europe					
Belgium	13	40	0	40	80
Denmark	25	29	18	29	26
Finland	19	24	24	24	33
France	4	24	6	67	26
French patients	18	52	n/a	30	n/a
Germany	1	10	17	65	41
German patients ($n = 593$)	9	73	n/a	18	n/a
Netherlands	4	15	52	26	63
Norway	22	33	0	44	33
Sweden	25	17	0	58	67
Switzerland	0	33	0	50	50
Southern Europe					
Greece	33	50	0	17	9
Italy	20	35	20	25	42
Portugal	46	27	27	0	18
Spain	27	27	10	31	41
Eastern Europe					
Czech Republic	23	31	9	33	10
Hungary	6	44	22	28	24
Poland	24	34	4	35	6
Russia	47	40	7	7	24
Near East					
Israel	52	14	29	0	44
Turkey	45	27	0	23	27
Latin America					
Argentina	7	36	7	36	79
Brazil	34	34	11	19	19
Chile	0	21	7	64	20
Colombia	21	28	21	29	31
Cuba	64	14	0	14	7
Mexico	32	27	14	21	33
Peru	15	54	0	31	14
Venezuela	38	38	0	25	38

Table 24. (continued)

Country	Perform PND without comment	Perform PND only with comment on possible negative consequences of her actions or requirement of psychological counseling	Refer	Refuse and not refer	Had similar request
Asia					
China	13	49	9	27	18
India	10	33	10	48	15
Japan	11	32	11	45	19
Thailand	0	52	12	28	0
Total geneticists	22	27	23	24	42
Total, excluding US	17	29	16	38	28

Table 25. Prenatal Diagnosis Solely for Maternal Anxiety. A 25-year-old woman with no first-hand knowledge of down syndrome requests PND. % would

Country	Perform PND (insurance should pay)	Perform PND only if she can pay out-of-pocket	Refer	Total would perform or refer 1985	Total would perform or refer 1995
English-speaking nations					
Australia	53	13	20	67	86
Canada	18	23	32	70	73
South Africa	44	50	6	—	100
United Kingdom	46	18	22	88	86
United States	30	58	8	88	96
US Primary Care Physicians	21	7	40	—	61
US Patients	78	—	—	—	—
Northern/Western Europe					
Belgium	53	27	7	—	87
Denmark	71	0	7	87	78
Finland	23	41	23	—	87
France	12	47	8	57	67
French patients (n = 383)	77	—	—	—	—
Germany	61	23	6	80	90
Netherlands	26	33	26	—	85
Norway	33	0	0	50	33
Sweden	50	25	0	91	75
Switzerland	0	83	0	100	83
Southern Europe					
Greece	18	64	0	43	82
Italy	14	62	5	82	81
Portugal	37	36	27	—	100
Spain	38	25	2	—	65

Table 25. (continued)

Country	Perform PND (insurance should pay)	Perform PND only if she can pay out-of-pocket	Refer	Total would perform or refer 1985	Total would perform or refer 1995
Eastern Europe					
Czech Republic	8	69	9	—	86
Hungary	9	53	9	20	71
Poland	21	40	8	—	69
Russia	14	50	0	—	64
Near East					
Israel	0	82	18	80	100
Turkey	23	41	9	20	73
Latin America					
Argentina	62	23	8	—	93
Brazil	20	42	11	44	73
Chile	27	0	7	—	34
Colombia	50	36	0	—	86
Cuba	43	14	0	—	57
Mexico	22	20	20	—	62
Peru	15	46	0	—	61
Venezuela	20	47	20	—	87
Asia					
China	6	21	12	—	39
India	4	30	17	56	51
Japan	4	33	12	18	49
Thailand	4	32	20	—	56
Total geneticists	27	40	10	73	77
Total, excluding US	26	32	12	61	71

Table 26. Who Should Pay? % thought public health insurance should pay for

PND for sex selection (couple with 4 daughters desires a son)	6
Preconception sex selection, in absence of X-linked disorders	5
PND for deaf couple wanting a deaf child	10
CF carrier testing	
Premarital	38
Preconception	48
In pregnancy	50
For everyone who asks	26
DNA banking on behalf of a person's descendants	54

Table 27. Late Abortion. An abnormality is diagnosed by ultrasound at 23 weeks. The woman cannot get a legal abortion at your institution. % would offer referral, if condition were

Country	Trisomy 18		Trisomy 21 (Down syndrome)		45,X (Turner syndrome)	
	In country/state	Outside country/state	In country/state	Outside country/state	In country/state	Outside country/state
English-speaking nations						
Australia	93	29	80	29	73	14
Canada	93	77	88	74	82	67
South Africa	93	64	93	64	87	50
United Kingdom	92	45	92	45	88	35
United States	98	59	97	58	93	55
US Primary Care Physicians[1]	n/a	93	n/a	n/a	n/a	n/a
US patients[2]	n/a	n/a	n/a	84	n/a	n/a
Northern/Western Europe						
Belgium	92	23	92	23	77	23
Denmark	92	39	92	39	72	22
Finland	52	10	48	10	38	10
France	97	48	93	46	62	25
Germany	64	39	56	36	33	24
Netherlands	72	57	64	48	60	48
Norway	88	29	75	14	50	14
Sweden	33	44	33	44	22	33
Switzerland	100	50	100	50	83	33
Southern Europe						
Greece	92	44	92	44	64	22
Italy	75	42	75	42	60	37
Portugal	91	64	91	64	78	60
Spain	61	41	56	43	33	32

Eastern Europe						
Czech Republic	89	34	88	31	46	22
Hungary	77	18	77	18	37	14
Poland	62	34	55	34	37	21
Russia	95	42	98	42	81	28
Near East						
Israel	86	39	95	44	91	37
Turkey	90	29	85	29	75	35
Latin America						
Argentina	42	46	36	36	18	18
Brazil	82	36	75	31	65	29
Chile	13	6	13	6	6	0
Colombia	69	27	54	18	33	25
Cuba	85	30	85	30	83	33
Mexico	71	31	58	29	30	16
Peru	46	25	36	13	27	0
Venezuela	69	43	69	29	27	25
Asia						
China	94	12	95	13	88	12
India	95	16	86	20	67	10
Japan	43	9	35	8	23	8
Thailand	80	24	70	12	24	16
Total geneticists	86	45	83	43	71	37
Total, excluding US	78	35	74	33	57	25

[1]Question described severe hydrocephalus and severe open spina bifida, not trisomy 18.
[2]Question described severe mental and physical disabilities, with normal lifespan, not trisomy 21.

III. RESULTS

Table 28. Abortion Laws

Country	Fetal abnormality	Social or maternal indications (except to save mother's life, rape or incest)
English-speaking nations		
Australia	Legal	
Canada	No national laws. May be difficult to obtain after 22 weeks	No national laws
South Africa	Legal, no time limit for severe disorders	Legal to 12 weeks; with a medical opinion to 20 weeks
United Kingdom	Legal	Legal
United States	Legal. No indications required before 24 weeks. Some states have no time limit; in others, difficult to obtain after 20–22 weeks	Legal. No indications required before 24 weeks. Difficult to obtain after 20–22 weeks
Northern/Western Europe		
Denmark	Legal without limit of time	Legal to 12 weeks
France	Legal without limit of time	Legal to 12 weeks
Germany	Illegal – no fetal indications	Legal
Netherlands	Legal	Legal
Norway	Legal without limit of time	Legal to 12 weeks
Sweden	Legal, no time limit. Requires special permission after 18 weeks	Legal to 18 weeks
Southern Europe		
Greece	Legal to 24 weeks	Legal to 12 weeks; 19 weeks for rape or incest
Italy	Legal to 24 weeks	Legal to 12 weeks
Portugal	Legal to 22 weeks. No time limit for anencephaly	Illegal, national referendum (49–51%)
Spain	Legal to 22 weeks	Legal for mother's health (no time limit); legal for rape to 12 weeks
Eastern Europe		
Czech Republic	Legal to 24 weeks; later if lethal defect	Legal to 12 weeks
Hungary	Legal to 20 weeks; no time limit for lethal disorders	Legal to 12 weeks
Poland	Legal for severe disorders	Illegal
Russia	Legal	Legal
Near East		
Israel	Legal, no time limit	Legal
Turkey	Legal, no time limit	Legal to 10th week; couple (not woman alone) must consent
Latin America	Illegal everywhere except Cuba	Illegal, except to save mother's life, or for rape, everywhere except Cuba. Illegal abortions are easily obtained
Cuba	Legal	Legal
Asia		
China	Legal, no time limit	Legal, no time limit
India	Legal to 20 weeks	Legal to 20 weeks
Japan	Illegal, easily obtained under maternal or social indications	Legal, no time limit for mother's health
Thailand	No laws	No laws

Table 29. Views on Abortion Laws. % agreeing with current laws in own country for

Country	Genetic abnormalities	Social indications
English-speaking nations		
Australia	73	80
Canada	86	78
South Africa	88	25
United Kingdom	78	84
United States	76	83
Northern / Western Europe		
Belgium	71	93
Denmark	89	100
Finland	100	95
France	86	85
Germany	85	54
Netherlands	77	96
Norway	100	75
Sweden	100	91
Switzerland	67	50
Southern Europe		
Greece	80	78
Italy	65	70
Portugal	0	10
Spain	57	43
Eastern Europe		
Czech Republic	87	90
Hungary	97	87
Poland	55	28
Russia	95	93
Near East		
Israel	91	55
Turkey	79	86
Latin America		
Argentina	14	21
Brazil	0	4
Chile	60	71
Colombia	0	0
Cuba	86	100
Mexico	12	15
Peru	18	36
Venezuela	7	13
Asia		
China	96	97
India	81	90
Japan	55	69
Thailand	33	17
Total geneticists	73	71
Total, excluding US	72	65

Table 30. Conditions for Which Abortion Should be Illegal. % thought abortion should be illegal for

Country	Fetus not the sex desired by parents		Severe untreatable obesity	Cleft lip/ plate	Predisposition to			HD	45,X
	1985	1995			Alcoholism	Alzheimer	Mental illness		
English-speaking nations									
Australia	100	93	27	13	21	0	0	0	0
Canada	66	58	17	22	20	15	13	4	4
South Africa	—	81	13	38	19	6	6	0	6
United Kingdom	83	67	25	18	19	12	10	0	3
United States	50	32	10	9	10	8	6	2	3
US Primary Care Physicians	—	47	25	22	21	19	15	7	12
US patients	—	53	25	25	26	27	6	13	25
US public	—	58	35	n/a	n/a	n/a	n/a	28	n/a
Northern/Western Europe									
Belgium	—	80	40	47	47	7	20	0	0
Denmark	86	63	35	30	30	7	7	0	7
Finland	—	91	33	29	29	14	24	0	0
France	100	92	53	65	66	36	32	14	22
Germany	98	90	30	52	44	31	31	8	18
Netherlands	—	85	44	42	36	12	23	4	0
Norway	100	88	50	38	63	50	13	0	13
Sweden	90	42	25	33	18	0	9	0	0
Switzerland	100	50	33	17	33	20	33	0	0
Southern Europe									
Greece	100	67	17	42	25	8	17	0	0
Italy	91	81	25	33	25	24	19	10	10
Portugal	—	100	64	55	55	40	40	9	18
Spain	—	73	60	49	55	43	38	26	34

Eastern Europe									
Czech Republic	—	84	17	25	28	9	15	4	7
Hungary	100	83	49	34	60	47	29	9	11
Poland	—	74	49	50	49	30	33	20	25
Russia	—	23	5	7	12	8	7	2	0
Near East									
Israel	86	52	19	0	10	14	9	0	0
Turkey	100	73	23	36	23	5	0	0	5
Latin America									
Argentina	—	93	79	85	71	64	71	46	33
Brazil	81	57	28	25	33	21	18	7	11
Chile	—	100	94	94	94	81	88	75	81
Colombia	—	57	36	50	43	36	43	29	29
Cuba	—	57	21	0	15	0	7	0	0
Mexico	—	65	48	47	46	29	27	18	28
Peru	—	69	69	69	69	54	62	39	62
Venezuela	—	82	67	58	58	42	46	21	29
Asia									
China	—	21	4	4	5	6	1	1	2
India	81	65	36	30	27	13	13	9	0
Japan	100	73	54	56	46	38	31	28	28
Thailand	—	92	84	84	72	44	56	28	36
Total geneticists	71	54	24	26	25	17	16	7	10
Total, excluding US	89	67	32	34	35	23	22	10	14

Table 31. Pregnant Women's Autonomy: Counseling for Maternal PKU. A pregnant woman with PKU is unwilling to stay on a special diet that may protect her fetus from retardation. % would

Country	Arrange involuntary hospitalization in first trimester and impose diet	Suggest abortion if she reaches second trimester without controlling diet (in parentheses: ask her spouse to tell her to abort)	Tell her about options of carrying to term or aborting	Offer to support whatever decision she makes
English-speaking nations				
Australia	20	40	13	47
Canada	15	7 (1)	23	69
South Africa	0	20 (7)	40	40
United Kingdom	9	4	25	70
United States	16	11 (1)	30	59
US Primary Care Physicians	21	18 (3)	30	39
US patients	21	43	—	—
US public	22	19	n/a	n/a
Northern/Western Europe				
Belgium	13	20	20	60
Denmark	4	36 (4)	14	50
Finland	5	0	18	77
France	3	33 (7)	39	24
French patients (n = 357)	40	62		
Germany	4	12 (2)	25	62
German patients (n = 593)	27	51		
Netherlands	0	4	20	68
Norway	11	22	56	22
Sweden	9	27	18	56
Switzerland	0	−17	0	83
Southern Europe				
Greece	25	58	17	25
Italy	22	15	5	80
Portugal	18	18	36	46
Spain	43	2	26	61
Eastern Europe				
Czech Republic	20	40 (6)	30	31
Hungary	26	12 (3)	23	63
Poland	31	11 (3)	20	54
Russia	38	49 (18)	27	24
Near East				
Israel	25	26 (4)	9	65
Turkey	46	50 (9)	4	46
Latin America				
Argentina	21	0	17	67
Brazil	21	14 (4)	24	60
Chile	47	0	13	6
Colombia	23	8	0	92
Cuba	43	36	7	57

Table 31. (continued)

Country	Arrange involuntary hospitalization in first trimester and impose diet	Suggest abortion if she reaches second trimester without controlling diet (in parentheses: ask her spouse to tell her to abort)	Tell her about options of carrying to term or aborting	Offer to support whatever decision she makes
Mexico	30	7	16	71
Peru	36	0	50	36
Venezuela	33	13	7	80
Asia				
China	78	92 (17)	2	6
India	57	65 (30)	17	17
Japan	28	39 (32)	20	40
Thailand	33	30 (8)	25	38
Total geneticists	23	22 (5)	24	52
Total, excluding US	27	29 (7)	20	47

Table 32. Reasoning in Maternal PKU Case. % reporting extremely or very important

Preventing harm to fetus	83
Professional responsibility to offer guidance	81
A child's right to be born healthy	77
A woman's right to control her own body	58
Avoiding forced medical/social control of pregnant women	51
Avoiding medical paternalism	48

Table 33. Parenthood for Persons with Disabilities at Risk for Having Children with Disabilities. % would support parental decision to take their chances of having a child with a disability

Country	Blind Couple, 100% chance of blindness in child	Woman with PKU, uncontrolled phenylalanine level in 3rd trimester, 100% chance of birth defects	Cystic fibrosis carrier–carrier couple, 25% chance of CF, without prenatal diagnosis	Woman with Fragile-X, living in an institution
English-speaking				
Australia	93	47	80	43
Canada	93	69	91	65
South Africa	88	40	75	55
United Kingdom	96	70	97	84
United States	94	59	89	53
US Primary Care Physicians	62	39	63	N/A
US patients	95	N/A	62	N/A
Northern/Western Europe				
Belgium	67	60	36	50
Denmark	85	50	54	50
Finland	86	81	91	71
France	75	28	68	45
Germany	90	62	65	57
Netherlands	78	76	70	65
Norway	78	22	100	38
Sweden	75	55	75	58
Switzerland	83	83	50	50
Southern Europe				
Greece	73	25	18	36
Italy	60	80	55	50
Portugal	73	46	64	56
Spain	63	73	48	66
Eastern Europe				
Czech Republic	37	31	17	32
Hungary	81	66	29	36

Poland	50	70	33	51
Russia	31	24	17	33
Near East				
Israel	74	65	68	78
Turkey	73	46	32	46
Latin America				
Argentina	93	83	93	71
Brazil	65	63	63	41
Chile	56	88	81	46
Colombia	71	92	79	50
Cuba	50	57	50	56
Mexico	60	77	65	36
Peru	43	50	58	33
Venezuela	73	80	63	88
Asia				
China	6	6	2	27
India	27	17	30	58
Japan	51	40	57	55
Thailand	44	46	29	20
Total geneticists	73	52	63	51
Total, excluding US	61	47	50	50

Table 34. What is Fair? % agreed that knowingly having a child with a serious genetic disorder is not fair

Country	To the child	To the child's siblings	To society ("socially irresponsible in an era of PND")
English-speaking nations			
Australia	27	7	7
Canada	40	20	18
South Africa	75	63	50
United Kingdom	36	19	21
United States	40	22	26
US Primary Care Physicians	52	32	55
US patients (n = 476)	46	25	26
US public	n/a	n/a	46
Northern/Western Europe			
Belgium	86	43	50
Denmark	67	22	19
Finland	36	27	10
France	51	40	30
French patients (n = 369)	77	n/a	56
Germany	18	10	8
German patients (n = 593)	48	21	27
Netherlands	33	16	11
Norway	56	44	11
Sweden	25	25	17
Switzerland	33	17	17
Southern Europe			
Greece	100	67	80
Italy	55	14	43
Portugal	72	27	55
Spain	58	40	45
Eastern Europe			
Czech Republic	77	43	65
Hungary	86	43	85
Poland	72	32	53
Russia	96	54	67
Near East			
Israel	68	59	68
Turkey	91	65	86
Latin America			
Argentina	29	50	14
Brazil	63	31	56
Chile	40	25	0
Colombia	64	33	50
Cuba	92	86	93
Mexico	82	73	78
Peru	86	71	79
Venezuela	94	63	56
Asia			
China	94	81	93
India	100	96	96
Japan	18	15	20
Thailand	76	13	60
Total geneticists	51	32	38
Total, excluding US	58	39	45

Table 35. Nontreatment of Genetically Handicapped Newborns. % would support parents' refusal of a lifesaving operation on a newborn with

Country	Trisomy 13	Severe, open spina bifida	Trisomy 21	45,X
English-speaking nations				
Australia	100	93	67	40
Canada	93	82	62	37
South Africa	100	94	69	27
United Kingdom	97	95	76	39
United States	98	79	56	39
US Primary Care Physicians	83	67	47	37
US patients*	70	25	33	N/A
Northern/Western Europe				
Belgium	100	100	80	33
Denmark	100	73	62	15
Finland	91	76	57	29
France	99	88	72	19
Germany	88	70	26	8
Netherlands	100	93	56	22
Norway	100	89	67	44
Sweden	83	67	67	33
Switzerland	100	67	50	33
Southern Europe				
Greece	75	75	58	25
Italy	76	52	29	14
Portugal	100	91	64	27
Spain	83	61	46	17
Eastern Europe				
Czech Republic	83	79	70	20
Hungary	88	85	41	15
Poland	62	63	30	25
Russia	88	88	75	35
Near East				
Israel	100	81	64	33
Turkey	82	73	50	27
Latin America				
Argentina	100	50	33	8
Brazil	76	59	39	19
Chile	56	31	13	6
Colombia	100	71	57	29
Cuba	64	43	43	29
Mexico	82	73	47	19
Peru	50	57	23	8
Venezuela	87	71	42	8
Asia				
China	68	68	65	50
India	61	56	35	35
Japan	77	55	32	27
Thailand	82	67	76	57
Total geneticists	87	74	52	31
Total, excluding US	82	72	50	27
Total geneticists would discuss giving up child for adoption	22	37	54	64
Total would discuss adoption, excluding US	14	23	39	50

*% agreeing with parents' decisions. Seventy-eight percent of patients also said doctor should *support* parents' decisions.

C. Directiveness in Counseling

Table 36. Preconception Counseling and Reproductive Alternatives. Both members of a couple carry a cystic fibrosis mutation. They oppose PND and selective abortion. % would provide information, unasked, about

Country	Adoption	Donor insemination	IVF with donated egg	Preimplantation diagnosis
English-speaking nations				
Australia	93	86	71	60
Canada	97	93	73	62
South Africa	87	60	53	50
United Kingdom	91	92	81	79
United States	95	96	87	84
US Primary Care Physicians	86	62	53	47
US patients	81	74	73	78
US public	71	54	n/a	n/a
Northern/Western Europe				
Belgium	85	92	85	70
Denmark	88	84	67	65
Finland	82	68	55	55
France	74	63	40	41
French patients	95	86	65	80
Germany	71	48	25	31
German patients ($n = 593$)	86	68	n/a	75
Netherlands	93	100	54	46
Norway	67	89	11	44
Sweden	50	50	17	50
Switzerland	83	83	33	17
Southern Europe				
Greece	60	73	46	50
Italy	71	50	41	37
Portugal	55	82	36	40
Spain	70	78	66	56
Eastern Europe				
Czech Republic	80	76	44	45
Hungary	77	82	42	41
Poland	68	45	36	38
Russia	37	71	44	49
Near East				
Israel	68	86	60	68
Turkey	57	10	10	43
Latin America				
Argentina	64	43	36	21
Brazil	70	49	44	49
Chile	63	19	13	0
Colombia	86	71	71	50
Cuba	73	82	82	60
Mexico	73	64	59	42
Peru	77	50	50	17
Venezuela	64	73	57	43
Asia				
China	41	20	19	22
India	64	46	32	59
Japan	46	23	20	28
Thailand	72	36	32	24
Total geneticists	77	69	56	55
Total, excluding US	68	56	41	4

Table 37. Views on Surrogacy. % would present option, unasked, to

Country	Woman with PKU who wants children	% thought surrogacy morally justified for autosomal dominant or X-linked disorders	CF carrier–carrier couple
English-speaking nations			
Australia	57	43	27
Canada	51	51	34
South Africa	60	25	27
United Kingdom	36	28	25
United States	78	74	58
US Primary Care Physicians	56	62	34
US patients	n/a	52	57
US public	n/a	n/a	46
Northern/Western Europe			
Belgium	14	14	8
Denmark	36	44	28
Finland	55	40	27
France	14	25	5
French patients	n/a	n/a	51
Germany*	26	25	18
German patients	n/a	n/a	55
Netherlands	46	33	19
Norway	11	11	11
Sweden	46	50	17
Switzerland	50	17	0
Southern Europe			
Greece	33	46	30
Italy	19	33	16
Portugal	70	30	18
Spain	52	34	28
Eastern Europe			
Czech Republic	44	64	25
Hungary	42	43	49
Poland	43	58	30
Russia	52	63	30
Near East			
Israel	47	58	29
Turkey	67	45	24
Latin America			
Argentina	15	29	8
Brazil	39	38	27
Chile	13	19	0
Colombia	82	43	36
Cuba	42	36	50
Mexico	63	42	35
Peru	67	46	8
Venezuela	63	50	46
Asia			
China	58	59	14
India	59	68	36
Japan	32	18	23
Thailand	52	40	12
Total geneticists	54	54	35
Total, excluding US	42	42	23

Table 38. Directiveness in Counseling After Prenatal Diagnosis for Conditions Appearing in Childhood, Part 1. % would provide unbiased counseling for

Country	Anencephaly	Severe, open spina bifida	Trisomy 13	Trisomy 21	CF	Achondroplasia	Sickle cell anemia	Child not of desired sex
English-speaking nations								
Australia	40	53	40	73	87	80	87	7
Canada	47	70	65	90	92	91	86	40
South Africa	0	31	25	56	50	67	73	13
United Kingdom	41	63	61	84	87	87	88	27
United States	51	70	63	86	88	88	88	46
US Primary Care Physicians	n/a	51	28	63	61	54	58	23
Northern/Western Europe								
Belgium	20	23	20	33	39	79	36	13
Denmark	31	32	39	54	58	69	68	27
Finland	38	52	48	71	71	81	71	10
France	3	15	10	28	32	43	50	3
Germany	16	49	38	68	70	69	69	11
Netherlands	39	46	44	72	69	70	80	22
Norway	44	67	56	78	67	89	78	44
Sweden	17	33	18	42	42	55	42	25
Switzerland	17	67	33	67	67	83	67	0
Southern Europe								
Greece	0	9	0	9	9	9	27	8
Italy	43	41	43	48	59	57	59	33
Portugal	36	36	36	46	55	46	55	46
Spain	23	36	35	35	38	23	36	12
Eastern Europe								
Czech Republic	6	7	6	5	12	23	31	9
Hungary	9	3	9	19	11	26	33	17

Poland	17	27	31	47	45	51	45	14
Russia	7	12	7	9	29	21	41	20
Near East								
Israel	27	23	27	27	54	46	59	36
Turkey	18	17	19	16	28	35	32	18
Latin America								
Argentina	23	46	25	54	62	54	46	8
Brazil	29	47	36	52	58	59	60	32
Chile	43	44	50	31	38	0	33	7
Colombia	27	46	42	50	50	50	50	25
Cuba	0	0	0	0	7	21	7	21
Mexico	8	10	13	16	21	24	18	15
Peru	0	7	0	7	15	7	15	0
Venezuela	13	20	19	29	33	43	29	7
Asia								
China	1	1	1	2	4	6	33	11
India	0	0	14	23	18	26	23	9
Japan	12	42	25	48	48	46	48	16
Thailand	4	8	8	4	16	20	20	0
Total geneticists	28	46	41	58	61	62	60	28
Total, excluding US	18	31	28	42	47	47	47	17

Table 39. Directiveness in Counseling after Prenatal Diagnosis for Conditions Appearing in Childhood, Part 2. % would provide unbiased counseling for

Country	NF	XXY	45,X	PKU in fetus	Severe obesity	Cleft lip/palate
English-speaking nations						
Australia	87	73	60	80	93	53
Canada	87	85	84	77	78	66
South Africa	81	75	75	50	75	56
United Kingdom	80	72	73	72	65	48
United States	89	84	81	77	81	67
US Primary Care Physicians	61	66	58	52	47	39
Northern/Western Europe						
Belgium	77	54	62	54	39	40
Denmark	69	81	85	58	69	65
Finland	67	62	67	67	52	48
France	56	34	39	23	20	8
Germany	64	42	41	46	56	32
Netherlands	72	67	67	67	57	48
Norway	78	100	100	56	67	56
Sweden	58	67	67	42	55	58
Switzerland	83	33	50	50	33	33
Southern Europe						
Greece	27	36	18	9	27	8
Italy	67	55	59	50	55	48
Portugal	73	73	82	55	55	55
Spain	32	19	17	32	26	25
Eastern Europe						
Czech Republic	31	36	44	30	51	42
Hungary	28	33	39	17	53	43
Poland	53	45	43	39	30	29
Russia	39	31	34	21	59	41
Near East						
Israel	77	73	71	50	67	91
Turkey	47	58	47	39	50	59
Latin America						
Argentina	39	15	23	15	31	8
Brazil	63	58	57	55	64	46
Chile	13	25	25	27	25	19
Colombia	42	42	42	42	33	25
Cuba	21	21	7	7	14	29
Mexico	23	24	24	16	18	13
Peru	0	7	7	0	0	23
Venezuela	27	29	27	14	43	6
Asia						
China	10	4	7	6	27	12
India	48	27	41	17	61	26
Japan	49	51	43	41	42	26
Thailand	32	20	21	4	12	8
Total geneticists	61	58	57	49	49	43
Total, excluding US	49	43	43	37	46	33

NF = neurofibromatosis; XXY = Klinefelter syndrome; 45,X = Turner syndrome.
Severe obesity = severe, untreatable obesity in absence of a known genetic syndrome.
Cleft lip/palate was asked twice, for a girl and a boy. Answers did not differ.

Table 40. Directiveness in Counseling after Prenatal Diagnosis for Adult-Onset Disorders. % would provide unbiased counseling for

Country	FH	HD	Predisposition to mental illness	Predisposition to alcoholism	Predisposition to Alzheimer disease
English-speaking nations					
Australia	80	80	87	87	80
Canada	81	88	86	81	82
South Africa	69	60	81	81	75
United Kingdom	76	87	79	76	83
United States	83	86	86	83	85
US Primary Care Physicians	50	61	65	57	60
Northern/Western Europe					
Belgium	64	57	77	64	64
Denmark	62	65	81	80	86
Finland	67	57	76	71	72
France	49	54	58	31	49
Germany	63	71	56	51	57
Netherlands	69	83	62	67	72
Norway	89	78	89	100	89
Sweden	50	58	75	92	68
Switzerland	67	80	67	33	50
Southern Europe					
Greece	36	9	36	36	27
Italy	55	64	73	73	68
Portugal	91	64	82	73	73
Spain	33	48	43	35	35
Eastern Europe					
Czech Republic	39	30	54	48	44
Hungary	42	25	53	61	44
Poland	49	51	54	51	49
Russia	65	33	58	50	42
Near East					
Israel	64	41	91	91	77
Turkey	70	40	56	45	42
Latin America					
Argentina	23	39	54	31	31
Brazil	59	68	65	72	71
Chile	27	38	25	19	25
Colombia	42	42	33	33	33
Cuba	31	14	23	23	17
Mexico	17	21	27	26	24
Peru	14	7	0	7	7
Venezuela	29	23	29	29	29
Asia					
China	8	3	13	19	21
India	33	26	46	41	39
Japan	56	54	69	57	60
Thailand	12	12	28	28	16
Total geneticists	59	63	65	63	64
Total, excluding US	48	49	53	51	51

FH = familial hypercholesterolemia, homozygous; HD = Huntington disease; Mental illness = schizophrenia or bipolar disorder.

Table 41. Directiveness in Counseling for Maternal Situations or Fetal Infections. % would provide unbiased counseling for

Country	Mother's life in danger	Pregnancy result of rape	HIV infection in fetus, first trimester	Toxoplasmosis in fetus, first trimester	Rubella in fetus, first trimester
English-speaking nations					
Australia	60	87	47	57	53
Canada	46	82	67	63	61
South Africa	31	63	33	50	44
United Kingdom	41	88	70	73	68
United States	44	85	69	75	73
US Primary Care Physicians	33	66	43	44	41
Northern/Western Europe					
Belgium	7	27	33	47	47
Denmark	12	60	27	46	35
Finland	24	62	43	57	52
France	34	53	40	49	34
Germany	29	77	63	59	54
Netherlands	41	59	67	42	39
Norway	33	78	44	56	56
Sweden	17	42	25	42	42
Switzerland	0	33	50	67	67
Southern Europe					
Greece	0	40	8	8	8
Italy	40	79	60	57	57
Portugal	36	64	27	46	36
Spain	46	59	41	34	34
Eastern Europe					
Czech Republic	12	30	10	17	11
Hungary	9	35	6	3	11
Poland	30	55	43	38	37
Russia	40	47	16	21	12
Near East					
Israel	14	38	27	27	27
Turkey	27	65	23	32	23
Latin America					
Argentina	23	36	58	25	39
Brazil	31	65	51	53	51
Chile	38	25	27	33	27
Colombia	39	55	42	33	33
Cuba	8	7	0	0	0
Mexico	17	29	13	16	14
Peru	7	25	15	8	8
Venezuela	13	50	27	14	21
Asia					
China	8	13	10	7	8
India	9	35	17	18	9
Japan	13	30	41	42	38
Thailand	4	36	12	12	8
Total geneticists	32	64	47	49	46
Total, excluding US	25	51	38	37	34

Table 42. Directively Pessimistic Counseling After Prenatal Diagnosis, Part 1. % would give pessimistically slanted information* (in parentheses and included in preceding figure): % would urge termination of pregnancy

Country	Severe, open spina bifida	Trisomy 21	Cystic fibrosis	Sickle cell anemia	Achondroplasia	XXY (Klinefelter syndrome)
English-speaking nations						
Australia	47 (13)	27	13	7	20	0
Canada	30 (7)	9 (2)	5	8 (1)	4 (1)	2 (2)
South Africa	69 (13)	44 (25)	44 (13)	13 (7)	25	13
United Kingdom	38 (3)	14 (2)	10	6	4	0
United States	28 (5)	13 (3)	9 (1)	6	6 (1)	5
US Primary Care Physicians	42 (20)	23 (10)	21 (8)	18 (5)	n/a	15 (6)
Northern/Western Europe						
Belgium	80 (33)	71 (14)	67 (20)	53 (13)	27 (7)	7 (7)
Denmark	68 (36)	42 (23)	42 (23)	24 (8)	27 (15)	8
Finland	48 (10)	24 (5)	29 (10)	39 (5)	14	5
France	82 (27)	66 (21)	59 (22)	38 (12)	20 (7)	1 (1)
Germany	49 (21)	23 (6)	23 (6)	15 (5)	13 (3)	3
Netherlands	58 (12)	19 (4)	19	11	15 (7)	0
Norway	33 (22)	22	22	22	11	0
Sweden	66 (8)	58 (8)	58	58 (8)	36 (9)	0
Switzerland	33 (17)	33 (17)	33 (17)	33 (17)	17	17 (17)
Southern Europe						
Greece	92 (50)	83 (50)	92 (50)	75 (33)	83 (50)	50 (17)
Italy	57 (19)	50 (15)	38 (10)	24 (10)	15	14
Portugal	64 (36)	46 (18)	36 (18)	36	55 (9)	27 (18)
Spain	55 (47)	40 (27)	36 (30)	27 (18)	25 (15)	13 (2)
Eastern Europe						
Czech Republic	91 (61)	92 (62)	86 (38)	63 (24)	74 (38)	45 (21)
Hungary	83 (50)	75 (42)	75 (36)	50 (11)	54 (29)	44 (8)
Poland	65 (32)	35 (11)	37 (9)	27 (5)	23 (7)	22 (6)
Russia	88 (67)	89 (68)	69 (29)	60 (21)	77 (30)	64 (38)

Table 42. (continued)

Country	Severe, open spina bifida	Trisomy 21	Cystic fibrosis	Sickle cell anemia	Achondroplasia	XXY (Klinefelter syndrome)
Near East						
Israel	77 (23)	73 (18)	45 (5)	41 (9)	55 (14)	18
Turkey	77 (27)	77 (32)	64 (23)	55 (23)	51 (5)	50 (14)
Latin America						
Argentina	23 (8)	23 (8)	0	8 (8)	0	0
Brazil	49 (25)	35 (15)	32 (10)	21 (3)	23 (10)	21 (7)
Chile	13	0	0	0	0	0
Colombia	46 (27)	50 (33)	25 (17)	0	17	17
Cuba	100 (71)	93 (50)	93 (64)	93 (64)	71 (36)	84 (29)
Mexico	82 (67)	59 (44)	47 (29)	35 (18)	19 (11)	16 (13)
Peru	79 (43)	29 (21)	31 (23)	31 (23)	14 (14)	0
Venezuela	67 (47)	50 (36)	47 (40)	50 (36)	21 (7)	7 (7)
Asia						
China	98 (89)	96 (90)	95 (82)	91 (67)	92 (77)	92 (73)
India	100 (74)	73 (59)	82 (55)	68 (36)	57 (22)	68 (32)
Japan	49 (16)	37 (15)	44 (12)	24 (5)	37 (20)	23 (9)
Thailand	84 (12)	76 (24)	76 (16)	72 (8)	56 (16)	44 (8)
Total geneticists	52 (25)	36 (20)	32 (15)	28 (12)	24 (12)	19 (10)
Total, excluding US	65 (36)	49 (29)	46 (24)	37 (17)	36 (20)	27 (16)

*On questionnaire: "Emphasize negative aspects so they will favor termination without suggesting it directly."

Table 43. Directively Pessimistic Counseling. Part 2: % would give pessimistically slanted information* (In Parentheses and included in preceding percent: % would urge termination of pregnancy)

Country	Anencephaly	Trisomy 13	Hurler syndrome	FH**	HD***	PKU in fetus****
English-speaking nations						
Australia	60 (33)	60 (27)	53 (7)	7	20 (7)	7
Canada	53 (18)	35 (8)	30 (6)	13 (2)	9 (1)	6
South Africa	94 (56)	75 (38)	56 (25)	25 (6)	40 (13)	38 (19)
United Kingdom	59 (14)	39 (3)	23 (1)	10	10 (4)	7 (1)
United States	49 (20)	37 (10)	27 (5)	11 (1)	10 (2)	7 (1)
US Primary Care Physicians	n/a	70 (47)	n/a	3 (1)	24 (10)	19 (7)
Northern/Western Europe						
Belgium	80 (47)	80 (40)	73 (13)	20	27	40
Denmark	69 (50)	62 (35)	52 (24)	19 (12)	31 (15)	15 (12)
Finland	62 (19)	52 (5)	38 (5)	19	38 (10)	29
France	95 (69)	89 (45)	70 (30)	18 (1)	25 (9)	13 (3)
Germany	74 (58)	61 (37)	42 (16)	9 (2)	16 (6)	6 (1)
Netherlands	62 (15)	56 (11)	31 (8)	15	12 (4)	15
Norway	56 (11)	44 (22)	44 (22)	11	22	33
Sweden	83 (50)	82 (18)	64 (9)	42	33	17
Switzerland	83 (33)	67 (17)	33 (17)	17	20 (20)	33 (17)
Southern Europe						
Greece	100 (83)	100 (67)	92 (58)	42 (8)	92 (50)	58 (17)
Italy	57 (38)	57 (24)	57 (24)	24 (5)	24	10
Portugal	64 (46)	64 (36)	64 (36)	0	36 (18)	18 (9)
Spain	68 (60)	58 (42)	51 (32)	21 (8)	25 (8)	23 (11)
Eastern Europe						
Czech Republic	94 (74)	94 (61)	88 (47)	51 (18)	68 (27)	54 (24)
Hungary	86 (71)	85 (47)	79 (27)	31 (8)	64 (14)	46 (14)
Poland	77 (52)	61 (31)	50 (20)	21 (2)	29 (8)	16 (5)
Russia	93 (84)	91 (74)	83 (54)	30 (12)	62 (33)	70 (49)
Near East						
Israel	73 (36)	73 (14)	73 (18)	27	59 (9)	27 (9)
Turkey	82 (46)	81 (33)	82 (36)	27 (9)	50 (23)	45 (23)
Latin America						
Argentina	54 (46)	58 (42)	39 (23)	15 (8)	23 (8)	0
Brazil	71 (49)	63 (32)	53 (17)	17 (4)	24 (10)	22 (8)
Chile	21	13	13	0	0	0
Colombia	73 (55)	58 (50)	58 (42)	8 (8)	17 (8)	33 (17)
Cuba	100 (71)	100 (79)	100 (79)	54 (31)	57 (43)	71 (36)
Mexico	89 (83)	83 (70)	69 (56)	29 (14)	35 (27)	25 (18)
Peru	85 (62)	69 (54)	69 (46)	14 (7)	36 (14)	21 (14)
Venezuela	88 (69)	81 (56)	68 (47)	29 (21)	39 (23)	21 (14)
Asia						
China	97 (93)	98 (91)	96 (85)	86 (56)	94 (73)	86 (68)
India	100 (83)	86 (57)	80 (50)	52 (29)	61 (35)	74 (44)
Japan	88 (60)	72 (29)	54 (22)	17 (6)	32 (6)	23 (8)
Thailand	92 (54)	84 (40)	77 (27)	64 (8)	76 (24)	60 (12)
Total geneticists	70 (47)	58 (31)	50 (24)	24 (9)	28 (13)	24 (12)
Total, excluding US	79 (58)	70 (42)	60 (32)	30 (13)	39 (20)	32 (17)

*On questionnaire: "Emphasize negative aspects so they will favor termination without suggesting it directly." FH = familial hypercholesterolemia, homozygous; HD = Huntington disease; PKU = phenylketonuria.

Table 44. Directively Pessimistic Counseling for Maternal Conditions or Infections. % would give pessimistically slanted information* (In parentheses: Urge termination)

Country	Mother's life in danger	Pregnancy result of rape	HIV infection in fetus	Toxoplasmosis in fetus, first trimester	Rubella, first trimester
English-speaking nations					
Australia	40 (33)	13	53 (27)	43 (21)	47 (20)
Canada	54 (34)	17 (7)	30 (11)	35 (12)	37 (12)
South Africa	69 (44)	38 (25)	67 (27)	50 (25)	56 (31)
United Kingdom	59 (33)	12 (2)	29 (4)	26 (4)	31 (4)
United States	56 (35)	14 (6)	30 (10)	24 (7)	27 (8)
US Primary Care Physicians	66 (50)	28 (17)	43 (25)	44 (21)	49 (25)
Northern/Western Europe					
Belgium	66 (39)	60 (20)	60 (13)	40 (7)	53 (20)
Denmark	89 (69)	40 (28)	65 (50)	50 (39)	62 (42)
Finland	86 (43)	29 (14)	52 (14)	34 (5)	43 (14)
France	64 (38)	34 (12)	49 (25)	29 (10)	56 (25)
Germany	69 (53)	17 (10)	25 (13)	23 (8)	40 (13)
Netherlands	56 (37)	41 (7)	33 (11)	50 (12)	54 (8)
Norway	67 (56)	22 (11)	44 (33)	44 (11)	44 (22)
Sweden	83 (50)	58 (8)	75 (8)	50 (17)	58 (17)
Switzerland	100 (83)	50	50 (17)	17 (17)	33 (17)
Southern Europe					
Greece	92 (75)	50 (30)	83 (83)	92 (67)	92 (67)
Italy	55 (25)	16 (11)	85 (5)	19 (5)	24 (5)
Portugal	64 (36)	27 (9)	73 (36)	55 (18)	64 (27)
Spain	40 (35)	22 (17)	30 (9)	43 (19)	47 (21)
Eastern Europe					
Czech Republic	88 (54)	66 (58)	88 (64)	80 (43)	88 (55)
Hungary	77 (49)	56 (24)	80 (34)	57 (31)	66 (34)
Poland	62 (36)	28 (17)	34 (17)	40 (14)	43 (19)
Russia	60 (36)	38 (18)	80 (73)	79 (56)	86 (63)
Near East					
Israel	82 (55)	62 (14)	73 (36)	68 (18)	73 (32)
Turkey	73 (41)	35 (10)	77 (27)	64 (23)	73 (23)
Latin America					
Argentina	69 (39)	18 (9)	25	25	31 (8)
Brazil	70 (40)	31 (20)	37 (20)	42 (18)	43 (17)
Chile	18	6	13	7	13
Colombia	54 (39)	27 (18)	58 (42)	58 (33)	58 (42)
Cuba	69 (46)	86 (64)	86 (43)	86 (43)	86 (43)
Mexico	75 (69)	50 (45)	68 (56)	67 (46)	68 (48)
Peru	79 (50)	41 (25)	46 (39)	46 (31)	62 (39)
Venezuela	80 (73)	50 (43)	47 (40)	43 (29)	50 (36)
Asia					
China	88 (77)	84 (67)	78 (62)	86 (61)	78 (57)
India	86 (59)	65 (35)	83 (61)	77 (41)	91 (55)
Japan	86 (60)	67 (47)	48 (19)	42 (13)	54 (23)
Thailand	88 (36)	48 (16)	80 (36)	80 (20)	80 (28)
Total geneticists	66 (44)	31 (19)	45 (24)	42 (19)	47 (22)
Total, excluding US	71 (50)	41 (27)	51 (30)	49 (25)	56 (29)

*On questionnaire: "Emphasize negative aspects so they will favor termination without suggesting it directly."

Table 45. Placement for Adoption. % would present option of carrying to term and giving up child for adoption

Condition or disorder	US	Outside US	Total
Pregnancy result of rape	86	46	62
Child not the sex desired by parents	74	38	53
Trisomy 21	71	16	38
Achondroplasia	64	15	36
45,X	63	19	37
XXY	63	19	37
Severe, open spina bifida	55	8	27
Cystic fibrosis	55	11	28
PKU in fetus	55	9	23
Neurofibromatosis	52	12	26
HIV infection in fetus	52	10	25
Cleft lip/palate in boy	52	19	33
Sickle cell anemia	51	11	25
Cleft lip/palate in girl	49	15	27
Rubella	48	9	23
Severe untreatable obesity	47	17	30
Toxoplasmosis in fetus	47	9	23
Familial hypercholesterolemia	43	12	23
Predisposition to schizophrenia or bipolar disorder	43	13	25
Predisposition to alcoholism	42	16	27
Huntington disease	42	13	25
Predisposition to Alzheimer disease	40	12	24
Mother's life in danger	17	6	10

Table 46. Personal Attitudes Toward Abortion for Conditions Appearing in Childhood, Part 1. % would abort in first trimester for

Country	Anencephaly	Severe, open spina bifida	Trisomy 13	Trisomy 21	CF**	Achondroplasia	NF***	Sickle cell anemia
English-speaking nations								
Australia	100	87	100	87	80	67	47	40
Canada	99	86	91	76	57	60	47	61
South Africa	100	100	100	100	88	50	50	50
United Kingdom	96	93	93	85	78	63	46	66
United States	94	90	93	80	62	57	51	52
US Primary Care Physicians	n/a	78	93	62	47	29	37	39
US patients* (*n* = 348)	88	63[1]	81[2]	36[3]	52	24	40	72
US public	n/a	41	48	22	n/a	n/a	n/a	n/a
Northern/Western Europe								
Belgium	93	92	93	92	83	71	69	92
Denmark	100	100	100	100	93	96	67	76
Finland	95	95	95	100	71	71	29	80
France	97	89	96	79	77	51	26	64
Germany	96	90	95	71	61	54	39	39
German patients (*n* = 593)	94	81	88	69	64	28	26	66
Netherlands	96	100	96	86	90	57	53	66
Norway	100	100	100	88	75	75	75	63
Sweden	100	100	100	100	92	75	50	83
Switzerland	83	83	83	50	67	67	33	67
Southern Europe								
Greece	100	100	100	100	91	100	82	82
Italy	71	73	33	57	50	50	36	41
Portugal	100	100	100	100	64	73	46	64
Spain	71	64	65	54	46	39	30	30

Eastern Europe								
Czech Republic	95	95	95	96	94	83	77	74
Hungary	97	94	94	89	83	71	66	53
Poland	83	80	76	62	54	44	55	41
Russia	100	98	98	98	93	95	91	84
Near East								
Israel	100	100	100	100	100	91	73	95
Turkey	100	100	96	90	90	80	79	82
Latin America								
Argentina	62	42	69	58	42	25	23	25
Brazil	92	81	89	73	57	49	52	42
Chile	33	20	13	6	6	6	0	0
Colombia	93	86	93	86	64	43	21	36
Cuba	100	100	100	100	100	100	93	100
Mexico	90	80	89	68	53	41	31	37
Peru	77	77	62	62	54	31	0	42
Venezuela	80	80	86	67	57	31	23	50
Asia								
China	99	99	98	94	98	96	86	92
India	96	96	96	78	86	74	73	61
Japan	97	80	87	65	67	59	37	36
Thailand	88	52	76	63	56	40	32	48
Total geneticists	93	89	91	79	68	62	53	56
Total, excluding US	93	88	90	79	72	64	53	58
Total geneticists for second trimester	91	85	88	74	61	54	45	50
Total for second trimester, excluding US	90	83	86	73	64	57	46	52

*% for patients based on those who think abortion should be legal for purposes other than saving mother's life, rape, or incest.
**CF = cystic fibrosis.
***NF = neurofibromatosis.
[1] Patient and public questionnaires: child would not be retarded but would be paralyzed from neck down, with nearly normal lifespan.
[2] Patient and public questionnaires: child would be severely retarded and would die within first few months of life.
[3] Patient and public questionnaires: child would be moderately retarded (could communicate, but not live independently).

Table 47. Personal Attitudes Toward First-trimester Abortion for Conditions Appearing in Childhood, Part II. % would abort in first trimester for

Country	XXY	Hurler Syndrome	45,X	PKU in fetus	Severe obesity**	Cleft lip/palate	Child is not desired sex
English-speaking nations							
Australia	47	87	40	60	33	13	0
Canada	43	92	38	42	24	17	3
South Africa	69	100	50	69	7	13	0
United Kingdom	37	87	36	42	14	2	2
United States	49	91	42	44	29	12	3
US Primary Care Physicians	48	n/a	37	45	13	10	3
US patients* (n = 348)	22	n/a	9	n/a	8	4	1
US public	n/a	n/a	n/a	n/a	16	n/a	7
Northern/Western Europe							
Belgium	39	93	36	71	14	31	7
Denmark	70	100	67	63	23	30	0
Finland	24	95	33	67	24	5	0
France	28	95	34	25	18	7	4
Germany	18	87	19	25	19	7	0
German patients (n = 593)	33	88	9	72	9	5	0
Netherlands	64	89	52	68	20	14	3
Norway	75	100	75	75	25	13	13
Sweden	50	100	55	50	17	17	25
Switzerland	33	83	17	80	17	17	0
Southern Europe							
Greece	82	100	91	73	27	18	0
Italy	19	67	25	36	24	14	5
Portugal	46	100	46	46	0	18	0
Spain	34	67	33	33	20	9	6
Eastern Europe							
Czech Republic	68	94	62	68	54	50	0
Hungary	57	97	63	54	29	29	9
Poland	41	73	36	26	12	16	2
Russia	91	100	91	88	79	64	14
Near East							
Israel	91	100	86	81	57	57	19
Turkey	94	96	95	80	60	47	25
Latin America							
Argentina	15	58	9	8	7	0	7
Brazil	57	88	54	50	22	30	7
Chile	0	13	12	0	0	0	0
Colombia	43	93	36	64	14	14	14
Cuba	100	100	93	93	43	69	0
Mexico	32	79	37	46	15	7	7
Peru	17	62	23	39	0	0	8
Venezuela	14	67	17	23	0	7	0

Table 47. (continued)

Country	XXY	Hurler syndrome	45,X	PKU in fetus	Severe obesity**	Cleft lip/palate	Child is not desired sex
Asia							
China	91	97	94	93	69	84	23
India	71	95	77	70	46	61	4
Japan	46	80	44	36	20	20	8
Thailand	28	79	28	36	8	4	4
Total geneticists	49	88	47	48	29	23	5
Total, excluding US	50	87	49	50	29	28	7
Total geneticists for second trimester	42	84	39	42	23	17	5
Total, 2nd trimester, excluding US	42	82	42	43	23	22	6

*% for patients based on those who think abortion should be legal for other purposes in addition to mother's life, rape, or incest.
**in absence of a known genetic syndrome.

Table 48. Personal Attitudes Toward First-trimester Abortion: Adult-Onset Disorders. % would abort for

Country	FH	HD	Predisposition to mental illness	Predisposition to alcoholism	Predisposition to Alzheimer disease
English-speaking nations					
Australia	67	87	36	7	29
Canada	54	61	14	6	12
South Africa	69	94	50	13	25
United Kingdom	42	71	26	10	31
United States	48	64	27	10	20
US Primary Care Physicians	10	50	24	8	16
US patients* (n = 348)	9	32	59	4	5
US public	n/a	21**	n/a	n/a	n/a
Northern/Western Europe					
Belgium	77	77	31	21	36
Denmark	82	93	48	19	41
Finland	52	73	10	5	14
France	36	46	21	7	14
Germany	30	48	12	3	11
German patients (n = 593)	11	33	72	5	6
Netherlands	59	79	35	15	36
Norway	75	86	25	25	25
Sweden	73	92	46	36	36
Switzerland	67	60	0	0	26

Table 48. (continued)

Country	FH	HD	Predisposition to mental illness	Predisposition to alcoholism	Predisposition to Alzheimer disease
Southern Europe					
Greece	55	100	73	27	73
Italy	30	46	36	20	23
Portugal	20	82	20	0	40
Spain	22	40	21	9	11
Eastern Europe					
Czech Republic	60	82	48	30	58
Hungary	37	77	37	17	35
Poland	31	51	25	9	23
Russia	77	90	70	48	75
Near East					
Israel	71	100	55	43	57
Turkey	42	88	84	44	75
Latin America					
Argentina	17	31	7	0	7
Brazil	39	50	35	17	38
Chile	0	0	6	6	6
Colombia	29	57	14	0	29
Cuba	71	79	71	23	85
Mexico	34	52	30	8	32
Peru	25	15	15	8	15
Venezuela	39	43	23	8	17
Asia					
China	86	94	81	64	68
India	59	78	57	36	61
Japan	32	47	38	25	33
Thailand	20	32	16	4	22
Total geneticists	48	64	33	17	29
Total, excluding US	48	64	37	21	34
Total geneticists for second trimester	41	57	28	14	23
Total, 2nd trimester, excluding US	40	58	31	18	29

*Patient % are based on those who thought abortion should be legal for purposes other than mother's life, rape, or incest.
**On public and patient questionnaires: severe incurable mental disease at age 40.
FH = familial hypercholesterolemia, homozygous.
HD = Huntington disease.

Table 49. Personal Attitudes Toward First-trimester Abortion for Maternal Situations or Fetal Infections. % would abort for

Country	Mother's life in danger	Pregnancy result of rape	HIV infection in fetus, first trimester	Toxoplasmosis in fetus, first trimester	Rubella in fetus, first trimester
English-speaking nations					
Australia	93	93	87	73	87
Canada	95	82	82	83	87
South Africa	100	93	94	100	100
United Kingdom	94	89	81	82	85
United States	96	87	85	79	84
US Primary Care Physicians	95	81	74	72	79
US patients* ($n = 348$)	71	75	64	20	68
US public	69	63	n/a	n/a	n/a
Northern/Western Europe					
Belgium	100	86	93	79	87
Denmark	100	96	96	93	100
Finland	100	81	81	76	86
France	81	79	71	58	84
Germany	94	72	65	58	81
German patients ($n = 593$)	87	80	68	31	84
Netherlands	100	88	76	82	85
Norway	75	100	88	88	88
Sweden	100	100	100	83	83
Switzerland	100	80	83	67	83
Southern Europe					
Greece	100	100	92	92	92
Italy	67	55	55	52	52
Portugal	100	90	73	91	100
Spain	56	48	49	51	57
Eastern Europe					
Czech Republic	94	83	93	86	93
Hungary	93	81	88	60	86
Poland	72	65	48	51	53
Russia	75	82	98	98	98
Near East					
Israel	96	100	100	96	100
Turkey	96	96	100	86	91
Latin America					
Argentina	71	54	55	50	50
Brazil	89	83	58	78	79
Chile	20	25	19	13	19
Colombia	77	92	86	86	79
Cuba	92	86	93	93	93
Mexico	81	71	73	82	82
Peru	69	46	69	69	77
Venezuela	93	73	60	57	71

Table 49. (continued)

Country	Mother's life in danger	Pregnancy result of rape	HIV infection in fetus, first trimester	Toxoplasmosis in fetus, first trimester	Rubella in fetus, first trimester
Asia					
China	97	91	88	92	89
India	91	91	87	86	91
Japan	100	93	69	62	76
Thailand	96	56	64	80	80
Total geneticists	92	83	78	75	82
Total, excluding US	89	80	75	74	81
Total geneticists for second trimester	87	69	73	68	73
Total, 2nd trimester, excluding US	83	66	70	65	71

*Patient %'s based on those who thought abortion should be legal for other reasons in addition to saving mother's life, rape, or incest.

Table 50. Geneticists' Dissatisfaction with Counseling. % feel

Country	Dissatisfied if patients refuse prenatal diagnosis if medically indicated	Dissatisfied if patients decide to carry a seriously affected fetus to term	Dissatisfied if legal abortion after PND not available	Satisfied if patients understand the medical genetic information	Satisfied if patients get emotional support, even if they forget information
English-speaking nations					
Australia	0	7	93	87	13
Canada	23	13	93	91	45
South Africa	50	38	88	87	56
United Kingdom	16	12	89	89	29
United States	27	15	91	97	37
Northern/Western Europe					
Belgium	50	50	93	100	50
Denmark	19	23	92	92	31
Finland	24	19	81	95	38
France	49	35	89	100	74
Germany	22	7	86	95	59
Netherlands	22	11	94	89	15
Norway	25	38	75	88	13
Sweden	17	8	84	100	17
Switzerland	17	33	100	100	83
Southern Europe					
Greece	100	75	92	100	58
Italy	30	25	80	95	50
Portugal	55	18	91	100	36
Spain	54	37	78	100	47
Eastern Europe					
Czech Republic	77	70	89	99	65
Hungary	88	75	94	100	36
Poland	51	18	78	94	70
Russia	84	81	93	98	60

Table 50. (continued)

Country	Dissatisfied if patients refuse prenatal diagnosis if medically indicated	Dissatisfied if patients decide to carry a seriously affected fetus to term	Dissatisfied if legal abortion after PND not available	Satisfied if patients understand the medical genetic information	Satisfied if patients get emotional support, even if they forget information
Near East					
Israel	43	43	100	100	62
Turkey	82	59	77	100	91
Latin America					
Argentina	21	14	64	100	31
Brazil	58	40	96	100	46
Chile	37	6	25	94	56
Colombia	50	43	72	100	40
Cuba	77	69	100	100	77
Mexico	58	46	91	98	42
Peru	85	39	69	100	77
Venezuela	40	47	88	100	67
Asia					
China	87	89	89	96	75
India	96	74	100	100	74
Japan	20	7	41	74	45
Thailand	20	36	40	96	56
Total geneticists	40	28	86	95	49
Total, excluding US	48	37	84	84	55

E. Eugenics

Table 51. Geneticists' Perceptions of Disability. % agreed

Country	Society will never provide adequate support	Some disabilities will never be overcome even with maximum social support	People with severe disabilities make society more rich and varied	Care usually falls on the mother
English-speaking nations				
Australia	68	100	13	87
Canada	62	82	29	85
South Africa	75	81	14	75
United Kingdom	80	91	36	93
United States	65	85	24	85
US Primary Care Physicians	49	90	19	78
US patients (n = 476)	56	60	19	55
Northern/Western Europe				
Belgium	53	87	0	86
Denmark	41	96	12	75
Finland	52	73	42	73
France	34	85	29	70
Germany	57	92	38	70
German patients (n = 593)	59	58	21	—
Netherlands	26	96	19	74
Norway	78	100	0	89
Sweden	63	92	17	67
Switzerland	50	83	50	50
Southern Europe				
Greece	42	91	0	100
Italy	52	65	25	67
Portugal	82	91	9	82
Spain	49	44	12	71
Eastern Europe				
Czech Republic	69	90	9	77
Hungary	40	94	9	77
Poland	80	86	37	80
Russia	57	86	12	63
Near East				
Israel	55	86	10	67
Turkey	29	77	10	36
Latin America				
Argentina	57	50	14	85
Brazil	58	82	4	77
Chile	38	53	44	67
Colombia	50	64	7	43
Cuba	29	85	7	50
Mexico	49	65	8	59
Peru	42	38	7	50
Venezuela	50	44	0	47
Asia				
China	18	79	2	11
India	52	91	14	65
Japan	30	65	49	31
Thailand	79	83	0	48
Total geneticists	56	83	23	70
Total, excluding US	51	81	22	62

Table 52. Responsible Parenthood. % agreed

Country	People who carry a recessive disorder should not mate with another carrier	People should know their own and their partner's carrier status before marriage	People at high risk should not have children unless they use PND and selective abortion	Women should have PND if indicated by age or family history
English-speaking nations				
Australia	0	64	0	13
Canada	3	45	8	44
South Africa	19	69	19	75
United Kingdom	6	34	3	30
United States	7	44	10	38
US Primary Care Physicians	38	65	34	78
US patients (n = 476)	5	66	40	81
US public	n/a	64	n/a	62
Northern/Western Europe				
Belgium	7	43	27	60
Denmark	4	33	11	85
Finland	9	23	5	59
France	17	58	18	75
French patients (n = 380)	7	70	46	93
Germany	3	23	13	34
German patients (n = 593)	3	49	40	80
Netherlands	0	22	4	30
Norway	0	22	9	56
Sweden	0	25	8	67
Switzerland	0	50	17	17
Southern Europe				
Greece	33	91	92	100
Italy	15	100	10	90
Portugal	9	73	27	91
Spain	26	88	33	77
Eastern Europe				
Czech Republic	52	93	57	95
Hungary	9	79	20	34
Poland	47	75	35	85
Russia	16	86	54	93
Near East				
Israel	14	73	38	68
Turkey	27	82	50	95
Latin America				
Argentina	7	71	29	72
Brazil	26	84	39	96
Chile	31	43	13	75
Colombia	50	71	43	79
Cuba	26	93	57	86
Mexico	31	85	75	95

Table 52. (continued)

Country	People who carry a recessive disorder should not mate with another carrier	People should know their own and their partner's carrier status before marriage	People at high risk should not have children unless they use PND and selective abortion	Women should have PND if indicated by age or family history
Peru	50	93	57	86
Venezuela	56	81	73	88
Asia				
China	92	92	95	91
India	70	87	78	100
Japan	19	22	6	25
Thailand	44	92	79	100
Total geneticists	23	55	27	58
Total, excluding US	31	62	37	67

Table 53. Effects of Social Class: 2 Cases of Blind Parents. % would try to persuade them not to have a child

Country	Case 1	Case 2
English-speaking nations		
Australia	20	33
Canada	12	24
South Africa	19	44
United Kingdom	5	24
United States	9	26
US Primary Care Physicians	30	69
US patients	28	75
Northern/Western Europe		
Belgium	60	87
Denmark	22	50
Finland	14	23
France	44	83
French patients (n = 363)	70	92
Germany	11	43
German patients (n = 593)	49	n/a
Netherlands	11	22
Norway	22	22
Sweden	25	67
Switzerland	33	50
Southern Europe		
Greece	36	92
Italy	45	62
Portugal	27	46
Spain	47	88

Table 53. (continued)

Country	Case 1	Case 2
Eastern Europe		
Czech Republic	70	87
Hungary	56	72
Poland	54	78
Russia	71	91
Near East		
Israel	17	41
Turkey	27	62
Latin America		
Argentina	21	62
Brazil	41	53
Chile	56	75
Colombia	36	50
Cuba	64	93
Mexico	47	80
Peru	64	86
Venezuela	33	88
Asia		
China	96	98
India	78	96
Japan	36	36
Thailand	52	64
Total geneticists	30	49
Total, excluding US	43	62

Case 1. A blind couple with the same autosomal recessive disorder have a 100% chance of having a blind child. They have considered and rejected adoption or donor gametes and want a biological child.
Case 2. A single blind woman on public welfare has 3 legally blind sons by 3 different fathers. She wants another child. There is a 50% risk that the child will be blind.

Table 54. Government Involvement in Prevention. % agreed governments should require

Country	Carrier tests for common genetic disorders before marriage
English-speaking nations	
Australia	0
Canada	6
South Africa	6
United Kingdom	4
United States	5
US Primary Care Physicians	36
US patients	31
Northern/Western Europe	
Belgium	7
Denmark	11
Finland	14
France	18
French patients (n = 379)	29
Germany	2

Table 54. (continued)

Country	Carrier tests for common genetic disorders before marriage
German patients (n = 593)	5
Netherlands	7
Norway	0
Sweden	0
Switzerland	0
Southern Europe	
Greece	67
Italy	33
Portugal	27
Spain	16
Eastern Europe	
Czech Republic	21
Hungary	43
Poland	14
Russia	54
Near East	
Egypt	100
Israel	5
Turkey	73
Latin America	
Argentina	7
Brazil	20
Chile	13
Colombia	29
Cuba	50
Mexico	39
Peru	50
Venezuela	25
Asia	
China	86
India	65
Japan	6
Thailand	68
Total geneticists	20
Total, excluding US	27

Table 55. Geneticists' Views on Sterilization. %would advise voluntary surgical sterilization for

Country	Woman with PKU	Cystic fibrosis carrier–carrier couple	Single blind woman on welfare, 50% risk of blindness in children ( ) = thought law should require sterilization	Woman with fragile-X living in institutional setting
English-speaking nations				
Australia	0	0	7	31
Canada	1	1	16 (1)	16
South Africa	0	0	38	64
United Kingdom	0	0	13	11
United States	1	3	16 (3)	23
US Primary Care Physicians	8	63	47 (15)	n/a
US patients (n = 476)	n/a	n/a	46 (21)	n/a
Northern/Western Europe				
Belgium	7	0	50	31
Denmark	4	4	31 (4)	15
Finland	9	5	14 (5)	24
France	0	10	20 (4)	20
French patients (n = 374)	n/a	n/a	57 (18)	n/a
Germany	0	5	15	14
Netherlands	0	0	11	0
Norway	0	0	11	13
Sweden	0	0	8 (8)	9
Switzerland	0	0	67	33
Southern Europe				
Greece	0	30	55 (19)	10
Italy	0	6	0	19
Portugal	10	9	36	33
Spain	6	33	85 (7)	61
Eastern Europe				
Czech Republic	3	27	55 (14)	39
Hungary	0	17	26 (3)	33
Poland	2	16	26 (11)	23
Russia	10	16	52 (23)	46
Near East				
Israel	0	9	27 (5)	0
Turkey	18	15	29 (6)	50
Latin America				
Argentina	0	36	0	20
Brazil	1	36	15 (4)	36
Chile	7	25	6 (19)	55
Colombia	0	64	17 (8)	33
Cuba	0	46	30 (14)	44
Mexico	5	67	22 (13)	72
Peru	0	58	27 (14)	63
Venezuela	11	62	9 (14)	67
Asia				
China	39	52	88 (82)	73
India	23	30	83 (50)	53
Japan	3	6	18 (6)	26
Thailand	21	38	64 (28)	57
Total geneticists	6	12	30 (12)	30
Total, excluding US	8	17	39 (17)	34

Table 56. "Cleaning up the Gene Pool". % agreed that "An Important goal of genetic counseling is to reduce the number of deleterious genes in the population"

Country	Percentages of agreement
English-speaking nations	
Australia	7
Canada	6
South Africa	13
United Kingdom	13
United States	5
Northern/Western Europe	
Belgium	13
Denmark	15
Finland	5
France	10
Germany	2
Netherlands	0
Norway	0
Sweden	0
Switzerland	0
Southern Europe	
Greece	58
Italy	14
Portugal	45
Spain	67
Eastern Europe	
Czech Republic	48
Hungary	43
Poland	66
Russia	58
Near East	
Israel	14
Turkey	73
Latin America	
Argentina	21
Brazil	40
Chile	31
Colombia	39
Cuba	57
Mexico	52
Peru	71
Venezuela	50
Asia	
China	100
India	87
Japan	12
Thailand	50
Total	25
Total, excluding US	37

F: Family Relationship

Table 57. A Woman Tests Positive for a Breast Cancer Mutation

% would	USA		Germany			Japan	Hungary
	M.S. Genetic counselors	M.D.–Ph.D. geneticists	Patients	Professionals	Patients		
Tell patient's sister that the sister is at risk, against the patient's wishes	25	46	n/a*	44	n/a**	44	48
Encourage patient's 20-year-old daughter to be tested	54	75	96	48	98	42	52
Encourage patient's 13-year-old daughter to be tested *now*	13	27	73	7	68	22	13
Suggest prophylactic double mastectomy for patient	11	18	n/a	4	n/a	0	4
Have counseled for breast cancer mutations	36	37	n/a	13	—	1	13

*91% of US patients said they were "very likely" or "extremely likely" to tell their sister.
**92% of German patients were "very likely" or "extremely likely" to tell their sister.

Table 58. Disclosure of a Cancer Mutation to Relatives at Genetic Risk, Against Patient's Wishes. % would

Country	Respect patient confidentiality	Tell relatives if they ask	Tell relatives unasked	Let referring doctor decide
English-speaking nations				
Australia	53	33	7	7
Canada	47	36	9	8
South Africa	13	40	20	27
United Kingdom	34	44	9	13
United States	47	30	12	11
US Primary Care Physicians	53	27	20	n/a
US patients*	40	47	13	n/a
Northern/Western Europe				
Belgium	29	50	14	7
Denmark	54	27	4	15
Finland	35	30	5	30
France	25	41	11	22
Germany	51	35	3	12
Netherlands	19	23	35	19
Norway	33	44	11	11
Sweden	58	25	8	8
Switzerland	25	50	0	25
Southern Europe				
Greece	20	10	40	30
Italy	41	12	23	24
Portugal	9	18	27	46
Spain	15	46	17	23
Eastern Europe				
Czech Republic	17	36	28	19
Hungary	17	41	0	41
Poland	23	34	17	26
Russia	16	21	48	16
Near East				
Israel	25	50	20	5
Turkey	30	35	15	15
Latin America				
Argentina	29	50	7	14
Brazil	18	44	21	17
Chile	27	53	7	13
Colombia	29	50	21	0
Cuba	0	43	43	14
Mexico	14	36	27	24
Peru	14	36	14	36
Venezuela	7	57	7	29
Asia				
China	10	13	60	17
India	5	23	14	60
Japan	40	37	5	11
Thailand	48	22	9	22
Total geneticists	36	32	17	15
Total, excluding US	29	33	21	18

*Patient questionnaire asked about disclosure of genetic risks for heart disease.

Table 59. Disclosure of a Huntington Disease Diagnosis to Relatives, Against Patient's Wishes. Trends, 1985–1995. % would

Country	Respect patient confidentiality		Tell relatives only if they ask		Tell relatives, unasked		Let referring doctor decide	
	1985	1995	1985*	1995	1985*	1995	1985	1995
English-speaking nations								
Australia	25	60	75	33	0	0	0	7
Canada	28	50	25	38	32	5	15	7
South Africa	—	13	—	53	—	7	—	27
United Kingdom	32	39	26	43	32	6	10	11
United States	40	53	28	32	24	6	8	9
US Primary Care Physicians	—	61	—	28	—	11	—	n/a
US patients (n = 476)	—	25	—	42	—	33	—	n/a
US public	—	44	—	22	—	19	—	n/a
Northern/Western Europe								
Belgium	—	53	—	40	—	0	—	7
Denmark	7	50	54	32	32	4	7	14
Finland	—	52	—	29	—	0	—	19
France	35	40	29	43	18	3	18	15
French patients (n = 391)	—	12	—	47	—	42	—	—
Germany**	20	59	65	30	9	0	6	11
German patients (n = 593)	—	24	—	42	—	35	—	n/a
Netherlands	—	29	—	42	—	21	—	8
Norway	0	56	33	44	50	0	17	0
Sweden	48	58	33	25	14	8	5	8
Switzerland	40	20	20	60	20	0	20	20
Southern Europe								
Greece	29	33	14	25	0	33	57	8
Italy	9	44	46	33	36	6	9	17
Portugal	—	18	—	46	—	9	—	27
Spain	—	22	—	41	—	14	—	24
Eastern Europe								
Czech Republic	—	24	—	39	—	20	—	18
Hungary	40	14	7	57	40	3	13	26

Poland	—	27	—	41	—	9	—	23
Russia	—	11	—	41	—	32	—	16
Near East								
Israel	20	35	46	45	27	10	7	10
Turkey	0	24	20	48	60	19	20	10
Latin America								
Argentina	—	29	—	64	—	0	—	7
Brazil	9	23	53	47	38	14	0	16
Chile	—	31	—	63	—	0	—	6
Colombia	—	21	—	57	—	21	—	0
Cuba	—	0	—	43	—	43	—	14
Mexico	—	17	—	40	—	25	—	18
Peru	—	21	—	50	—	7	—	21
Venezuela	—	7	—	47	—	13	—	33
Asia								
China	—	11	—	22	—	54	—	13
India	26	26	15	26	33	13	26	35
Japan	37	53	37	38	16	3	10	6
Thailand	—	46	—	27	—	5	—	23
Total geneticists	32	41	34	35	25	12	10	12
Total, excluding US	26	34	28	37	25	15	11	14

*1985 survey said "siblings" instead of "relatives."
**1985 figures are for West Germany.

Table 60. Family Decisions: Whose View Prevails. % responding

		World totals
Decisions are shared by husband and wife		75
Husband's view prevails (countries with 19% or more)		8
Brazil	39	
Chile	25	
Colombia	39	
Greece	27	
India	35	
Mexico	62	
Peru	50	
South Africa	39	
Turkey	19	
Wife's view prevails (countries with more than 19%)		13
Chile	25	
Hungary	50	
Norway	22	
Portugal	27	
Russia	19	
Venezuela	27	

Table 61. Family Access to an Individual's Cells. % would

Country	Collect cells from mentally incompetent man with HD, against his legal guardian's wishes, on behalf of his minor children	Use patient's cells to test daughter but not son, according to patient's wishes	Require pretest agreements allowing access for relatives	Have counseled people who refused family access
English-speaking nations				
Australia	64	79	33	67
Canada	74	75	37	39
South Africa	93	73	31	31
United Kingdom	81	66	26	59
United States	81	85	27	32
Northern/Western Europe				
Belgium	100	54	40	27
Denmark	82	62	32	43
Finland	68	71	29	30
France	88	46	72	32
Germany	74	68	25	22
Netherlands	86	68	22	56
Norway	89	67	11	56
Sweden	100	90	8	33
Switzerland	83	50	17	67
Southern Europe				
Greece	100	91	64	58
Italy	85	41	50	53
Portugal	100	82	27	9
Spain	100	68	42	26

Table 61. (continued)

Country	Collect cells from mentally incompetent man with HD, against his legal guardian's wishes, on behalf of his minor children	Use patient's cells to test daughter but not son, according to patient's wishes	Require pretest agreements allowing access for relatives	Have counseled people who refused family access
Eastern Europe				
Czech Republic	98	89	63	10
Hungary	94	58	66	6
Poland	88	67	45	3
Russia	88	76	65	0
Near East				
Israel	100	67	41	17
Turkey	100	59	36	4
Latin America				
Argentina	100	92	50	21
Brazil	95	49	60	3
Chile	100	40	38	0
Colombia	93	70	57	21
Cuba	100	82	86	29
Mexico	93	77	56	11
Peru	86	82	57	0
Venezuela	81	67	75	19
Asia				
China	95	37	73	15
India	96	67	62	14
Japan	68	53	42	3
Thailand	91	88	72	28
Total	85	70	41	24
Total, excluding US	87	62	48	21

Table 62. Spouse/Partner's Access to Information. % thought spouse/partner should have access without consent

Geographical area	Translocation Down*	HD	CF	CF carrier	Predisposition to Schizophrenia	Predisposition to alcoholism
United States	20	9	6	3	4	4
US Primary Care Physicians	31	15	15	15	14	15
US patients	43	—	30	27	31	23
Other English-speaking nations	17	11	6	4	5	7
Northwestern Europe	17	6	6	3	3	3
German patients	48	n/a	25	28	21	20
Southern Europe	39	32	27	23	24	19
Eastern Europe	44	30	24	20	25	26
Near East	40	26	21	11	23	23
Asia	39	37	34	24	31	33
Latin America	56	37	32	26	31	30
Total geneticists	29	18	15	12	15	14
Total, excluding US	34	24	21	16	20	19

HD = Huntington disease; CF = cystic fibrosis.
*May lead to the couple's giving birth to a child with Down syndrome.

Table 63. Spouse/Partner's Rights

Country	Would tell prenatal test results of XXY to separated husband, in process of divorce, against wife's wishes	Would tell both spouses, unasked, which one carried a balanced translocation that caused Down syndrome in their child		Thought father of fetus should be able to require mother to have prenatal diagnosis
		1985	1995	
English-speaking nations				
Australia	21	67	67	0
Canada	15	60	84	7
South Africa	25	—	93	19
United Kingdom	12	41	76	4
United States	14	62	79	5
US Primary Care physicians	17	—	38	18
US patients	—	—	—	13
Northern/Western Europe				
Belgium	43	—	73	29
Denmark	28	67	96	19
Finland	9	—	68	18
France	22	69	43	20
French patients (n = 375)	n/a	n/a	n/a	25
Germany*	13	38	45	6
German patients (n = 593)	n/a	n/a	n/a	20

Table 63. (continued)

Country	Would tell prenatal test results of XXY to separated husband, in process of divorce, against wife's wishes	Would tell both spouses, unasked, which one carried a balanced translocation that caused Down syndrome in their child		Thought father of fetus should be able to require mother to have prenatal diagnosis
		1985	1995	
Netherlands	35	—	85	11
Norway	0	67	88	0
Sweden	8	29	83	8
Switzerland	17	100	67	0
Southern Europe				
Greece	46	72	80	42
Italy	33	64	71	38
Portugal	73	—	70	36
Spain	54	—	89	49
Eastern Europe				
Czech Republic	30	—	55	46
Hungary	61	27	21	40
Poland	31	—	50	28
Russia	46	—	36	56
Near East				
Israel	46	71	84	55
Turkey	50	80	35	77
Latin America				
Argentina	83	—	100	29
Brazil	39	63	85	26
Chile	38	—	73	44
Colombia	69	—	71	57
Cuba	86	—	79	71
Mexico	49	—	74	62
Peru	50	—	58	71
Venezuela	57	—	93	50
Asia				
China	70	—	58	59
India	39	44	56	41
Japan	30	6	23	30
Thailand	36	—	67	76
Total geneticists	29	54	66	24
Total, excluding US	36	46	60	32

*1985 figures are for West Germany.

Table 64. Approaches to Accidental Findings of Nonpaternity. % would

Country	Warn mother alone, before test, so she can withdraw from testing	Make couple aware, in informed consent before test, that finding may appear	Disclose finding to mother only	Urge mother to tell husband	Tell husband who asks directly
English-speaking nations					
Australia	33	47	67	13	53
Canada	49	78	47	34	24
South Africa	53	47	63	67	25
United Kingdom	38	32	44	28	11
United States	65	68	50	42	36
US Primary Care Physicians	72	63	31	74	42
US Patients	n/a	n/a	81	54	74
US Public	n/a	n/a	54	32	n/a
Northern/Western Europe					
Belgium	46	50	57	29	43
Denmark	37	48	48	16	27
Finland	36	50	33	18	23
France	62	14	82	27	10
French patients	n/a	n/a	84	n/a	60
Germany	52	35	58	8	18
Netherlands	31	30	30	15	23
Norway	44	44	33	0	0
Sweden	50	42	42	17	8
Switzerland	67	17	67	33	20
Southern Europe					
Greece	42	42	58	33	25
Italy	57	24	76	10	19
Portugal	64	46	27	55	36
Spain	67	40	61	31	38

Table 64. (continued)

Country	Warn mother alone, before test, so she can withdraw from testing	Make couple aware, in informed consent before test, that finding may appear	Disclose finding to mother only	Urge mother to tell husband	Tell husband who asks directly
Eastern Europe					
Czech Republic	56	19	71	16	15
Hungary	58	43	66	26	17
Poland	77	28	76	21	14
Russia	68	17	95	24	21
Near East					
Israel	55	23	68	29	18
Turkey	43	38	43	9	18
Latin America					
Argentina	75	57	43	43	50
Brazil	64	64	58	41	15
Chile	56	31	50	31	19
Colombia	86	31	57	21	43
Cuba	86	50	64	50	15
Mexiço	89	40	61	33	17
Peru	79	54	39	69	23
Venezuela	83	62	64	48	38
Asia					
China	37	39	74	61	25
India	78	27	78	36	18
Japan	65	51	21	45	24
Thailand	80	52	80	36	42
Total geneticists	60	50	56	35	26
Total, excluding US	56	39	58	31	21

Table 65. Testing Children. % who thought parents should be able to have minor children tested for

Country	Huntington disease	Alzheimer disease	Alcoholism (predisposition)	Cancer mutations	Familial hypercholesterolemia	Had requests to test children for adult-onset disorders	Would tell a minor results of Alzheimer test
English-speaking nations							
(totals)	9	9	19	62	69	65	16
Australia	13	13	7	64	73	93	0
Canada	12	11	26	63	71	56	22
South Africa	25	25	67	81	81	50	0
United Kingdom	2	2	4	58	67	76	12
United States	27	25	48	70	81	38	18
US Primary Care Physicians	66	58	76	81	89	20	44
US patients	n/a	61	n/a	93	92	n/a	32
US public	n/a	54	n/a	85	82	n/a	n/a
Northern/Western Europe							
(totals)	16	14	27	51	63	49	12
Belgium	15	15	50	62	92	57	17
Denmark	33	33	33	56	67	37	14
Finland	18	18	23	41	64	41	13
France	21	19	37	72	85	51	9
French patients	45	n/a	n/a	97	97	n/a	n/a
Germany	14	10	24	45	56	44	12
German patients (n = 593)	37	n/a	n/a	86	87	n/a	n/a
Netherlands	11	11	18	36	57	84	13
Norway	11	11	11	56	78	67	0
Sweden	8	8	25	50	42	58	0
Switzerland	40	20	33	80	67	83	20

Table 65. (continued)

Country	Huntington disease	Alzheimer disease	Alcoholism (predisposition)	Cancer mutations	Familial hypercholesterolemia	Had requests to test children for adult-onset disorders	Would tell a minor results of Alzheimer test
Southern Europe							
(total)	66	62	66	80	89	34	1
Greece	90	70	70	80	90	40	0
Italy	48	48	57	76	91	38	0
Portugal	64	64	64	82	91	27	0
Spain	70	66	70	82	88	33	2
Eastern Europe							
(total)	73	70	78	88	93	16	5
Czech Republic	73	70	78	86	95	30	5
Hungary	58	56	58	75	81	17	0
Russia	86	80	93	91	93	12	7
Near East							
Israel	41	38	59	82	91	77	15
Turkey	68	68	91	86	91	18	20
Latin America							
Argentina	69	69	85	85	85	23	0
Brazil	69	66	80	89	90	8	15
Chile	73	75	81	93	94	19	6
Colombia	64	64	79	71	79	21	8
Cuba	100	100	100	100	100	0	15
Mexico	82	77	77	84	85	32	4
Peru	100	100	93	100	100	7	0
Venezuela	73	67	73	100	93	25	15
Asia							
China	85	61	69	85	84	18	6
India	74	70	54	82	87	14	9
Japan	58	55	59	74	73	13	25
Thailand	84	84	92	96	96	4	16
Total geneticists	40	35	50	71	79	35	14
Total, excluding US	48	42	51	72	77	34	11

Table 66. The Huntington Disease Exclusionary Test: Presymptomatic Testing Prenatally. A man at 50% risk for Huntington disease does not wish to know his own genetic status, but requests prenatal testing. The test will tell only whether the fetus has the same 50% risk as the father. If a child at 50% risk is carried to term and the father later develops HD, the child will know that she/he has the Huntington mutation. % would

Country	Perform test as requested	Perform test only if couple agree, in advance, to abort an at-risk fetus	Refuse to perform test unless father agrees to learn own genetic status
English-speaking nations			
Australia	43	50	7
Canada	61	21	18
South Africa	54	31	15
United Kingdom	44	51	5
United States*	n/a	n/a	n/a
Northern/Western Europe			
Belgium	50	7	43
Denmark	69	31	0
Finland	73	9	18
France	46	17	37
Germany*	n/a	n/a	n/a
Netherlands	46	42	12
Norway	67	0	33
Sweden	46	27	27
Switzerland	80	0	20
Southern Europe			
Greece	45	55	0
Italy	47	0	26
Portugal	60	10	30
Spain	53	5	42
Eastern Europe			
Czech Republic	70	14	17
Hungary*	n/a	n/a	n/a
Poland	72	1	26
Russia	40	25	35
Near East			
Israel	67	11	22
Turkey	46	27	27
Latin America			
Argentina	46	0	54
Brazil	62	2	36
Chile	62	0	38
Colombia	54	0	46
Cuba	92	0	8
Mexico	38	2	60
Peru	39	15	46
Venezuela			
Asia			
China	59	19	22
India	30	45	25
Japan*	n/a	n/a	n/a
Thailand	52	27	31
Total	57	18	25

*In the United States, Germany, Hungary, and Japan a question about breast cancer (Table 58) was substituted.

Table 67. Adoptees' and Birthparents' Rights to Genetic Information. % agreeing: adopted children have the right to know, when they reach age of legal adulthood, their biological parents'

Country	Medical history up to time of adoption	Health status until parents' death	Cause of death	Names	A still fertile birth mother should be told if a child given up for adoption develops X-linked retinitis pigmentosa
English-speaking nations					
Australia	100	93	100	8	100
Canada	83	78	80	6	90
South Africa	88	67	67	29	94
United Kingdom	63	57	70	31	90
United States	91	78	81	10	90
US Primary Care Physicians	85	76	82	11	94
US patients	93	90	92	23	93
US public	n/a	76	n/a	n/a	n/a
Northern/Western Europe					
Belgium	39	54	50	23	85
Denmark	22	31	32	8	52
Finland	27	27	33	27	43
France	30	32	43	11	69
French patients (n = 340)	69	—	—	—	95
Germany	51	59	67	40	65
German patients (n = 593)	—	—	—	—	96
Netherlands	58	62	72	35	81
Norway	11	22	22	33	33
Sweden	36	36	55	20	58
Switzerland	50	83	83	17	67
Southern Europe					
Greece	82	80	70	50	100
Italy	69	72	69	0	71
Portugal	90	100	100	40	91
Spain	65	72	80	19	90
Eastern Europe					
Czech Republic	60	69	75	13	95
Hungary	74	72	78	13	77
Poland	71	69	71	38	85
Russia	62	71	77	25	96
Near East					
Israel	91	87	96	26	91
Turkey	86	91	86	41	95
Latin America					
Argentina	92	92	100	57	77
Brazil	76	73	82	27	89
Chile	75	63	67	27	75
Colombia	67	79	92	42	100
Cuba	92	85	92	42	93

Table 67. (continued)

Country	Medical history up to time of adoption	Health status until parents' death	Cause of death	Names	A still fertile birth mother should be told if a child given up for adoption develops X-linked retinitis pigmentosa
Mexico	74	88	90	32	90
Peru	82	85	100	46	100
Venezuela	79	86	100	36	94
Asia					
China	81	75	82	22	95
India	73	68	76	25	91
Japan	41	40	43	9	58
Thailand	84	76	80	44	83
Total geneticists	73	70	75	21	84
Total, excluding US	65	66	72	26	81

H. Privacy and Discrimination

Table 68. Who Should Have Access to Genetic Information (total percents). Access for

Condition	Adult patient		Blood relatives		Employer		Life insurer		School		Public Health Dept. (without individual identifiers)
	If wants to know	Even if does not want to know	With consent	Without consent	With consent	Not at all	With consent	Not at all	With consent	Not at all	
CF carrier status	79	21	82	14	39	59	46	48	41	58	79
CF	70	30	81	16	59	38	61	30	69	27	84
HD	80	20	71	19	57	39	60	32	52	46	82
Familial polyposis coli	68	32	74	22	50	48	60	32	48	49	82
HIV	45	55	74	16	54	38	59	30	57	30	88
Translocation that may cause Down syndrome in person's children	58	42	74	22	36	61	45	50	38	60	76
Risk for schizophrenia	83	17	78	14	46	49	50	43	45	50	76
Risk for alcoholism	76	24	78	10	46	49	49	44	43	52	75

Table 69. Access to Information from DNA on File in a Medical Record*. % agreed should be access without consent for

Country	Law enforcement agencies	Spouse/partner	Blood relatives	Employers	Employers if involves public safety	Life insurers	Health insurers	Registry of motor vehicles
English-speaking nations								
Australia	13	33	33	0	20	0	0	0
Canada	24	14	19	1	24	2	3	9
South Africa	13	56	25	0	31	6	13	13
United Kingdom	17	31	42	0	13	0	0	4
United States	26	22	21	1	21	3	4	5
US Primary Care Physicians	25	25	18	2	31	11	13	14
US patients	19	53	44	3	28	8	9	8
US public	n/a	20	9	2	n/a	5	5	n/a
Northern/Western Europe								
Belgium	7	33	53	0	0	0	0	0
Denmark	23	23	35	0	19	0	0	0
Finland	22	11	11	0	16	0	11	0
France	18	34	46	0	4	0	0	0
French patients (n = 391)	n/a	55	41	n/a	n/a	n/a	n/a	6
Germany	34	18	16	0.4	12	1	1	2
German patients (n = 593)	10	42	32	0.3	19	1	6	3
Netherlands	16	4	16	0	0	0	0	0
Norway	0	33	44	0	0	0	0	0
Sweden	33	42	42	8	8	8	8	8
Switzerland	17	0	0	0	0	0	0	0

Table 69. (continued)

Country	Law enforcement agencies	Spouse/partner	Blood relatives	Employers	Employers if involves public safety	Life insurers	Health insurers	Registry of motor vehicles
Southern Europe								
Greece	25	44	44	0	33	11	11	11
Italy	12	33	22	0	12	6	6	17
Portugal	20	60	70	0	40	0	0	0
Spain	19	47	42	2	25	9	9	2
Eastern Europe								
Czech Republic	18	64	70	0	12	15	19	6
Hungary	13	41	31	9	16	0	3	13
Poland	41	31	22	2	35	11	17	12
Russia	70	50	70	3	67	28	43	29
Near East								
Israel	21	53	39	5	33	5	32	17
Turkey	26	38	24	0	35	14	14	10
Latin America								
Argentina	8	46	42	15	31	7	8	15
Brazil	44	55	63	12	38	9	14	12
Chile	14	43	36	0	39	0	14	14
Colombia	39	39	54	0	23	0	0	31
Cuba	21	64	71	7	36	29	50	21
Mexico	27	66	64	3	48	15	27	12
Peru	23	46	46	15	33	8	17	0
Venezuela	29	69	67	0	50	0	0	29
Asia								
China	72	70	72	10	62	70	83	42
India	57	87	65	18	77	50	59	64
Japan	59	37	17	0	12	3	7	4
Thailand	42	63	42	17	46	33	33	22
Total geneticists	33	34	33	3	26	11	15	10
Total, excluding US	36	41	41	4	29	16	21	13

*On public questionnaire, "test results."

Table 70. Telling a School System a Diagnosis of XYY. *Parents of a mildly retarded 6-year-old son have received a questionnaire from the school system asking whether their son has had genetic testing and its results. The boy is XYY. % would counsel parents to

Country	Not answer questionnaire	Answer it, do not mention test results	Disclose symptoms but not XYY diagnosis	Disclose diagnosis
English-speaking nations				
Australia	0	13	33	40
Canada	5	15	47	28
South Africa	0	6	50	25
United Kingdom	7	17	40	26
United States	—	—	—	—
US Primary Care Physicians	4	13	29	53
US patients	3	1	n/a	96
Northern/Western Europe				
Belgium	13	27	33	13
Denmark	4	19	26	37
Finland	5	9	41	32
France	16	47	24	3
French patients (n = 383)	8	n/a	n/a	92
Germany	16	21	46	7
Netherlands	4	27	35	23
Norway	0	11	33	33
Sweden	9	0	46	18
Switzerland	0	17	33	33
Southern Europe				
Greece	0	0	58	33
Italy	5	14	29	33
Portugal	9	18	46	18
Spain	0	10	55	24
Eastern Europe				
Czech Republic	5	22	53	12
Hungary	9	15	50	24
Poland	7	12	49	17
Russia	13	16	40	4
Near East				
Israel	5	9	55	27
Turkey	0	5	32	32
Latin America				
Argentina	0	8	39	54
Brazil	0	10	44	44
Chile	13	19	25	25
Colombia	0	21	29	36
Cuba	0	0	0	50
Mexico	2	5	24	53
Peru	0	7	29	43
Venezuela	13	20	47	7
Asia				
China	3	10	24	61
India	0	13	22	48
Japan	12	17	12	12
Thailand	4	4	24	56
Total geneticists	6	15	43	28
Total, excluding US	7	16	37	27

*8% said this case could not happen in their country.

Table 71. DNA Fingerprinting. % thought DNA fingerprinting should be required for

Country	Persons convicted of sex crimes	Persons convicted of other serious crimes	Persons charged with sex crimes	Persons charged with other serious crimes	Members of armed forces, to identify casualties	Newborns, to prevent mix-ups in hospitals	Applicants for passports
English-speaking nations							
Australia	100	100	50	50	57	0	7
Canada	93	86	71	62	56	21	17
South Africa	93	80	73	53	80	47	27
United Kingdom	96	82	79	65	47	26	15
United States	97	91	67	59	80	38	20
US Primary Care Physicians	96	89	69	61	73	42	26
US patients	97	95	77	72	86	73	n/a applicants for credit cards 34
Northern/Western Europe							
Belgium	93	87	93	93	39	7	14
Denmark	96	92	80	72	46	8	4
Finland	96	91	68	64	62	29	19
France	92	87	87	83	59	17	4
French patients	93	88	80	79	60	47	12
Germany	86	81	89	84	35	12	4
German patients (n = 593)	94	83	82	71	62	31	10
Netherlands	93	89	85	81	54	4	4
Norway	100	88	63	38	38	38	0
Sweden	93	92	83	67	42	8	17
Switzerland	100	100	100	100	50	0	0
Southern Europe							
Greece	91	91	82	82	36	18	9
Italy	91	86	81	76	24	33	10

Portugal	100	100	89	89	60	40	0
Spain	98	96	96	94	81	51	17
Eastern Europe							
Czech Republic	100	95	89	87	52	2	4
Hungary	85	85	71	74	44	32	6
Poland	99	100	93	81	68	39	13
Russia	100	95	98	88	86	54	21
Near East							
Israel	91	75	96	68	79	33	10
Turkey	96	95	96	96	76	68	23
Latin America							
Argentina	93	93	100	100	79	71	50
Brazil	89	85	82	76	75	55	28
Chile	93	80	87	73	53	40	33
Colombia	100	100	93	93	86	71	57
Cuba	100	100	100	100	92	85	31
Mexico	95	94	94	92	75	65	33
Peru	93	86	93	79	79	71	39
Venezuela	100	100	100	100	93	81	33
Asia							
China	98	87	96	81	80	73	56
India	100	96	96	81	64	41	36
Japan	79	74	66	59	83	40	19
Thailand	100	100	92	92	84	52	48
Total geneticists would fingerprint	95	89	79	71	69	38	21
Total would fingerprint, excluding US	93	87	86	78	63	38	21
Total geneticists would keep DNA on file	89	84	58	53	55	29	19
Total would keep DNA on file, excluding US	90	90	70	66	57	37	27

Table 72. Screening in the Workplace. Assume that there is an accurate, simple, reliable mass screening test for alpha-l-antitrypsin deficiency that could screen factory workers exposed to dust and smoke. % thought test should be

Country	Offered at all	Voluntary		Offered with financial incentives to employees	Offered with penalties for refusal	Reimbursed by public insurance
		1985	1995			
English-speaking nations						
Australia	92	92	100	20	20	87
Canada	99	73	98	5	4	86
South Africa	100	—	100	6	13	81
United Kingdom	99	87	99	4	4	87
United States	98	77	94	14	6	78
US Primary Care Physicians	96	—	88	20	14	52
US patients	n/a	—	51	n/a	n/a	n/a
US public	n/a	—	39	n/a	n/a	n/a
Northern/Western Europe						
Belgium	100	—	100	0	0	93
Denmark	89	73	100	19	0	92
Finland	100	—	100	17	0	77
France	86	53	97	7	0	85
Germany*	92	94	96	8	3	75
German patients ($n = 593$)	—	—	61	—	—	—
Netherlands	93	—	100	20	0	88
Norway	78	84	89	0	0	75
Sweden	82	91	83	22	0	75
Switzerland	100	100	100	33	0	67
Southern Europe						
Greece	100	83	91	46	0	100
Italy	79	82	100	12	6	84
Portugal	100	—	90	30	0	100
Spain	90	—	95	21	3	91

Eastern Europe						
Czech Republic	94	—	72	52	19	—
Hungary	94	27	81	48	12	93
Poland	93	—	83	26	13	90
Russia	97	—	46	14	14	84
Near East						
Israel	100	73	85	18	9	86
Turkey	90	20	88	68	14	81
Latin America						
Argentina	92	—	93	30	0	69
Brazil	99	34	77	19	0	76
Chile	81	—	100	20	7	93
Colombia	90	—	75	20	0	85
Cuba	100	—	80	25	0	92
Mexico	85	—	78	30	5	81
Peru	100	—	73	63	29	88
Venezuela	92	—	79	58	0	75
Asia						
China	95	—	58	85	25	93
India	96	26	96	52	27	82
Japan	68	63	96	31	6	43
Thailand	83	—	84	54	22	87
Total geneticists	94	72	88	25	8	81
Total, excluding US	92	68	85	31	9	82

*1985 figures are for West Germany.

Table 73. Access to Results of Employment-Related Tests. % thought should be access for

Country	Employer				Government Health Department						Union		
	With worker's consent		No access at all		With worker's consent		Without consent		No access at all		With worker's consent	Without consent	No access at all
	1985	1995	1985	1995	1985	1995	1985	1995	1985	1995			
English-speaking nations													
Australia	100	80	0	20	25	73	33	0	42	27	67	0	33
Canada	70	65	21	34	55	54	19	6	26	41	63	2	35
South Africa	—	75	—	19	—	75	—	6	—	19	81	6	13
United Kingdom	74	78	10	22	62	81	19	3	19	16	77	0	23
United States	64	76	12	21	51	69	22	4	27	27	74	2	24
US Primary Care Physicians	—	84	—	8	—	72	—	9	—	19	78	5	17
US patients	—	78	—	5	—	n/a	—	n/a	—	n/a	66	5	29
Northern/Western Europe													
Belgium	—	20	—	80	—	47	—	13	—	40	40	13	47
Denmark	53	65	47	35	57	73	14	8	29	19	65	0	35
Finland	—	67	—	33	—	62	—	0	—	38	53	0	47
France	23	55	77	44	37	38	50	4	13	58	14	0	86
Germany*	32	44	64	56	28	43	28	3	44	54	56	3	41
German patients (n = 593)	—	71	—	22	—	—	—	—	—	—	—	—	—
Netherlands	—	54	—	46	—	39	—	4	—	58	48	0	52
Norway	17	44	83	44	34	67	33	11	33	22	44	11	44
Sweden	65	46	35	54	53	55	21	0	26	46	46	0	54
Switzerland	40	67	60	33	40	33	0	0	60	67	50	0	50

Southern Europe													
Greece	50	78	0	11	17	78	83	22	0	0	89	11	0
Italy	72	45	27	40	27	50	64	25	9	25	44	31	25
Portugal	—	64	—	27	—	46	—	27	—	27	82	9	9
Spain	—	73	—	24	—	70	—	23	—	6	43	0	57
Eastern Europe													
Czech Republic	—	70	—	23	—	68	—	21	—	11	62	23	15
Hungary	50	81	29	16	22	63	71	19	7	19	68	10	23
Poland	—	69	—	28	—	59	—	30	—	11	66	28	6
Russia	—	65	—	5	—	53	—	47	—	0	52	46	3
Near East													
Israel	93	73	7	23	50	64	29	5	21	32	65	0	35
Turkey	20	65	20	35	0	70	100	25	0	5	81	10	10
Latin America													
Argentina	—	50	—	43	—	79	—	7	—	14	50	7	43
Brazil	58	77	10	16	29	67	71	26	10	7	81	13	6
Chile	—	60	—	27	—	60	—	27	—	13	69	15	15
Colombia	—	62	—	23	—	62	—	39	—	0	38	8	54
Cuba	—	64	—	27	—	46	—	46	—	8	54	15	31
Mexico	—	68	—	16	—	48	—	46	—	5	54	7	39
Peru	—	78	—	11	—	33	—	67	—	0	67	0	33
Venezuela	—	93	—	0	—	69	—	25	—	6	53	0	47
Asia													
China	—	67	—	3	—	41	—	58	—	2	47	50	3
India	27	68	14	9	23	65	68	30	9	4	52	48	0
Japan	54	66	23	27	27	70	64	7	9	23	73	5	22
Thailand	—	76		14	—	33	—	48	—	19	83	17	0
Total geneticists	59	68	22	25	43	60	32	16	25	24	64	11	25
Total, excluding US	55	64	30	28	37	55	40	22	23	22	58	16	26

*1985 figures are for West Germany.

Table 74. The Susceptible Worker. If a worker is found susceptible to work-related disease, and the job cannot be made safer, employer should (% agreeing)

Country	Shift worker to less whether to hazardous job that pays the same	Let worker decide stay in same job or move	If there is no other job to which worker can be shifted, let worker stay on, after explaining risks
English-speaking nations			
Australia	33	67	100
Canada	21	79	94
South Africa	25	75	100
United Kingdom	23	77	98
United States	25	75	98
US Primary Care Physicians	44	56	91
US patients	30	70	90
Northern/Western Europe			
Belgium	29	71	100
Denmark	19	81	100
Finland	23	77	100
France	35	65	61
French patients	40	57	—
Germany	24	76	97
German patients	28	72	—
Netherlands	78	22	30
Norway	38	62	88
Sweden	17	83	100
Switzerland	20	80	100
Southern Europe			
Greece	45	55	100
Italy	45	55	88
Portugal	64	36	100
Spain	75	25	89
Eastern Europe			
Czech Republic	33	67	86
Hungary	50	50	94
Poland	43	57	96
Russia	40	60	97
Near East			
Israel	32	68	76
Turkey	40	60	77
Latin America			
Argentina	100	0	92
Brazil	67	33	66
Chile	62	38	93
Colombia	n/a	n/a	100
Cuba	100	0	92
Mexico	n/a	n/a	97
Peru	0	100	100
Venezuela	n/a	n/a	94
Asia			
China	74	26	78
India	48	52	96
Japan	29	71	94
Thailand	72	28	83
Total geneticists	36	64	91
Total, excluding US	42	58	88

Table 75. Privacy Versus Public Safety. A public bus driver with familial hypercholesterolemia is at high risk of heart attack or stroke while working. He has a blood cholesterol of over 400 and high blood pressure. He refuses to tell his employer or to retire on disability. % would

Country	Preserve patient confidentiality	Tell transportation department about patient	Warn transportation dept. that passengers are at risk because of medical status of a driver, and request complete physical examinations of all drivers	If dept. will not do this, tell them about patient
English-speaking nations				
Australia	53	20	27	0
Canada	51	20	24	6
South Africa	31	6	56	6
United Kingdom	66	9	20	5
United States	62	4	29	5
US Primary Care Physicians	38	13	34	16
US patients*	20	69	n/a	n/a
US public* (n = 476)	27	58	n/a	n/a
Northern/Western Europe				
Belgium	64	0	29	7
Denmark	75	11	14	0
Finland	77	5	13	5
France	50	3	46	1
French patients (n = 390)	6	61	n/a	n/a
Germany	75	7	17	1
German patients (n = 593)	11	38	n/a	n/a
Netherlands	85	0	15	0
Norway	63	0	25	12
Sweden	83	0	17	0
Switzerland	75	25		
Southern Europe				
Greece	42	0	50	8
Italy	48	0	43	10
Portugal	46	0	46	9
Spain	49	12	35	4

Table 75. (continued)

Country	Preserve patient confidentiality	Tell transportation department about patient	Warn transportation dept. that passengers are at risk because of medical status of a driver, and request complete physical examinations of all drivers	If dept. will not do this, tell them about patient
Eastern Europe				
Czech Republic	34	19	28	19
Hungary	6	6	69	20
Poland	22	5	49	24
Russia	26	0	45	29
Near East				
Israel	55	9	32	5
Turkey	27	14	46	14
Latin America				
Argentina	71	7	21	0
Brazil	27	10	36	27
Chile	27	0	53	20
Colombia	57	7	29	7
Cuba	43	21	14	21
Mexico	23	13	39	25
Peru	31	8	39	23
Venezuela	27	7	53	13
Asia				
China	8	23	27	43
India	22	9	26	44
Japan	38	3	50	9
Thailand	20	16	32	32
Total geneticists	49	8	31	12
Total, excluding US	42	10	32	16

*Patient and public questionnaires described a school bus driver, not a public bus driver.

Table 76. "Genetic Discrimination": Provider Reports

	% of geneticists reporting any refusals at all	Number of individuals or family members refused*
Reported refusals of life insurance because of carrier status		
Australia	13	—
Canada	17	8
United Kingdom	17	14
United States	12	237
Netherlands	17	7
Germany	3	7
3 other countries	—	4
Total worldwide	7	277
Life insurance because of genetic predisposition in absence of symptoms		
Australia	47	40
Canada	25	29
United Kingdom	38	66
United States	14	260
Netherlands	20	6
3 other countries	—	13
Total worldwide	9	414
Refusals of employment because of carrier status		
United States	4	20
Belgium, France, Italy, UK, Germany	0.5–3	7
Total worldwide	2	27
Refusals of employment because of genetic predisposition, in absence of symptoms		
United Kingdom	11	5
United States	3	23
Germany	2	4
Spain	2	20
5 other countries	—	9
Worldwide total	3	61
Total reported refusals of employment or insurance		779
In USA		540*
Outside USA		239

*These figures are incomplete. This column lists only numbers actually reported. In USA, 62 geneticists who reported refusals did not report the number of individuals refused life insurance on the basis of predisposition, 49 did not report the number refused life insurance because of carrier status, 22 did not report the number of refusals of employment on the basis of carrier status, and 20 did not report the number of refusals of employment because of predisposition, for a total of 153 reports of refusals unaccompanied by numbers. Even if each had only one refusal, the total for the US would be 693.

Table 77. Warning about Risks to Insurance and Employment. A woman age 35 wants to be tested for autosomal dominant polycystic kidney disease (ADPKD) to see whether she can be a kidney donor for her brother, who has ADPKD. % would tell her that if the test shows that she has a mutation for ADPKD, it might be difficult to

Country	Get life insurance	Get employment, if employers can access her records
English-speaking nations		
Australia	80	33
Canada	66	42
South Africa	80	60
United Kingdom	77	56
United States	80	63
US Primary Care Physicians	86	48
Northern/Western Europe		
Belgium	40	33
Denmark	63	44
Finland	32	14
France	22	11
Germany	40	27
Netherlands	82	56
Norway	33	33
Sweden	18	18
Switzerland	17	0
Southern Europe		
Greece	46	36
Italy	5	5
Portugal	26	36
Spain	27	8
Eastern Europe		
Czech Republic	17	15
Hungary	60	65
Poland	36	35
Russia	33	42
Near East		
Israel	27	29
Turkey	46	41
Latin America		
Argentina	29	33
Brazil	66	65
Chile	60	47
Colombia	50	75
Cuba	23	17
Mexico	53	45
Peru	50	36
Venezuela	38	27
Asia		
China	27	27
India	78	57
Japan	21	22
Thailand	44	36
Total geneticists	54	43
Total, excluding US	42	23

Table 78. Disability Laws and Genetics. % said if laws exist they cover

Country	Country has laws prohibiting employment discrimination against persons with disabilities	Asymptomatic persons with ADPKD mutations	People with genetic predispositions to disorders such as schizophrenia	CF carriers
English-speaking nations				
Australia	93	77	69	69
Canada	93	68	69	68
South Africa	20	100	100	100
United Kingdom	81	34	36	33
United States	97	56	52	50
US Primary Care Physicians	n/a	59	56	55
Northern/Western Europe				
Belgium	67	75	75	75
Denmark	46	80	80	80
Finland	57	73	64	64
France	72	53	53	48
Germany	85	42	41	40
Netherlands	81	75	75	69
Norway	100	57	57	57
Sweden	83	88	88	88
Switzerland	0	n/a	n/a	n/a
Southern Europe				
Greece	27	100	100	100
Italy	33	40	33	22
Portugal	55	67	67	67
Spain	40	72	77	77
Eastern Europe				
Czech Republic	43	53	51	50
Hungary	67	8	8	8
Poland	43	42	43	41
Russia	33	31	29	29
Near East				
Israel	56	67	50	58
Turkey	59	75	75	75
Latin America				
Argentina	8	100	100	100
Brazil	57	22	22	22
Chile	13	50	50	50
Colombia	50	83	83	83
Cuba	23	80	80	80
Mexico	5	25	25	25
Peru	14	33	33	33
Venezuela	0	0	0	0
Asia				
China	36	43	50	44
India	35	0	11	0
Japan	78	57	64	57
Thailand	4	25	25	25
Total geneticists	69	52	50	48
Total, excluding US	56	48	49	47

I. Screening and Public Health

Table 79. Newborn Screening: Blood Spots

Country	If DNA automation techniques are available on blood spots throughout, classical PKU screening programs, would screen for				Blood spots should be kept indefinitely	Children should have access to blood spots at age of legal majority
	CF*	FH**	Fragile-X	Thalassemia		
English-speaking nations						
Australia	86	79	71	86	86	67
Canada	74	70	68	79	52	71
South Africa	81	75	81	88	44	56
United Kingdom	86	69	70	80	86	78
United States	81	78	66	82	51	75
Northern/Western Europe						
Belgium	100	87	87	87	60	73
Denmark	73	65	46	69	62	65
Finland	73	68	59	62	48	76
France	91	70	72	93	80	79
Germany	67	63	38	54	35	73
Netherlands	71	61	46	65	46	78
Norway	75	75	75	75	63	57
Sweden	75	67	58	83	91	58
Switzerland	83	50	17	83	33	67
Southern Europe						
Greece	100	91	91	100	73	91
Italy	95	91	86	95	62	81
Portugal	100	91	73	91	46	90
Spain	86	81	83	89	55	80

Eastern Europe						
Czech Republic	98	88	77	65	55	77
Hungary	88	85	97	91	74	85
Poland	91	89	82	83	51	82
Russia	98	83	93	88	37	71
Near East						
Israel	86	86	76	81	42	94
Turkey	100	91	91	96	55	86
Latin America						
Argentina	92	93	93	92	69	100
Brazil	89	87	80	83	48	83
Chile	88	81	80	73	29	69
Colombia	100	86	100	92	79	86
Cuba	100	100	100	100	31	85
Mexico	98	91	94	94	50	86
Peru	79	83	92	100	67	75
Venezuela	86	86	86	86	47	93
Asia						
China	88	82	82	93	39	66
India	96	91	91	100	41	55
Japan	47	56	45	47	11	46
Thailand	92	92	84	96	68	76
Total	82	77	69	79	50	74
Total, excluding US	82	77	71	78	49	74

*CF = cystic fibrosis.
**FH = familial hypercholesterolemia.

Table 80. Newborn Screening: Goals. % agreed that

Country	Governments should require newborn screening for Duchenne muscular dystrophy	An important goal of newborn screening is to identify and counsel parental carriers before next pregnancy
English-speaking nations		
Australia	0	80
Canada	15	74
South Africa	13	88
United Kingdom	13	86
United States	11	73
US primary care physicians	n/a	90
US patients	27	71
Northern/Western Europe		
Belgium	13	93
Denmark	7	78
Finland	5	68
France	21	81
Germany	9	52
Netherlands	7	67
Norway	0	56
Sweden	0	58
Switzerland	17	50
Southern Europe		
Greece	50	83
Italy	19	81
Portugal	55	100
Spain	29	81
Eastern Europe		
Czech Republic	38	83
Hungary	17	69
Poland	28	99
Russia	54	91
Near East		
Israel	32	86
Turkey	48	95
Latin America		
Argentina	14	92
Brazil	20	88
Chile	19	69
Colombia	43	93
Cuba	36	100
Mexico	34	94
Peru	39	100
Venezuela	21	100
Asia		
China	71	65
India	65	87
Japan	13	40
Thailand	62	68
Total geneticists	23	74
Total, excluding US	29	75

Table 81. Cystic Fibrosis Carrier Testing*. % thought CF carrier testing should be offered, routinely, *without a family history of* CF

Country	Persons considering having children	Pregnant women	Persons considering marriage	Newborns		Children under 13	Teenagers 13–17	Only high-risk ethnic groups	Everyone
				1985	1995				
English-speaking nations									
Australia	93	71	87	59	20	7	36	33	43
Canada	77	66	63	27	22	6	30	63	23
South Africa	80	53	73	—	40	7	13	67	7
United Kingdom	91	77	73	30	26	6	42	27	46
United States	75	72	60	22	32	14	23	57	37
US Primary Care Physicians	68	72	78	—	57	38	37	44	n/a
US patients	87	77	59	—	63	n/a	n/a	15	59
Northern/Western Europe									
Belgium	93	67	85	—	33	10	25	25	25
Denmark	48	70	36	20	16	0	4	33	42
Finland	75	50	50	—	0	0	0	100	25
France	96	77	71	12	35	11	17	42	43
Germany**	52	44	27	30	21	7	6	25	42
Netherlands	92	46	50	—	17	4	9	35	54
Norway	44	56	33	33	22	11	11	56	33
Sweden	60	55	36	29	17	9	0	64	17
Switzerland	83	67	83	80	17	0	0	0	50
Southern Europe									
Greece	80	80	70	0	30	11	13	50	50
Italy	72	60	61	0	32	11	32	50	13
Portugal	70	70	70	—	30	10	13	50	44
Spain	86	71	48	—	62	35	19	77	35

Table 81. (continued)

Country	Persons considering having children	Pregnant women	Persons considering marriage	Newborns		Children under 13	Teenagers 13–17	Only high-risk ethnic groups	Everyone
				1985	1995				
Eastern Europe									
Czech Republic	84	80	60	—	60	16	25	73	34
Hungary	69	78	53	7	50	16	33	34	41
Poland	75	71	58	—	62	32	24	68	27
Russia	83	55	62	—	78	19	15	67	52
Near East									
Israel	70	61	44	7	14	5	14	81	27
Turkey	60	50	39	0	57	15	0	100	23
Latin America									
Argentina	70	36	55	—	50	20	27	90	33
Brazil	60	39	39	16	60	29	23	75	28
Chile	25	25	25	—	60	25	0	75	25
Colombia	88	57	88	—	88	38	38	38	38
Cuba	90	70	50	—	56	0	20	100	0
Mexico	83	59	44	—	75	31	9	88	14
Peru	100	50	100	—	75	100	50	100	50
Venezuela	100	100	0	—	100	50	0	67	33
Asia									
China***	n/a	n/a	n/a	—	n/a	n/a	n/a	n/a	n/a
India	75	83	75	31	80	50	25	100	25
Japan	63	71	44	51	36	36	41	65	79
Thailand	65	25	54	—	22	14	9	83	13
Total geneticists	74	66	55	25	36	14	21	54	36
Total, excluding US	73	61	52	28	38	15	19	52	36

*Question assumes test identifies 90% or more in high risk ethnic groups (persons of European descent), but far fewer in other groups, and costs US$ 30–50.
**West Germany in 1985 survey.
***Cystic fibrosis is extremely rare in China.

Table 82. Regulating DNA Banks. % thought DNA banks should

Country	Be permitted to share anonymous DNA with researchers, without depositor's consent	Be required to get depositor's permission each time a sample is used in research, even if anonymous	Give depositors a right to share in profits from commercialization of such research	Require depositors to agree, in advance, to allow access for at-risk relatives	Be supported by public agencies, if done for a person's descendants
English-speaking nations					
Australia	21	36	7	73	93
Canada	55	25	19	53	81
South Africa	60	33	13	67	60
United Kingdom	57	29	19	65	91
United States	51	35	21	44	25
Northern/Western Europe					
Belgium	87	0	7	53	79
Denmark	48	23	4	64	86
Finland	59	18	9	50	82
France	75	28	4	74	75
Germany	56	33	13	38	54
Netherlands	46	31	24	40	70
Norway	88	13	0	50	89
Sweden	67	25	0	50	92
Switzerland	83	33	40	17	50
Southern Europe					
Greece	50	20	30	60	83
Italy	67	43	20	62	73
Portugal	40	20	20	56	82
Spain	53	41	23	70	84
Eastern Europe					
Czech Republic	52	47	46	68	79
Hungary	55	42	18	91	69
Poland	69	25	27	50	70
Russia	41	45	74	71	81

Table 82. (continued)

Country	Be permitted to share anonymous DNA with researchers, without depositor's consent	Be required to get depositor's permission each time a sample is used in research, even if anonymous	Give depositors a right to share in profits from commercialization of such research	Require depositors to agree, in advance, to allow access for at-risk relatives	Be supported by public agencies, if done for a person's descendants
Near East					
Israel	73	18	18	59	82
Turkey	41	41	29	65	82
Latin America					
Argentina	64	14	21	64	54
Brazil	47	42	36	64	51
Chile	44	56	33	69	69
Colombia	43	50	50	64	61
Cuba	54	39	62	77	100
Mexico	67	33	44	94	52
Peru	36	57	57	92	77
Venezuela	50	50	57	86	67
Asia					
China	55	37	55	74	69
India	48	48	44	74	46
Japan	23	57	22	52	34
Thailand	56	72	68	92	64
Total	53	36	26	55	54
Total, excluding US	55	36	30	62	68

J. Ethical Reasoning

Table 83. Questions that Respondents Found Most Difficult

	% listing
1. Which disorders to abort	41
2. Sex selection by prenatal diagnosis	29
3. Parenthood for persons with disabilities	28
4. Telling blood relatives against a patient's wishes	25
5. False paternity	24
6. Third party access to genetic information (employers, insurers, schools, etc.)	17

Table 84. Overview of Reasons for Choices (n = 44, 178 qualitative responses in respondents' own words). % of responses

	US geneticists	US patients (n = 3,102 responses)	Geneticists outside US	Total geneticists
Basis of Reasoning				
Principles	9	9	9	9
Action rules (practical short-cuts for principles)	70	61	47	54
Experience	2	2	1	1
None given	20	28	43	36
Consequences mentioned or implied (notable exception: China 59%)	27	35	33	31
Relationships mentioned or implied				
None	84	75	84	84
With patient	7	1	4	5
Patient's own relationships, interfamilial	4	6	3	3
Patient's own relations, marital	1	2	2	2
With patient's family	2	4	2	2
With society (China = 10%)	1	4	3	2
With own professional integrity	1	n/a	2	2
Principles				
Autonomy (total responses)	44	35	34	38
Respect for persons	22	11	13	16
Right to know	6	4	6	6
Diminished autonomy	4	3	3	3
Need information for decision-making	7	11	7	7
Right to privacy	6	5	4	5
Truth telling	2	1	1	1
Prevention of harm (total)	20	21	26	24
Deserts (special characteristics making people more or less deserving) total	12	15	13	13
Depends on patient characteristics	7	9	7	8
Perceptions of disability	3	4	5	5
Beneficence (China = 24)	4	11	11	9
Prepare for future	1	2	2	2
Improvement of life	3	9	9	7

Table 84. (continued)

	US geneticists	US patients (n = 3102 responses)	Geneticists outside US	Total geneticists
Justice (total)	10	4	7	8
Allocation of resources	5	1	5	5
Fairness	2	1	2	2
Utility	3	2	1	1
Nonmoral answers	8	14	9	8
Profit-making by geneticists	—	3	—	—

Table 85. Ethical Principles Behind Qualitative Responses (n = 44, 146 written-in responses giving reasons for choices). % responses

Country	Autonomy	Non-maleficence	Deserts	Beneficence	Justice	Non-moral reasons
English-speaking nations						
Australia	39	29	10	3	10	10
Canada	38	27	10	4	10	10
South Africa	42	24	13	9	4	10
United Kingdom	42	21	12	4	7	13
United States	44	20	12	4	10	8
Northern/Western Europe						
Belgium	48	26	13	5	7	8
Denmark	39	24	13	4	6	12
Finland	45	26	8	4	9	7
France	31	30	14	6	8	11
Germany	39	26	16	5	5	9
Netherlands	38	25	13	5	10	11
Norway	31	29	17	9	6	5
Sweden	37	24	16	7	12	14
Switzerland	28	29	14	7	10	12
Southern Europe						
Greece	30	34	21	6	5	4
Italy	36	29	13	5	11	6
Portugal	41	28	12	5	4	9
Spain	41	24	13	7	8	8
Eastern Europe						
Czech Republic	34	26	17	9	7	8
Hungary	34	21	24	8	6	7
Poland	39	28	14	8	8	6
Russia	31	28	21	10	6	5
Near East						
Israel	54	23	12	4	7	6
Turkey	32	29	14	10	8	6
Latin America						
Argentina	43	21	13	7	10	6
Brazil	42	24	12	7	8	7
Chile	37	24	10	6	14	8
Colombia	50	20	10	5	9	5

Table 85. (continued)

Country	Autonomy	Non-maleficence	Deserts	Beneficence	Justice	Non-moral reasons
Cuba	40	24	12	13	7	4
Mexico	37	27	11	12	8	6
Peru	36	30	9	8	10	7
Venezuela	38	29	9	8	11	6
Asia						
China	20	26	13	27	6	8
India	25	31	15	15	7	9
Japan	41	25	13	6	8	10
Thailand	40	26	8	10	8	8
Total	38	24	13	9	8	8

Table 86. Gender Differences in Controversial Uses of Prenatal Diagnosis (%)

Case	US M.D.s		P	Non-US M.D.s		P	US genetic counselors (n = 555)
	Men n = 186	Women n = 152		Men n = 640	Women n = 647		
PND for sex selection, (average of 5 cases)*			0.010			0.001	
Perform PND	30	33		24	29		32
Offer referral	27	40		8	11		46
Refuse PND	43	27		68	60		22
Attempt to dissuade	30	28	ns	53	55	ns	20
PND for a deaf couple who will abort a hearing fetus			0.050			ns	
Perform PND	27	29		13	14		42
Refuse PND	46	34		81	78		23
Offer referral	27	37		6	8		35
Perform prenatal paternity testing for woman involved with 2 men	57	55	ns	40	47	ns	54
PND for maternal anxiety	79	87	ns	54	61	0.050	92
PND provided regardless of use	43	52	0.020	30	29	ns	47
Patients are entitled to any service they can pay for	26	35	0.001	30	31	ns	43
Withholding any service is paternalistic	47	60	0.020	41	38	0.020	59

*The 5 different case vignettes on sex selection are described in the text. As differences in responses to the various cases were small, these have been averaged.

Table 87. Gender Differences in Nondirectiveness in Counseling (%)

	US M.D.s		P	Non-US M.D.s		P	US genetic counselors $n = 555$
	Men n = 186	Women $n = 152$		Men $n = 640$	Women $n = 647$		
Counseling after prenatal diagnosis							
Mother's life in danger urge termination	59	76	0.010	72	71	ns	51
Would be nondirective about							
Trisomy 21	71	85	0.020	38	40	ns	94
XXY (Klinefelter syndrome)	78	80	ns	41	38	ns	90
Cleft lip/palate in a boy	72	61	0.050	35	31	ns	71
Cleft lip/palate in a girl	67	61	ns	33	26	0.001	70

Table 88. Gender Differences Regarding Privacy and Disclosure (%)

Case	US M.D.s		P	Non-US M.D.s		P	US genetic counselors $n = 555$
	Men $n = 186$	Women $n = 152$		Men $n = 640$	Women $n = 647$		
1. Disclose genetic information to relatives at genetic risk, against a patient's wishes*			0.020			<0.001	
• Maintain patient confidentiality	56	50		38	25		56
• Tell relatives, unasked	14	7		17	20		6
• Tell relatives if they ask	23	30		29	39		30
• Refer decision to family physician	7	13		16	16		8
2. Disclose genetic risk of schizophrenia to relatives, against patient's wishes	29	27	ns	32	50	<0.001	29
3. Disclose nonpaternity							
• Tell mother alone before test that the test may reveal nonpaternity	72	61	0.001	55	58	ns	58
• Disclose to husband							
Who does not ask	2	1	ns	3	2	ns	3
Who asks directly	37	39	ns	25	17	0.001	35

Table 88. (continued)

Case	US M.D.s		P	Non-US M.D.s		P	US genetic counselors
	Men $n = 186$	Women $n = 152$		Men $n = 640$	Women $n = 647$		$n = 555$
4. Access to genetic information for life insurers**			0.03			.03	
• With consent	59	46		49	44		43
• Without consent	0	0		6	11		0
• No access at all	40	54		45	45		57
5. Access to genetic information (HD) without consent for							
• employers, schools, government	1	3	ns	4	6	ns	1
• spouses	13	9	ns	19	30	<0.001	4
• blood relatives	11	15	ns	19	27	<0.001	7
• patient's minor children	8	7	ns	12	14	ns	3

*Average of responses to 4 genetic conditions; Huntington disease, Familial hypercholesterolemia, Hemophilia A, Li–Fraumeni syndrome (familial cancer). Responses to all conditions varied by less than 3% and have been averaged.
**Average of 3 cases: CF carrier status, translocation Down, predisposition to alcoholism.

Table 89. Sociodemographic Characteristics of Genetics Professionals, by Gender

Characteristic	US M.D.s		P	Non-US M.D.s		P	US genetic counselors
	Men $n = 186$	Women $n = 152$		Men $n = 640$	Women $n = 647$		(%) $n = 555$
Years in genetics (median)	14	9	0.002	10	9	<0.001	7
>50% time in direct patient care	37%	58%	0.010	25%	35%	<0.001	70%
>25% time in research	37%	18%	0.004	33%	36%	ns	10%
Based at university	55%	40%	0.030	41%	34%	0.002	40%
Age (median)	47	42	<0.001	46	41	<0.001	34
Married	87%	76%	0.050	88%	69%	<0.001	64%
No children	5%	15%	0.040	5%	12%	<0.001	41%
Income >US$ 100,000	71%	50%	0.001	14%	2%	<0.001	4%
Income <US$ 50,000	2%	11%	0.001	42%	87%	<0.001	81%

Table 90. Differences between US Geneticists, Genetic Counselors, and Primary Care Physicians

Area	M.D./Ph.D. geneticists (n = 529)	Genetic counselors (n = 555)	Primary care physicians (n = 499)
1. Autonomy			
Patients entitled to any service they can pay for	30	43	26
Women should have unqualified right to abortion before viability	84	87	63
Provider who refuses sex selection should offer referral	49	62	41
After taking a test, patients have right not to know results	83	82	62
A woman's decision about abortion should be her own, without outside intervention	81	89	65
Disclose colleague of disagreement about meaning of test results	82	83	97
PND for sex selection			
Perform	31	32	34
Refer	33	46	34
Refuse	36	22	32
PND for deaf couple desiring deaf child			
Perform	28	42	21
Refer	30	35	29
Refuse	41	23	51
Preconception sex selection should be available to all	38	38	24
Prenatal paternity testing			
Perform, no comment	36	53	21
Perform with negative comments	22	26	41
Support decision of pregnant woman with PKU	54	66	39
Support parenthood for a woman with fragile X	44	62	n/a
Blind woman on welfare			
Persuade her not to have children	36	15	69
Suggest sterilization	25	10	47
Support decision of			
(a) blind couple to have a blind child	88	98	62
(b) CF carrier couple to take chances of having child with CF	79	96	63
Bringing a child into the world with serious disabilities is			
Unfair to child	48	26	52
Unfair to sibs	28	15	32
Socially irresponsible in an era of PND	38	16	55

Table 90. (continued)

Area	M.D./Ph.D. geneticists (n = 529)	Genetic counselors (n = 555)	Primary care physicians (n = 499)
2. Counseling			
Would present option of surrogacy to			
CF carrier couple	60	57	34
PKU mother	78	79	50
Counseling after PND as unbiased as possible for			
CF	82	95	61
NF	84	94	61
Sickle cell anemia	82	94	58
Achondroplasia	83	93	
Trisomy 21	77	94	63
HD	78	93	61
Rape	78	93	66
XXY	78	90	66
Familial hypercholesterolemia	76	90	50
Schizophrenia/bipolar disorder	86	87	65
45,X	75	87	58
Toxoplasmosis	64	85	44
Rubella in fetus	60	85	41
PKU in fetus	71	83	52
Severe spina bifida	61	79	51
HIV	61	78	43
Trisomy 13	51	75	28
Mother's life in danger	38	50	33
Cleft lip/palate	65	70	39
Predisposition to Alzheimer	85	86	60
Predisposition to alcoholism	83	83	57
Obesity	80	83	47
Pessimistic counseling for			
Trisomy 13	49 (17)	25 (4)	70 (47)
Rubella	40 (14)	15 (3)	49 (25)
Rape	21 (11)	7 (2)	28 (17)
Severe spina bifida	39 (8)	21 (2)	42 (20)
Toxoplasmosis	34 (12)	13 (2)	44 (21)
CF	15 (2)	3	21 (8)
Sickle cell anemia	10 (1)	2	18 (5)
FH	19 (3)	3	3 (1)
HD	17 (3)	4	24 (10)
Trisomy 21	21 (5)	5	23 (10)
HIV in fetus	39 (15)	21 (5)	43 (25)
Optimistic counseling for			
Cleft lip/palate	33	30	57
Predisposition to Alzheimer	10	13	33
Alcoholism	14	17	39
Schizophrenia/bipolar disorder	9	12	27
45,X	18	13	28
XXY	14	9	19
NF	9	4	30
Familial hypercholesterolemia	5	7	45
Severe obesity	18	17	52

Table 90. (continued)

Area	M.D./Ph.D. geneticists (n = 529)	Genetic counselors (n = 555)	Primary care physicians (n = 499)
Achondroplasia	8	5	41
"Wrong sex"	53	47	77
PKU in fetus	18	14	30
A woman should have PND if medically indicated	58	19	78
Would themselves abort for			
CF	72	52	47
HD	67	58	50
FH	60	37	10
45,X	47	37	37
Trisomy 21	84	76	62
XXY	54	39	48
PKU	49	38	45
Rape	88	87	81
Trisomy 13	94	93	93
Cleft lip/palate	17	7	10
Severe spina bifida	91	89	78
Sickle cell	61	43	39
Predisposition to			
Alzheimer	26	14	16
Alcoholism	14	6	8
Schizophrenia	33	21	24
NF	54	49	37
Severe obesity	31	27	13
Achondroplasia	58	55	29
Wrong sex	4	1	3
HIV	87	83	74
Toxoplasmosis	84	75	72
Rubella	86	82	79
3. *Family Relationships*			
Maintain confidentiality of patient with HD (do not tell relatives at risk)	50	57	61
Accidental findings of nonpaternity			
Disclose to mother only	53	48	31
Urge her to tell husband	39	45	74
Test children for			
HD	37	16	66
Alzheimer	34	15	58
Alcoholism	52	41	76
16 year old should be able to refuse test for treatable disorder	77	79	28
4. Privacy			
Tell bus driver's employer about his FH	36	12	13
5. Payment			
National public health insurance should pay for:			
DNA banking for benefit of future generations	33	18	13
Newborn screening for sickle cell anemia	59	66	53

Table 90. (continued)

Area	M.D./Ph.D. geneticists (n = 529)	Genetic counselors (n = 555)	Primary care physicians (n = 499)
CF carrier tests for:			
High-risk ethnic groups	73	70	25
Newborns	83	85	29
Would perform PND without medical indications	32	28	42
For sex selection	31	34	34
Workplace screening should be voluntary			
Alpha-1-antitrypsin deficiency	75	81	52
Heart disease, cancer, diabetes (executives only)	58	68	43

Table 91. US Patients and Professionals: Areas of Agreement. %'s agreeing

	Patients	M.D./Ph.D. geneticists	Genetic counselors	Primary care physicians
1. NO access to genetic information without consent, for insurers and employers	97	98	99	95
2. Genetic information about birthparents should be transmitted on anonymous basis to adoptees, if the adoptee wishes it, when adult				
Medical info up to time of adoption	93	88	94	85
Medical Info up to time of parents' deaths	90	78	90	76
Causes of death	92	79	83	82
Birth parents' names	23	13	6	11
3. Full disclosure of ambiguous information	95	93	96	n/a
4. Provider should perform PND or offer referral for sex selection	62	64	79	68
5. Provider should perform PND or offer referral for a deaf couple who want a deaf child	57	59	76	50
6. Provider should perform or refer for PND for paternity testing, for a woman with two boyfriends	87	93	96	91
7. DNA fingerprinting should be used for persons				
Convicted of sex crimes	97	98	98	96
Charged with other serious crimes	72	59	60	61
Members of armed forces	86	81	78	73
Newborns	73	34	40	42
Passport applicants	50	19	21	26

Table 92. US Patients and Professionals: Areas of Difference, % Agreeing

	Patients	M.D./Ph.D. geneticists	Genetic counselors	Primary care physicians
Individual Autonomy				
Patients are entitled to any service they request and can pay for out of pocket	59	30	38	26
Withholding any requested service is paternalistic	69	53	57	57
After taking a test, people have a right NOT to know results	41	81	82	62
2. *Uses of PND* PND should be:				
Provided regardless of its intended use	80	48	47	44
Done to relieve maternal anxiety, without medical indications	75	83	92	28
3. *Personal Willingness to Abort (examples)*				
Condition				
Trisomy 21	36	85	76	62
Achondroplasia	24	56	55	29
Obesity only	8	31	27	13
Huntington Disease	32	72	58	50
XXY	22	54	43	48
4. *Views on disability*				
Some disabilities will never be overcome, even with maximum social support	61	95	83	90
Most of the care for a child with a disability falls on mother	55	85	86	78
People with severe disabilities enrich society	19	21	29	19
Not fair to bring a child into world with serious genetic disorder if birth could be prevented	44	48	30	52
5. *Genetic testing*				
Couples should know carrier status before marriage	66	53	36	65
Women with family history of genetic disorder should use PND	81	54	19	78
People at high risk should not have children unless they use PND	40	12	5	34

Table 92. (continued)

	Patients	M.D./Ph.D. geneticists	Genetic counselors	Primary care physicians
6. *Directive counseling in some situations*				
Case: A blind woman on welfare has 3 blind sons. She wants another child, with 50% risk of being blind.				
Dissuade her from having a child	75	36	15	69
Suggest voluntary sterilization	46	25	10	47
Thought law should require sterilization	21	3	2	15
7. *Testing children* % favored testing children, at parents' request for predisposition to				
Alzheimer disease	61	34	15	58
Breast cancer	93	74	65	81
8. *Telling relatives about a genetic diagnosis of HD without patient's consent*				
%would tell relatives	75	35	38	39
9. *Access to DNA in a medical record* %favored automatic access for				
Spouse	74	24	18	25
Blood relatives	53	21	17	18
10. *Non-paternity*				
%Thought, provider should tell husband about accidental finding of nonpaternity, if he asks directly.	74	38	35	42
11. *Spouse's access* %thought spouse/partner should have automatic access to information, without consent for				
Down translocation	42	24	15	31
Risk for schizophrenia	32	7	1	14
CF carrier status	26	5	1	15
CF	29	9	1	15
Risk for alcoholism	24	6	1	15
DNA in medical record	54	27	18	25
12. *Telling a school system about a diagnosis of XYY*				
%would disclose diagnosis	96	33	26	53
13. *Confidentiality versus public safety*				
%would preserve confidentiality of a school bus driver with familial hypercholesterolemia, at high risk of stroke or heart attack while on the job	20	62	65	38

Table 93. International Comparisons of Patients' Views, % Agreeing

	US (n = 476)	Germany (n = 593)	France (n = 394)
1. *Autonomy*			
Parents should be told all information relative to health of fetus	98	99	92
Parents have right to know fetus' sex	92	91	86
If doctor refuses a procedure on moral grounds, owes referral	86	86	88
Patients have right not to know test results	40	35	35
People have a right to whatever services they can pay for	60	48	54
Counselors should provide support for patients' decisions	80	92	75
Parents have right to select sex	13	4	4
Doctor who refuses sex selection owes a referral	50	29	26
If law forbids procedure, doctor owes referral outside country	50	29	33
Withholding any service is paternalistic and a denial of patients' rights	69	46	48
A women's decision about abortion should be her own, no interference by anyone	60	52	39
2. *Prenatal diagnosis*			
Doctor should tell pregnant woman that blood test may lead to decision about abortion	72	85	835
Doctor should do PND for paternity test (in absence of rape or incest)			
Without comment	31	9	18
With negative comment	22	27	25
Only if woman consents to counseling	34	47	27
Not at all	13	18	30
Parents should be told fetal sex if they ask	92	92	86
Women at high risk should have PND	81	80	93
Women should abort if PND indicates serious disorder	21	43	67
Women are under pressure to have PND	30	29	n/a
Parents have right to choose quality of their children	37	4	54
PND should be available to all women on request	80	94	68
Useless to provide PND unless treatment available	15	18	19
Father of fetus should be able to make mother have PND	13	20	25
Counseling should precede PND	78	98	94

Table 93. (continued)

	US (n = 476)	Germany (n = 593)	France (n = 394)
3. Preconception sex selection			
Should be			
Available to all, no restrictions	35	11	22
Only to married couples	16	1	3
Available only if family has 3+ children	13	11	26
Not available at all	34	71	37
Would use it myself	26	10	20
Would benefit society	11	4	13
Insurance should pay	15	9	19
4. Disability			
Genetically impaired newborn needs a life-saving operation			
Respect parents' decision not to have operation	78	68	90
Do operation anyway	10	6	6
Put child up for adoption	23	13	12
Society will never give enough support for people with disabilities	58	59	—
Disability will never be overcome, even with social support	59	58	—
Unfair to bring child into world with a serious disability, if birth could have been prevented	44	48	77
Unfair to sibs	24	21	—
Most care falls on mother	59	80	—
People with severe disabilities make society richer	21	21	—
Socially irresponsible knowingly to bring child into world with serious disability	26	27	56
5. Parenthood for people with disabilities			
A. Blind couple whose child will be blind want a child			
Doctor should:			
Try to talk them out of it	25	49	69
Suggest sterilization	9	23	19
Suggest they go ahead	16	14	5
Say will support their decision	95	77	87
Decision may be best for couple	70	41	52
Would do same in their situation	25	12	11
Unfair to child	27	41	57
Unfair to society	12	5	22
B. Blind woman on welfare wants a child			
Dissuade her	74	n/a	92
Suggest sterilization	45	n/a	57
Law should require sterilization	21	n/a	18
C. Pregnant woman with PKU			
Doctor should			
Recommend voluntary hospitalization	92	95	92
Arrange forced hospitalization	21	27	40
Offer abortion	43	51	62
Charge her with child abuse	19	15	9

Table 93. (continued)

	US (n = 476)	Germany (n = 593)	France (n = 394)
D. Decision of a CF carrier–carrier couple to take their chances of having a child with CF is			
Best for parents	57	35	31
I would do likewise	21	19	12
Unfair to child	56	41	64
Unfair to society	17	7	27
6. Prevention			
People should tell partners their genetic information	84	89	78
People should know their genetic status before marriage	64	49	70
People at high risk shouldn't marry each other	5	4	7
People at high risk should not have children unless they use PND	41	40	46
States should require carrier tests before marriage	31	5	29
7. Telling relatives			
People should tell relatives their genetic information	89	79	84
A patient refuses to tell blood relatives about a diagnosis of HD	25	24	12
Doctor should:			
Maintain confidentiality	25	24	12
Tell relatives who ask	42	42	47
Find relatives and tell them even if they don't ask	33	35	42
Respondents would themselves want to know diagnosis of fatal, untreatable genetic disease			
Yes	85	68	n/a
No	4	19	n/a
Don't know	11	13	n/a
Doctor should tell them even if they don't want to know			
Yes	48	35	n/a
No	36	55	n/a
Don't know	17	10	n/a
Tell birth mother about adopted child's illness	93	96	95
Parents should be able to test children for			
Breast cancer	93	88	97
Heart disease	92	87	97
Diseases that are neither preventable nor treatable	61	34	45
A person should be able to make a relative have a blood test	41	n/a	40
Doctor should disclose nonpaternity to a man who asks	74	n/a	60
8. Privacy			
Access to DNA in a Medical record, without patient's consent for			
Law enforcement agencies	19	8	—
Life insurers	8	1	—

Table 93. (continued)

	US (n = 476)	Germany (n = 593)	France (n = 394)
Blood relatives	44	32	41
Employer	3	0.3	—
Family doctor	54	43	59
Spouse	53	42	55
Employer, if public safety job	28	19	—
Government agencies, such as Registry of Motor Vehicles	8	3	6
Would tell a school system a diagnosis of their son's XYY, which sometimes leads to antisocial behavior	96	n/a	92
DNA fingerprinting should be required for			
Convicted sex offenders	97	94	93
Persons convicted of other serious crimes	95	83	88
Persons charged with sex offenses	77	82	80
Persons charged with other serious crimes	72	71	79
Members of armed forces	86	62	60
Newborns	73	31	47
Passport applicants	50	10	12
People on welfare	47	24 (asylum seekers)	7
Credit card applicants	34	n/a	5
9. Employment			
Screening of workers should be voluntary, as opposed to required, for			
Genetic risk for lung disease	51	61	—
Executives, for genetic risk for cancer, heart disease, diabetes	81	96	—
Access to test results for:			
Employee	99	98	—
For workers, employer (without consent)	17	7	8
(with consent)	78	71	—
For executives, employer (with consent)	81	60	—
Life insurer (with consent)	66	35	—
Let genetically susceptible worker decide whether to stay in hazardous job	70	72	57
Genetic tests should be included in pre-employment physicals	7	4	0.6
10. Public safety			
Bus driver at high risk of heart attack on job. Doctor should			
Maintain patient confidentiality	20	11	6
Tell employer	69	38	61
Other	11	50	33
11. Newborn screening			
Would want to know if future child would have incurable disease	72	79	n/a
State should require newborn tests for untreatable genetic disorders	26	62	n/a

WHO/HGN/GL/ETH/98.1
ENGLISH ONLY
Distr.: GENERAL

PROPOSED INTERNATIONAL GUIDELINES ON ETHICAL ISSUES IN MEDICAL GENETICS AND GENETIC SERVICES

WORLD HEALTH ORGANIZATION

HUMAN GENETICS PROGRAMME

1998

PROPOSED INTERNATIONAL GUIDELINES
ON
ETHICAL ISSUES IN MEDICAL GENETICS
AND GENETIC SERVICES

Report of a WHO Meeting on Ethical Issues in Medical Genetics

Geneva, 15–16 December 1997

PREFACE

The following rationale for preparing guidelines on ethical issues in medical genetics and genetic services appears in a document of the World Health Organization on Ethics and Health at the Global Level: WHO's Role and Involvement[1]:

"A great number of international committees, commissions and advisory groups and many reports and guidelines have been produced on specific biomedical issues. In dealing with ethics, WHO will not duplicate what has been done or is being done in existing national or regional forums. WHO's involvement in ethics will be directly derived from its global mission, its inclusive vision of health, and its responsibility for coordinating international health action.

Cutting across specific health issues and disciplines, WHO's main aim must be to enhance the integration of ethics in overall public health policies and practices as well as in international health cooperation. Perhaps the most pressing reason for this is to promote greater equity in access and use of health services by all individuals and in all countries. WHO will contribute to coordinating national and regional approaches, identifying gaps and workable solutions, and promoting harmonization of standards and practices at a global level."

To this end, the WHO Human Genetics Programme developed the draft document "Guidelines on Ethical Issues in Medical Genetics and the Provision of Genetics Services"[2]. This document was circulated worldwide, and comments were received from all Regions and WHO staff. The draft document and all responses received formed the background information for a WHO Meeting on "Ethical Issues in Medical Genetics" held in Geneva from 15 to 16 December 1997. The participants at the meeting were experts in this field from both developing and developed countries.

The purpose of the meeting was to review ethical issues in medical genetics and to propose international guidelines on ethical issues in medical genetics and genetic services. These proposed guidelines, which were unanimously adopted by the participants in this meeting, appear below.

LIST OF EXPERTS

Professor O.O. Akinyanju, College of Medicine, University of Lagos, P.M.B. 12003, Lagos, Nigeria

Professor K. Berg, Director, Institute of Medical Genetics, University of Oslo; and Director, Department of Medical Genetics, Ulleval University Hospital, P.O. Box 1036 Blindern, N-0315 Oslo, Norway (**Chairman**)

Dr J.M. Cantú Garza, Chief, Genetics Division, Centro de Investigación Biomédica de Occidente, Instituto Mexicana del Seguro Social, Siena No. 1068 Lomas de Providencia, Guadalajara, Jalisco, Mexico

Professor M.A.F. El-Hazmi, Department of Medical Biochemistry, College of Medicine and King Khalid University Hospital, P.O. Box 2925, Riyadh 11461, Saudi Arabia

Professor D.D. Farhud, Head, Unit of Human Genetics & Anthropology, University of Teheran, School of Public Health & Institute of Public Health Research, P.O. Box 1310, Teheran, Islamic Republic of Iran

Professor J.C. Fletcher, Professor of Biomedical Ethics, The Center for Biomedical Ethics, Box 348, Health Sciences Center, University of Virginia, Charlottesville, VA 22908, USA (**Co-Rapporteur**)

Professor N. Fujiki, Professor Emeritus, Fukui Medical School, Shimoaizuki, Matsuoka-cho, Fukui Prefecture, 910-11 Japan

Professor H. Hamamy, Professor of Medical Genetics, Mustansiriya College of Medicine, Baghdad, Iraq

Professor V.I. Ivanov, Director, National Research Centre for Medical Genetics, Moskvorechie str., Moscow 115478, The Russian Federation

Professor B.M. Knoppers, Centre de recherche en droit public, Faculté de droit, Université de Montréal, C.P. 6128, succursale A, Montréal, Québec, Canada H3C 3J7

Dr Xin Mao, Division of Genetics, Department of Psychiatry, West China University of Medical Sciences, Chengdu 610041, China *and* Section of Molecular Carcinogenesis, Haddow Laboratories, Institute of Cancer Research, 15 Cotswold Road, Sutton, Surrey SM2 5NG, UK

Professor J.-F. Mattei, Centre de Génétique Médicale, Hôpital d'Enfants de la Timone, F-13385 Marseille Cedex 5, France

Professor V.B. Penchaszadeh, Director, Division of Medical Genetics, Beth Israel Medical Center, First Avenue at 16th Street, New York, NY 10003, USA

Professor I.C. Verma, Head, Department of Medical Genetics, Sir Ganga Ram Hospital, Rajinder Nagar, New Delhi 110060, India

Professor D.C. Wertz, Division of Social Science, Ethics and Law, Eunice Kennedy Shriver Center for Mental Retardation, Inc., 200 Trapelo Road, Waltham, MA 02254, USA (**Co-Rapporteur**)

Professor R. Williamson, Director, The Murdoch Institute, Royal Children's Hospital, Flemington Road, Parkville, Melbourne, VC 3052, Australia

Council for International Organizations of Medical Sciences (CIOMS)

Dr J. Gallagher, Consultant

WHO Secretariat

Dr Hu Ching-Li, Chairman, WHO Steering Group on Ethics and Health
Dr B.G. Mansourian, Director, Office of Research Policy and Strategy Coordination
Dr N. Biros, Director General's Executive Secretariat
Dr V. Boulyjenkov, Human Genetics, Division of Noncommunicable Diseases (**Secretary**)
Mr S.S. Fluss, Office of Health Policy in Development
Miss G. Pinet, Health Legislation

EXECUTIVE SUMMARY

The contents of this document were unanimously agreed upon by the experts present at the meeting. The rapid developments in the area of genetics make it necessary to continue discussions as new developments take place. For this reason, it is likely that the proposed guidelines will need modifications at certain intervals. The group felt that it is important for WHO to provide leadership in the international debate on issues related to ethics and the provision of medical genetics services.

The recommendations that follow are designed to assist decision-makers at both national and international levels to protect people and families with genetic disabilities, to recognize the great potential of advances in human and medical genetics for public health, and to develop policies and practices that will ensure that these applications can become accessible to all and are provided with due regard to ethics and justice worldwide.

The advances in human genetics that have occurred during the past twenty years have revolutionized knowledge of the role of inheritance in health and disease. We now know that our DNA determines not only the cause of catastrophic single-gene disorders, which affect millions of persons worldwide, but also predisposition to cancer, heart disease, psychiatric disorders and even to some infectious diseases.

When used properly, this knowledge will be extremely important in helping to achieve better health for people in all countries. However, these advances will only be acceptable if their application is carried out ethically, with regard to autonomy, justice, education and the beliefs and laws of each nation and community.

Genetics teaches that there is no such thing as a "superior" or "inferior" genome; humankind depends for its richness and its survival on the interaction of its complex genetic diversity with the environment.

General ethical considerations. The medical application of genetic knowledge must be carried out with due regard to the general principles of medical ethics; doing good to individuals and families, not doing harm, offering autonomy of choice after information is given, and facilitating personal and social justice. These principles, which have often been outlined by WHO and other groups, are the same as in other medical fields.

The proper use of genetic data. Since we inherit our genes from our parents, pass them on to our children, and share them with our close and distant relatives, every genetic diagnosis, test and procedure involves many people.

It is ethically imperative that genetic data should only be used to the advantage of members of a family or ethnic group, and never to stigmatize or discriminate against them.

Voluntary use of genetic screening and testing. There shall be no compulsory genetic testing of adult individuals or populations.

Every test shall be offered in such a way that individuals and families are free to refuse or accept according to their wishes and moral beliefs.

All testing should be preceded by adequate information about the purpose and possible outcomes of the test and potential choices that may arise.

Children shall only be tested when it is for the purpose of better medical care, as in the case of newborn screening when early treatment will be of benefit to the child.

Prenatal testing. Prenatal diagnosis should be offered to those who need it, but there must be no pressure on couples to accept such testing, nor to use the results of the test to compel either continuing or terminating a pregnancy when the fetus is affected with a genetic disorder.

Decisions in the context of reproduction should rest with those being tested, not with physicians or the government.

The woman should be an important decision-maker in all matters related to reproduction.

Prenatal diagnosis should be done only to give parents and physicians information about the health of the fetus; its use for paternity testing, except in cases of rape or incest, or for gender selection, apart from sex-linked disorders, is not acceptable.

Justice demands equitable access to services. Genetic services for the prevention, diagnosis and treatment of disease should be available to all, without regard to ability to pay, and should be provided first to those whose needs are greatest.

Genetic data should be treated as confidential at all times. Genetic data should only be used to advantage and empower an individual or family, and for better treatment or prevention of disease. Data relevant to health care should be collected and kept by medical geneticists in secure confidential files.

Genetic data should not be given out to insurance companies, employers, schools or governments, other than after the full informed consent of the person tested. In some countries it may be possible, or necessary, to protect both confidentiality and non-discrimination through legal means.

Genetic counselling is the provision of accurate, full and unbiased information in a caring, professional relationship that offers guidance, but allows individuals and families to come to their own decisions. Counselling is essential before any genetic testing is carried out, and should continue afterwards if the results entail choices for the person and family tested.

Genetic counselling should be available to all, and should be as non-directive as possible.

Education about genetics for the public and health care professionals is of paramount importance.

Genetics is playing an increasingly important part in medical practice, and many people are concerned about possible abuse of this new knowledge. It is important that education about genetic principles relevant to human health be emphasized appropriately for all people in all cultures.

Education is a two-way process, and geneticists and other health care professionals have much to learn from support and advocacy groups representing those with genetic disorders. Such groups are an integral part of genetic services, and should be guaranteed a voice in policy and education.

In conclusion, WHO and its constituents are urged to consider these issues as a matter of priority, to facilitate global consultation and consensus on them, and to ensure that ethical codes are put in place nationally and internationally so as to guarantee that advances in genetics can benefit the community in an ethical way. It was agreed that WHO and many other international bodies are correct in rejecting any intentional step towards human cloning as unsafe, ethically unacceptable and a distraction from the more crucial issues dealt with in these guidelines.

1. INTRODUCTION

The proposed guidelines are intended to protect people and families with genetic conditions and to inform health policy officials at the highest level of government, public health officials and workers, and physicians and other health care professionals.

The document should also be seen as a contribution to WHO's consideration of "Health Policy for the 21st Century," especially since it concerns global issues of equity in access to medical genetics services by all individuals and in all countries. One of WHO's main aims must be to further the integration of ethics in overall health policies and practices as well as in international health cooperation.

The primary purpose of these proposed guidelines is to assist policy-makers, officials, practitioners and other health workers in the Member States of WHO in ensuring that genetic information and genetic services are introduced into the broader medical practice of the nations in ethically acceptable ways. A secondary purpose is to allay fears and to reassure the public that adequate controls exist in member countries to prevent abuses of genetic information and unacceptable practices. Member States will develop their own policies and practices, using the broader principles and guidelines of this document as points of departure.

The knowledge gained from discoveries in human genetics has the potential for making a significant improvement in the health of the world's people when applied properly at the individual, family and community levels. If such discoveries and information are integrated into primary health care in ethically acceptable ways and with respect for cultural and religious diversity, the increased benefits for diagnosis, treatment and prevention of human genetic conditions will be appreciable in both developed and developing countries.

Genetics and biomedical technology open vast new avenues for research and can provide humankind with much needed therapeutic tools. But, where human life and dignity are at stake, technology cannot be left on its own to govern ethics, nor can health technology, practices and procedures be left to the vagaries of economic forces and personal interests, fears or vulnerabilities. The profound economic and technological inequalities that exist between nations and population groups have an impact on biomedical research and health care practice. Efforts must be made worldwide to mitigate such inequalities, so that the safety and rights of all individuals and communities are adequately protected and a proper balance is achieved between their rights and responsibilities.

In pursuing this task, it was recognized that some ethical problems of human genetics are highly debatable and – at this time in history – are issues beyond the reach of practical consensus among countries. The meeting was convened to explore the possibility of reaching moral consensus on such issues, but it could not produce consensus where this does not exist in the international community. We recognize that the laws of nations differ with respect to these particular issues, and that laws are subject to debate, evolution and change.

In the face of such controversial issues, our view is not that WHO is obliged to resolve these controversies, but that WHO can call for international unity among leaders and peoples as regards the task of including genetic knowledge and its discoveries within the goals of medicine. We recognize and respect the diversities of culture, religion and social structure that shape the public's moral attitudes in different countries.

We also call for mutual respect and continued dialogue on such issues among scientifically informed members of the world's cultural and religious traditions, and for an exchange of views between developed and developing countries.

Educating the public should be a primary goal of WHO, and education is a prerequisite to informed discussion of the issues in this document. WHO should play a leading role in providing a platform for public discussion of ethics and genetics worldwide.

2. ETHICAL PRINCIPLES IN MEDICINE

The traditional sources of ethical guidelines in medicine apply also to medical genetics, which is a field of medicine (Table 1). Medical genetics' main concerns, however, extend beyond those of the traditional structure of medicine and the physician–patient relationship. For example: (a) genetic information may affect an entire family, rather than only the individual; (b) genetic discoveries may be predictive of future adverse events in an individual's or family member's health; (c) genetic information and the choices of the present may affect future generations; and (d) medical genetics has a tradition of non-directiveness in counselling.

The principle of respect for autonomy includes: (a) respecting the self-determination and choices of autonomous persons, and (b) protecting persons with diminished autonomy, e.g., children and persons with mental impairments.

The principle of beneficence (L. "bene" = good) is the source of physicians' obligation to give highest loyalty to the welfare of individuals and families. Beneficence also bears upon a goal of medicine to improve the health of populations with the voluntary cooperation of the populations involved.

Non-maleficence (L. "male" = evil, harm) is the source of the traditional medical norm of "do no harm", meaning a duty to prevent harm altogether, or, if harm cannot be avoided, to minimize harm to individuals and families.

The goals of justice can be described somewhat differently: treating persons fairly, giving persons what they deserve, or giving persons that to which they are entitled. The term "distributive" (or social) justice means to allocate benefits (e.g., property) and burdens (e.g., taxation) fairly and with equity, in order to enhance social harmony and cooperation. Distributing the benefits (e.g., of diagnosis and treatment) and the burdens (e.g., of rationing of expensive care or of research risks) of health care ought to be governed by ethically justified rules such as: to each according to need, to each according to an equal share or opportunity, etc.

At present, the principles laid out in Table 1 are not applied with equal force around the world, especially respect for persons. Health professionals need to pay

Table 1. Relevant Ethical Principles in Medicine

Respect for the autonomy of persons: respecting the self-determination of individuals and protecting those with diminished autonomy;
Beneficence: giving highest priority to the welfare of persons and maximizing benefits to their health;
Non-maleficence: avoiding and preventing harm to persons or, at least, minimizing harm;
Justice: treating persons with fairness and equity, and distributing the benefits and burdens of health care as fairly as possible in society.

special attention to these principles in areas of the world where they are unfamiliar or infrequently used.

It is a common misconception that prevention and care of genetic disorders and birth defects concern only people living in industrialized countries. Genetic conditions occur with similar frequencies in different nations and irrespective of the socioeconomic status of individuals. In fact, at all levels of society, children born with genetic disadvantages have higher risks of getting sick and dying of environmental causes such as infections and malnutrition. A meaningful right to health care must include access to services for the diagnosis, treatment and prevention of genetic disorders. The priority assigned to genetic services with respect to other health services is a matter of public health policy in each country.

WHO Member States should be encouraged to draw up public health policies that include standards for genetic services along the lines recommended in the Report of a WHO Scientific Group on Control of Hereditary Diseases[3]. People have the right to equitable access to genetic services according to the standard of care that exists in each country, according to need and irrespective of the ability to pay. Also, some parties, such as women, children and people with disabilities, are especially disadvantaged and vulnerable in some societies and deserve special consideration. Professionals should help to protect such persons wherever they are at risk of harm.

Within genetic services, priority should be given to programmes that address the heaviest burdens and needs of the majority of the population. In particular, efforts should be directed towards extending the reach of genetic services at the primary care level, with the utilization of technologies and personnel that are appropriate to the needs, expectations and beliefs of the community. On the other hand, it is an inequitable use of scarce resources to develop expensive high technology services that cater only to the wealthier sectors of society while being largely inaccessible to the majority.

The principle of distributive justice should ensure that scarce resources are utilized equitably on the basis of need, and thus would oppose catering to consumers' requests for genetic services (e.g., prenatal diagnosis) to gratify cultural or personal desires rather than for medical reasons.

3. GOALS AND PRACTICES OF MEDICAL GENETICS

Medical genetics is the field of medicine that is most centrally involved in providing services to persons with genetic conditions and their families. The goals of medical genetics services are to help people with a genetic disadvantage and their families to live and reproduce as normally as possible, to make informed choices in reproductive and health matters, to assist people to obtain access to relevant medical services (diagnostic, therapeutic, rehabilitative or preventive) or social support systems, to help them adapt to their unique situation, and to become informed on relevant new developments.

Conditions studied by medical geneticists include diseases caused by defects in single genes (e.g., haemophilia, sickle cell anaemia, neurofibromatosis, cystic fibrosis), disorders caused by interaction between several genes and environmental factors (e.g., common congenital malformations, diabetes, hypertension, cardiovascular disease, breast cancer, mental disorders), and conditions caused by chromosomal anomalies (e.g., Down syndrome). Diagnostic work in medical genetics includes laboratory work at the DNA, protein and chromosome levels as well as clinical observation of disorders, including birth defects.

Whereas single-gene disorders are rare, conditions caused by an interaction between genes and environmental factors are frequent and include disorders such as cardiovascular diseases, several cancers, asthma, diabetes mellitus and mental disorders. Preventive aspects of work in medical genetics include identification of high-risk individuals with respect to common disorders for the purpose of preventing disease (e.g., heart disease) or securing early diagnosis and treatment (several cancers). At present there are significant research efforts aimed at developing somatic cell gene therapies or therapies to improve or block the function of genes.

Medical genetics services should be organized at all levels of medical care and should be directed by specially trained physicians. Actions may be conducted by a variety of health personnel according to the level of care and the particular organization of health delivery in each society. The different members of the genetics team may include Ph.D. geneticists, nurses, primary care physicians, other health professionals, specially trained health care workers or genetic counsellors, social workers and laboratory personnel.

4. APPLICATION OF ETHICAL PRINCIPLES TO GENETIC SERVICES

The application of the above-mentioned ethical principles to genetic services is illustrated in Table 2.

Non-discriminatory language emphasizes the personhood of those with genetic conditions. Thus, for example, someone with Down syndrome is best described as a "person (or child) with Down syndrome" rather than a "Down syndrome child"

Table 2. Ethical Principles Applied to Genetic Services

- Fair allocation of public resources to those who most need them (justice).
- Freedom of choice in all matters relevant to genetics. The woman should be an important decision-maker in reproductive matters (autonomy).
- Voluntary approach necessary in services, including approaches to testing and treatment; avoidance of coercion by government, society or physicians (autonomy).
- Respect for human diversity and for those whose views are in the minority (autonomy, non-maleficence).
- Respect for people's basic intelligence, regardless of their knowledge (autonomy).
- Education about genetics for the public, medical and other health professionals, teachers, clergy and other persons who are sources of religious information (beneficence).
- Close cooperation with patient and parent organizations, if such organizations exist (autonomy).
- Prevention of unfair discrimination or favouritism in employment, insurance or schooling based on genetic information (non-maleficence).
- Teamwork with other professionals through a network of referrals. When possible, help individuals and families to become informed members of the team (beneficence, autonomy).
- Use of non-discriminatory language that respects individuals as persons (autonomy).
- Timely provision of indicated services or follow-up treatment (non-maleficence, beneficence).
- Refraining from providing tests or procedures not medically indicated (non-maleficence).
- Providing ongoing quality control of services, including laboratory procedures (non-maleficence).

or "Down syndrome case." Words that dehumanize persons with disabilities or stigmatize them should be avoided.

5. GENETIC COUNSELLING

Non-directive counselling has two major elements. The first is the provision of accurate, full and unbiased information that individuals and families may use in making decisions. The second is an understanding, empathic relationship that offers guidance and helps people to work towards their own decisions. In non-directive counselling, the professional avoids purposely slanting information that may lead people to do what the counsellor thinks best. Individuals and families must depend on the counsellor as a source of accurate information, and usually have no way of discovering when information is biased. Non-directive counselling does not mean presenting information and then abandoning individuals and families to make their own decisions without help. Most people may want to talk with someone who will listen to their concerns, help them to express and understand their own values, and help them to work toward their own decisions. Non-directive counsellors do not tell people what to do; decisions are those of the individuals and the families. The counsellor should, as much as possible, support all decisions.

One factor in favour of non-directive counselling was that genetics evolved as a largely diagnostic speciality with little treatment. As more treatments become available, and as susceptibility testing for common multifactorial diseases may suggest lifestyle changes that could benefit the individual's health, the counselling approach may become similar to approaches in general medicine, where the doctor may recommend beneficial treatment or lifestyle changes. Counselling related to reproductive choices should remain non-directive. The ethical principles that underlie genetic counselling and their applications are outlined in Table 3.

Full disclosure of test results includes ambiguous test results, new and controversial interpretations, and differences among professional colleagues in regard to test interpretation.

Re-contact means keeping abreast of new developments and re-contacting individuals or families on a timely basis regarding any new developments relevant to their health or reproduction, unless otherwise instructed by the individual or family.

6. GENETIC SCREENING AND TESTING

Genetic screening refers to tests offered to a population group to identify asymptomatic people at an increased risk from a particular adverse outcome. Examples are phenylalanine screening for phenylketonuria in newborn babies, or the use of maternal serum biochemical markers in pregnant women to screen for fetuses with Down syndrome. In all cases, individuals whose screens indicate that they are at higher risk must be offered a definitive diagnostic test.

Genetic testing is the analysis of the status of a particular gene. A genetic test may establish: (a) a specific diagnosis of a genetic condition in a symptomatic individual, (b) the certainty that a particular condition will develop in an individual who is

Table 3. Ethical Principles Applied to Genetic Counselling

- ! Respect for persons and families, including full disclosure, respect for people's decisions, and accurate and unbiased information (autonomy).
- ! Preservation of family integrity (autonomy, non-maleficence).
- ! Full disclosure to individuals and families of all information relevant to health (non-maleficence, autonomy).
- ! Protection of the privacy of individuals and families from unjustified intrusions by employers, insurers, and schools (non-maleficence).
- ! Information to individuals and families about possible misuses of genetic information by institutional third parties (non-maleficence).
- ! Informing individuals that it is the individual's ethical duty to tell blood relatives that the relatives may be at genetic risk (non-maleficence).
- ! Informing individuals about the wisdom of disclosing their carrier status to a spouse/partner if children are intended, and the possibility of harmful effects on the marriage from disclosure (non-maleficence).
- ! Informing people of their moral duties to disclose a genetic status that may affect public safety (non-maleficence).
- ! Unbiased presentation of information, insofar as this is possible (autonomy).
- ! Non-directive approach, except when treatment is available (autonomy, beneficence).
- ! Children and adolescents to be involved in decisions affecting them, whenever possible (autonomy).
- ! Duty to re-contact if appropriate and desired (non-maleficence, beneficence, autonomy).

asymptomatic at the time of the testing (presymptomatic diagnosis), or (c) the presence of a genetic predisposition to develop a particular complex disease such as cancer or cardiovascular disease.

The main objective of genetic screening and testing is to prevent disease or secure early diagnosis and treatment.

Ordinarily, population screening programmes are offered only when proven methods of treatment or prevention are available. In selecting population groups to be screened because they are thought to have higher than average risks, it is important to avoid the possibility of stigmatizing the entire group. Anonymous screening for epidemiological purposes may be conducted after notification of the population to be screened, in the absence of preventive or therapeutic options for the individuals screened. Screening programmes are usually better received if they work in cooperation with community leaders in the group to be screened. Screening should be preceded by educational programmes for the group.

If screening is provided for newborn babies, there is an obligation on health care providers to make sure that appropriate and timely treatment is provided. Suggested ethical guidelines for screening and testing are listed in Table 4.

7. INFORMED CONSENT AND GENETIC TESTING

The ethical requirement of informed consent differs between a research setting and a clinical situation. International ethical guidelines for research[4] are applicable to efforts to establish new genetic tests or to develop quality control of tests[5], as well as to other research-related situations. Informed consent for any study to validate a genetic test must be obtained if the specimen can be linked to the person from whom it was taken. People should be informed of possible future uses of the specimen, whether identifiers will be retained, and if so, whether individuals will be re-contacted about new developments concerning their health care.

Table 4. Proposed Ethical Guidelines for Genetic Screening and Testing

! Genetic screening and testing should be voluntary, not mandatory, with the exception noted in the last point below (autonomy);
! Genetic screening and testing should be preceded by adequate information about the purpose and possible outcomes of the screen or test and potential choices to be made (autonomy, non-maleficence);
! Anonymous screening for epidemiological purposes may be conducted after notification of the population to be screened (autonomy);
! Results should not be disclosed to employers, insurers, schools or others without the individual's consent, in order to avoid possible discrimination (autonomy, non-maleficence);
! In rare cases where disclosure may be in the best interests of the individual or of public safety, the health provider may work with the individual towards a decision by him or her (beneficence, non-maleficence, justice);
! Test results should be followed by genetic counselling, particularly when they are unfavourable (autonomy, beneficence);
! If treatment or prevention exists or is available, this should be offered with a minimum of delay (beneficence, non-maleficence);
! Newborn screening should be mandatory and free of charge if early diagnosis and treatment will benefit the newborn (beneficence, justice).

If a child or adolescent is tested as part of a research project, the child's assent should be sought. An adequate explanation for a child's assent should describe the potential harms and benefits of testing in a simple manner appropriate to the child's age.

Table 5 illustrates ethical requirements of informed consent applicable to testing in clinical practice and in research situations, respectively.

8. PRESYMPTOMATIC AND SUSCEPTIBILITY TESTING

Presymptomatic testing refers to identification of healthy individuals who may have inherited a gene for a late-onset disease, and if so will develop the disorder if they live long enough (e.g., Huntington disease). Susceptibility testing identifies healthy individuals who may have inherited a genetic predisposition that puts them at increased risk of developing a multifactorial disease, such as heart disease, Alzheimer disease or cancer, but who, even so, may never develop the disease in question.

Presymptomatic testing in the absence of therapeutic options should be available if the following conditions are met:

a. The information provided by testing will be used to prevent harm to the person tested, or to spouse, family, prospective children or others.
b. The person is fully informed about the limitations of testing, including possibilities of uninformative results, and inability to predict exact age of onset or (sometimes) severity of symptoms.
c. The person (or the legally authorized representative) is mentally capable of giving consent.
d. Testing is accompanied by a counselling programme of appropriate length and intensity for the disorder.

In regard to requests for testing children, in the absence of medical benefit through prevention or treatment, presymptomatic or susceptibility tests for adult-onset

Table 5. Proposed Ethical Guidelines Concerning Autonomy and Informed Consent

A. *Applicable to clinical practice*:
Genetic testing in clinical practice should be voluntary and should occur in the context of a comprehensive genetic service and valid process of informed consent, with an explanation of the following elements:

! the purpose of the test,
! the chance that it will give a correct prediction,
! the implications of the test results for the individual and family,
! the tested person's options and alternatives,
! the test's potential benefits and risks, including social and psychological,
! that social risks include discrimination by insurers and employers (even though this may be illegal), and
! that whatever decision individuals and families make, their care will not be jeopardized.

B. *Applicable to research and quality control*:
Elements of a valid informed consent process include an explanation of:

! the experimental nature and purpose of the study,
! why the individual is invited to participate, and that participation is voluntary,
! the procedure,
! the discomforts and risks (if any) of the test to both the individual and the family,
! the uncertainty of the results of the test for prediction and accurate genetic counselling,
! the possible benefits to others and to science,
! the confidentiality of records identifying the tested individual,
! whom to contact for questions about research or in the event of a research injury,
! the right of the individual to withdraw at any time, and
! the right of the individual and family to unrestricted health care, even if the individual withdraws.

Table 6. Proposed Ethical Guidelines for Presymptomatic and Susceptibility Testing

! Genetic susceptibility testing of persons with a family history of heart disease, cancer or other common diseases of possible genetic origin should be encouraged, provided that information from the test can be used effectively for prevention or treatment (beneficence).
! All susceptibility testing should be voluntary, preceded by adequate information and based on informed consent (autonomy).
! Presymptomatic testing should be available for adults at risk who want it, even in the absence of treatment, after proper counselling and informed consent (autonomy).
! Testing of children or adolescents should be carried out only if there are potential medical benefits to the child or adolescent (autonomy, beneficence, non-maleficence).
! Employers, insurers, schools, government agencies or other institutional third parties should not be given access to test results (non-maleficence).

disorders are usually best postponed until adulthood, when the young adult can make her/his own decision. In counselling, geneticists need to explain to parents the potential benefits and potential harms of testing children.

Proposed ethical guidelines for presymptomatic and susceptibility testing are presented in Table 6.

9. DISCLOSURE AND CONFIDENTIALITY

Disclosure and confidentiality issues are some of the most frequent ethical problems appearing in medical genetics. Because of the possibility of harm from disclosure to institutional third parties, utmost care must be taken to protect confidentiality. However, a genetic diagnosis in an individual may indicate genetic risks in his/her relatives. In those circumstances, the genetic service provider should encourage the individual to ask the relatives to seek genetic counselling. If the individual refuses, especially in cases where effective and affordable treatment or preventive measures are available, the counsellor may ethically make direct contact with the relatives, bearing in mind that the information provided should concern only their own genetic risks, not the genetic status nor the identity of the relative who refused to inform them. Counsellors should also make sure that adequate follow-up takes place.

Proposed guidelines on issues of disclosure and confidentiality are listed in Table 7.

10. PRENATAL DIAGNOSIS

Prenatal diagnosis of genetic disorders and fetal anomalies has expanded significantly for hundreds of conditions through DNA analysis of fetal cells, and the increased use of ultrasound and maternal serum biochemical screening (amniocentesis). The purpose of prenatal diagnosis is to rule out the presence in the fetus of a particular medical condition for which the pregnancy is at an increased risk. This information is provided to the couple to assist in their decision-making process regarding the available options, such as carrying the pregnancy to term, preparing for a difficult delivery and for special newborn care, or terminating the pregnancy. Genetic counselling is particularly important prior to prenatal diagnosis and, after a result indicating an affected fetus, to secure fully informed choices. Information about the purposes, benefits and limitations of maternal serum biochemical screening must be given when

Table 7. Proposed Ethical Guidelines Concerning Disclosure and Confidentiality

! Professionals should disclose to tested individuals all test results relevant to their health or the health of a fetus. Adequate information is a prerequisite for free choice and is necessary to the open communication and trust that should mark the relationship between the provider and the person counselled.
! Test results, including normal results, should be communicated to the tested person without undue delay.
! Test results not directly relevant to health, such as non-paternity, or the sex of the fetus in the absence of X-linked disorder, may be withheld if this appears necessary to protect a vulnerable party or if prescribed by national law.
! The wish of individuals and families not to know genetic information, including test results, should be respected, except in testing of newborn babies or children for treatable conditions.
! Information that could cause grave psychological or social harm may be temporarily withheld. Within the general duty of disclosure, the counsellor may exercise judgement about when a tested person is ready to receive information.
! If a couple intends to have children, individuals should be encouraged to share genetic information with their partners.
! Where appropriate, as part of their general duty to educate, counsellors should inform people that genetic information may be useful to their relatives and may invite individuals to ask the relatives to seek genetic counselling.
! The provision of genetic information to relatives about the family so as to learn their own genetic risks should be possible, especially when a serious burden can be avoided.
! Results of carrier tests, presymptomatic tests, susceptibility tests and prenatal tests should be kept confidential from employers, health insurers, schools and government agencies. People should not be penalized or rewarded for their genetic constitutions. Information about a symptomatic condition may be disclosed as part of general medical information, in accordance with laws and practices in different countries.
! Registries (if any) should be protected by the strictest standards of confidentiality.

Table 8. Proposed Ethical Guidelines for Prenatal Diagnosis

! Equitable distribution of genetics services, including prenatal diagnosis, is owed first to those with the greatest medical need, regardless of ability to pay or any other considerations (justice).
! Prenatal diagnosis should be voluntary in nature. The prospective parents should decide whether a genetic disorder warrants prenatal diagnosis or termination of a pregnancy with an affected fetus (autonomy).
! If prenatal diagnosis is medically indicated, it should be available regardless of a couple's stated views on abortion. Prenatal diagnosis may, in some cases, be used to prepare for the birth of a child with a disorder (autonomy).
! Prenatal diagnosis is carried out only to give parents and physicians information about the health of the fetus. The use of prenatal diagnosis for paternity testing, except in cases of rape or incest, or for gender selection, apart from sex-linked disorders, is not acceptable (non-maleficence).
! Prenatal diagnosis solely for relief of maternal anxiety, in the absence of medical indications, should have lower priority in the allocation of resources than prenatal diagnosis with medical indications (justice).
! Counselling should precede prenatal diagnosis (non-maleficence).
! Physicians should disclose all clinically relevant findings to the woman or couple, including the full range of variability in the manifestations of the condition under discussion (autonomy).
! The woman's and/or the couple's choices in a pregnancy with an affected fetus should be respected and protected, within the framework of the family and of the laws, culture and social structure of the country. The couple, not the health professional, should make the choice (autonomy).

offering the test, including the fact that any abnormal screening result will need confirmatory testing by invasive prenatal diagnosis and may potentially lead to a decision about abortion.[1]

Cultures, religions and national laws differ with regard to abortion of an affected fetus after prenatal diagnosis. WHO cannot resolve these differences, but can suggest general guidelines for the provision of prenatal diagnosis, subject to the framework of the law in each individual country. Proposed ethical guidelines for the provision of prenatal diagnosis and of counselling about it are given in Tables 8 and 9.

Pre-test counselling makes post-test counselling (for those with an affected fetus) much less difficult because prospective parents are better prepared. Counselling should include the items in Table 9, as a minimum.

11. BANKED DNA

Stored DNA in tissue or blood samples may provide useful information for examination of genetic disorders in families or for research. Information from DNA specimens may be of importance for relatives and not only for the person from whom DNA originates. Therefore, access to stored DNA by family members needs to be considered.

[1] References in this document to abortion as a choice available to individuals and couples following a prenatal diagnosis assumes that any such abortion is not prohibited by law. In this respect, WHO refers to the Official United Nations Report of the International Conference on Population and Development (Cairo, 5–13 September 1994) which states, in part, (para 8.25): "In no case should abortion be promoted as a method of family planning.... Women who have unwanted pregnancies should have ready access to reliable information and compassionate counselling. Any measures or changes related to abortion within the health system can only be determined at the national or local level according to the national legislative process."

Table 9. Proposed Counselling Points Prior to Prenatal Diagnosis

- ! Name(s) and general characteristics of the major disorder(s) that the test may identify. The list of disorders need not be exhaustive. The characteristics of the disorder(s) should be described also in terms of their effects on the future child, on the parents and on family life.
- ! Possibilities for treatment of the disorder(s) after birth and availability of supportive care.
- ! Description of the likelihood (risk) that the fetus may have the disorder(s). Risks should be expressed in several ways (as a percentage, as a proportion and verbally).
- ! The possibility of unfavourable test results or of fortuitous or unexpected findings.
- ! Alternatives available for those with an affected fetus, for example, carrying the fetus to term and caring for the child at home; placing the child in an institutional setting, if available; placing the child for adoption; termination of pregnancy; prenatal treatment of the fetus or early treatment after birth.
- ! The possibility of ambiguous laboratory or ultrasonography results.
- ! Information that, because most conditions diagnosed in the fetus cannot be treated before birth, knowing about the existence of a condition may not help the fetus.
- ! Information that the test does not guarantee a healthy baby, because there are many disorders that cannot be identified before birth, or professionals may not know that a family is at risk of a specific disorder (in addition to the disorder that motivated the examination).
- ! The medical risks to fetus and mother posed by the testing procedure.
- ! Non-medical risks, if any (e.g., to parental employment or health care, where applicable).
- ! Information that non-invasive screens used early in pregnancy, such as maternal serum alpha fetoprotein screening, may be the first step on the road to prenatal diagnosis and a possible decision about abortion.
- ! Costs of the test and sources of reimbursement for the mother or couple, if applicable.
- ! Names and addresses of genetic support groups or organizations for persons with genetic disorders that people can contact if they wish.

Existing stored specimens or samples, such as those in university or hospital departments or collections of blood spots, should not be subject to new rules for consent or re-contact that may be established in the future.

In developing policies about samples to be collected in the future, it is helpful to keep the following issues in mind:

- protection of individuals from possible discrimination by employers and insurers, etc.;
- possible benefits to the individuals from research findings;
- the possibility of multiple uses of the same sample in different and unforeseen research projects;
- possible sharing of samples among collaborators, including international collaborators and commercial entities;
- advantages and disadvantages for individuals and researchers of removing all identifiers (including coded numbers) from a sample.

A blanket informed consent that would allow use of a sample for genetic research in general, including future as yet unspecified projects, appears to be the most efficient and economical approach, avoiding costly re-contact before each new research project. The consent should specify that family members may request access to a sample to learn their own genetic status but not that of the donor. While spouses may not have such a right of access, their concerns should be considered. All samples should be used with appropriate regard for confidentiality. Proposed guidelines for access to banked DNA are given in Table 10.

Biomedical research in human genetics can lead to the development of diagnostic and pharmaceutical products. Patents may be necessary to raise funding to develop such products commercially, but gene sequences without proven utility should not be

Table 10. Proposed Ethical Guidelines for Access to Banked DNA

- ! A blanket informed consent that would allow use of a sample in future projects is the most efficient approach.
- ! Control of DNA may be familial, not only individual. Blood relatives may have access to stored DNA for purposes of learning their own genetic status, but not for purposes of learning the donor's status.
- ! Family members should have access regardless of whether they contributed financially to the banking of the DNA.
- ! DNA should be stored as long as it could be of benefit to living or future relatives or fetuses.
- ! Attempts should be made to inform families, at regular intervals, of new developments in testing and treatment. Donors should inform DNA banks of current addresses for follow-up.
- ! After all relatives have died or all attempts to contact survivors have failed, DNA may be destroyed.
- ! Spouses should not have access to DNA banks without the donor's consent, but may be informed that DNA has been banked. If a couple is considering having children, it is the moral obligation of the party whose DNA has been banked to provide the spouse with any relevant information.
- ! Except for forensic purposes or instances when the information is directly relevant to public safety, there should be no access for institutions without the donor's consent. Insurance companies, employers, schools, government agencies and other institutional third parties that may be able to coerce consent should not be allowed access, even with the individual's consent.
- ! Qualified researchers should have access if identifying characteristics are removed.
- ! Potentially valuable specimens that could be useful to concerned families in the future should be saved and should be available.

granted patents. Patenting has the potential to impede international collaboration, especially between developing and developed countries, to the ultimate detriment of service delivery to those with genetic disorders. Genetics differs from many areas of research in that important new knowledge can come from a family, or an ethnic group, with a particular genetic variant. If this leads to the development of a diagnostic test or new therapies, equity requires that the donors, or the community generally, should receive some benefit.

12. ASSISTED REPRODUCTION AND MEDICAL GENETICS

Although not directly related to medical genetics, various types of assisted reproduction are often discussed in connection with genetic counselling. Couples who are at risk of having a child with a genetic disorder may choose alternative options. These may include egg or sperm or embryo donation, or surrogacy. Countries have legitimate wide differences in their beliefs about the acceptability of each of these practices. In addition, these alternatives are often expensive in health resources. Whichever reproductive alternatives are offered must be consistent not only with the cultural traditions and beliefs of each country, but also with overall respect for the autonomy of individuals and families. In this context, reproductive cloning (the creation of a fetus whose genome is entirely derived from another individual) has been rejected by many international bodies, including WHO, has aroused fears in many societies, and is not in accord with currently accepted international ethical standards.

13. ACKNOWLEDGEMENTS

The participants at the meeting would like to express their thanks to those who sent their comments on the distributed draft document "Guidelines on Ethical Issues in

Medical Genetics and the Provision of Genetics Services"[2] which was used as background information for the purpose of this meeting. An updated version of the draft document, where all comments received have been considered, will be published in 1998 as a "Review of Ethical Issues in Medical Genetics and Genetic Services."

14. REFERENCES

1. Ethics and Health at the Global Level: WHO's Role and Involvement. WHO Executive Board Meeting – Information Document: EB95/INF.DOC./20, 23 January 1995.
2. Guidelines on Ethical Issues in Medical Genetics and the Provision of Genetics Services. Unofficial WHO document, WHO/HDP/GL/ETH/95.1.
3. Control of Hereditary Diseases. Report of a WHO Scientific Group, WHO Technical Report Series No. 865, Geneva, 1996.
4. International Ethical Guidelines for Biomedical Research Involving Human Subjects. CIOMS/WHO, Geneva, 1993.
5. Neil A. Holtzman and Michael S. Watson (eds), Promoting safe and effective genetic testing in the United States. Report of the Joint DOE-NIH ELSI Task Force on Genetic Testing. Baltimore: Johns Hopkins University Press, 1998.

INDEX

International Library of Ethics, Law, and the New Medicine

1. L. Nordenfelt: Action, *Ability and Health.* Essays in the Philosophy of Action and Welfare. 2000 ISBN 0-7923-6206-3
2. J. Bergsma and D.C. Thomasma: *Autonomy and Clinical Medicine.* Renewing the Health Professional Relation with the Patient. 2000 ISBN 0-7923-6207-1
3. S. Rinken: *The AIDS Crisis and the Modern Self.* Biographical Self-Construction in the Awareness of Finitude. 2000 ISBN 0-7923-6371-X
4. M. Verweij: *Preventive Medicine Between Obligation and Aspiration.* 2000 ISBN 0-7923-6691-3
5. F. Svenaeus: *The Hermeneutics of Medicine and the Phenomenology of Health.* Steps Towards a Philosophy of Medical Practice. 2001 ISBN 0-7923-6757-X
6. D.M. Vukadinovich and S.L. Krinsky: *Ethics and Law in Modern Medicine.* Hypothetical Case Studies. 2001 ISBN 1-4020-0088-X
7. D.C. Thomasma, D.N. Weisstub and C. Hervé (eds.): *Personhood and Health Care.* 2001 ISBN 1-4020-0098-7
8. H. ten Have and B. Gordijn (eds.): *Bioethics in a European Perspective.* 2001 ISBN 1-4020-0126-6
9. P.-A. Tengland: *Mental Health.* A Philosophical Analysis. 2001 ISBN 1-4020-0179-7
10. D.N. Weisstub, D.C. Thomasma, S. Gauthier and G.F. Tomossy (eds.): *Aging: Culture, Health, and Social Change.* 2001 ISBN 1-4020-0180-0
11. D.N. Weisstub, D.C. Thomasma, S. Gauthier and G.F. Tomossy (eds.): *Aging: Caring for our Elders.* 2001 ISBN 1-4020-0181-9
12. D.N. Weisstub, D.C. Thomasma, S. Gauthier and G.F. Tomossy (eds.): *Aging: Decisions at the End of Life.* 2001 ISBN 1-4020-0182-7
    (Set ISBN for Vols. 10-12: 1-4020-0183-5)
13. M.J. Commers: *Determinants of Health: Theory, Understanding, Portrayal, Policy.* 2002 ISBN 1-4020-0809-0
14. I.N. Olver: *Is Death Ever Preferable to Life?* 2002 ISBN 1-4020-1029-X
15. C. Kopp: *The New Era of AIDS.* HIV and Medicine in Times of Transition. 2003 ISBN 1-4020-1048-6
16. R.L. Sturman: *Six Lives in Jerusalem.* End-of-Life Decisions in Jerusalem – Cultural, Medical, Ethical and Legal Considerations. 2003 ISBN 1-4020-1725-1

KLUWER ACADEMIC PUBLISHERS – DORDRECHT / BOSTON / LONDON